Studies in Big Data

Volume 45

Series editor

Janusz Kacprzyk, Polish Academy of Sciences, Warsaw, Poland
e-mail: kacprzyk@ibspan.waw.pl

The series "Studies in Big Data" (SBD) publishes new developments and advances in the various areas of Big Data- quickly and with a high quality. The intent is to cover the theory, research, development, and applications of Big Data, as embedded in the fields of engineering, computer science, physics, economics and life sciences. The books of the series refer to the analysis and understanding of large, complex, and/or distributed data sets generated from recent digital sources coming from sensors or other physical instruments as well as simulations, crowd sourcing, social networks or other internet transactions, such as emails or video click streams and others. The series contains monographs, lecture notes and edited volumes in Big Data spanning the areas of computational intelligence including neural networks, evolutionary computation, soft computing, fuzzy systems, as well as artificial intelligence, data mining, modern statistics and operations research, as well as self-organizing systems. Of particular value to both the contributors and the readership are the short publication timeframe and the world-wide distribution, which enable both wide and rapid dissemination of research output.

**Indexing: The books of this series are submitted to ISI Web of Science, DBLP, Ulrichs, MathSciNet, Current Mathematical Publications, Mathematical Reviews, Zentralblatt Math: MetaPress and Springerlink.

More information about this series at http://www.springer.com/series/11970

Taeho Jo

Text Mining

Concepts, Implementation,
and Big Data Challenge

 Springer

Taeho Jo
School of Game
Hongik University
Seoul, Korea (Republic of)

ISSN 2197-6503   ISSN 2197-6511   (electronic)
Studies in Big Data
ISBN 978-3-030-06302-3   ISBN 978-3-319-91815-0   (eBook)
https://doi.org/10.1007/978-3-319-91815-0

Printed on acid-free paper

This Springer imprint is published by the registered company Springer International Publishing AG part
of Springer Nature.
The registered company address is: Gewerbestrasse 11, 6330 Cham, Switzerland

Preface

This book is concerned with the concept, the theories, and the implementations of text mining. In Part I, we provide the fundamental knowledge about the text mining tasks, such as text preprocessing and text association. In Parts II and III, we describe the representative approaches to text mining tasks, provide guides for implementing the text mining systems by presenting the source codes in Java, and explain the schemes of evaluating them. In Part IV, we cover other text mining tasks, such as text summarization and segmentation, and the composite text mining tasks. Overall, in this book, we explain the text mining tasks in the functional view, describe the main approaches, and provide guidance for implementing systems, mainly about text categorization and clustering.

There are a few factors that provided motivation for writing this book. Textual data including web documents were already dominant over relational data items in information systems, strongly, since the 1990s. So we need techniques of mining important knowledge from textual data as well as relational data. Once obtaining the techniques, we need to provide guides for implementing text mining systems for system developers. Therefore, in this book, we set the functional views of the text mining tasks, the approaches to them, and the implementation of the text mining systems as the scope of this book.

This book consists of four parts: foundation (Chaps. 1–4), text categorization (Chaps. 5–8), text clustering (Chaps. 9–12), and advanced topics (Chaps. 13–16). In the first part of this book, we provide the introduction to text mining and the schemes of preprocessing texts before performing the text mining tasks. In the second part, we cover types of text categorization, the approaches to the task, and the schemes of implementing and evaluating the text categorization systems. In the third part, we describe text clustering with respect to its types, approaches, implementations, and evaluation schemes. In the last part, we consider other text mining tasks such as text summarization and segmentation, and combinations of the text mining tasks.

This book is intended for the three groups of readers: students, professors, and researchers. Senior undergraduate and graduate students are able to study the contents of text mining by themselves with this book. For lecturers and professors

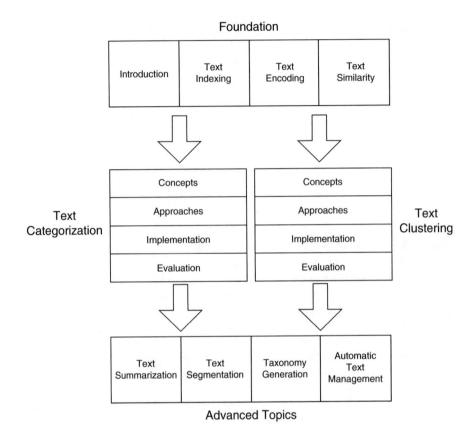

Fig. 1 Organization of four parts

in universities, this book may be used as the main material for providing lectures on text mining. For researchers and system developers in industrial organizations, it may be used as the guide book for developing text mining systems. For understanding this book, it requires only elementary level of the three disciplines of mathematics: probabilities, linear algebra, and vector calculus.

As illustrated in Fig. 1, this book is organized with the four parts. Part I deals with the four aspects: introduction, text indexing, text encoding, and text association. Part II focuses on the text categorization with respect to its concepts, approaches, implementations, and evaluation schemes. Part III mention the text clustering with respect to its four aspects. Part IV is concerned with other text mining tasks and the hybrid ones, as the advanced topics.

Sejong, Korea   Taeho Jo

Contents

Part I
Foundation

Part I is concerned with the foundation of text mining and text processing as the preparation for its main tasks. Because textual data are dominant over numerical one in the real world, we need to realize the importance of text mining and explore its basic concepts. In this part, we mention text indexing and encoding as the main preprocessing before performing the main text mining tasks. We study the schemes of computing the similarities between texts and extracting association rules from textual data as the basic data mining task. Therefore, in this section, we mention what concerns the four chapters involved in this chapter.

Chapter 1 is concerned with the introduction to text mining. We will mention text categorization, text clustering, text summarization, and text segmentation as the typical instances of text mining. Machine learning algorithms will be mentioned as the main approaches to the text mining tasks, rather than other kinds of ones. We explore other areas which are related mainly to the text mining, comparing it with them. Therefore, in this book, the kinds of approaches are restricted to the machine learning algorithms.

Chapter 2 is concerned with the process of indexing a text into a list of words. The tokenization and its implementation will be mentioned as the process of segmenting a text into tokens. We mention stemming as the process of mapping a varied form of words into their root forms, and present its implementation, together with its explanations. We will present the process, the implementation, and examples of the process, stop-word removal. We also mention the additional steps of text indexing to the three basic steps.

Chapter 3 is concerned with the process of encoding texts into their structured forms, which are numerical vectors. We explore the criteria for selecting some feature candidates which are extracted from texts through the process which is covered in Chap. 2. We describe the process of computing the feature values for representing texts into numerical vectors. We provide also the scheme of computing similarity between texts and between text groups. Although numerical vectors are popular representations of texts, we need to consider the issues in encoding texts to them.

Chapter 4 is concerned with the scheme of extracting the association rules from text. We mention data association as the primitive data mining tasks and introduce the Apriori algorithm as its tool. The data association is specialized into the word association which extracts the association rules from a text collection. We deal also with the text association which does the rules from it. The output from the task which is covered in this chapter is given as symbolic rules as the form of if-then.

Chapter 1
Introduction

This chapter is concerned with the introduction to the text mining and its overview is provided in Sect. 1.1. In Sect. 1.2, we explain the texts which are source from which the text mining is performed with respect to their structures and formats. In Sect. 1.3, we describe the tasks of data mining, such as the classification, the clustering, and the association. In Sect. 1.4, we cover other types of data mining than text mining, in order to provide its background. Therefore, this chapter is intended to provide the basic concepts of text mining and the background for understanding it in the subsequent chapters.

1.1 Definition of Text Mining

Text mining is defined as the process of extracting the implicit knowledge from textual data [18]. Because the implicit knowledge which is the output of text mining does not exist in the given storage, it should be distinguished from the information which is retrieved from the storage. The text classification, clustering, and association are the typical tasks of text mining, and they are covered in the subsequent chapters, in detail. Text mining is the special type of data mining, and other types such as relational data mining, web mining, and big data mining are explained in Sect. 1.4. Therefore, this section is intended to explore the overview of text mining, before mentioning the tasks of text mining and the types of data mining.

Text is defined as the unstructured data which consists of strings which are called words [82]. Even if the collection of strings belongs to text in the broad view, it requires the meanings of individual strings and the combination of them by rules, called grammars, for making the text. Here, the scope of text is restricted to the article which consists of paragraphs and is written in a natural language. We assume that a paragraph is referred to an organized group of sentences and a

© Springer International Publishing AG, part of Springer Nature 2019
T. Jo, *Text Mining*, Studies in Big Data 45,
https://doi.org/10.1007/978-3-319-91815-0_1

text is an ordered set of paragraphs. Note that what consists of words written in an artificial language such as source code or mathematical equations is excluded from the text scope.

Text mining is regarded as the special type of data mining, as mentioned above, and we need to explore the data mining conceptually, in order to understand it. The data mining is referred to the process of getting the implicit knowledge from any kind of data in the broad view. However, in the traditional data mining, the type of data which plays roles of source is restricted to the relational data. Classification, regression, clustering, and association are typical main tasks of data mining [91] and will be mentioned in Sect. 1.3. In Sect. 1.4, we will consider web mining and big data mining as the other types of text mining.

We already mentioned classification, regression, clustering, and association, as the main tasks of data mining. Classification is referred to the process of classifying data items into their own categories, and Regression is done to that of estimating an output value or output values to each data item. Clustering is regarded as the process of segmenting a group of various data items into several subgroups of similar ones, which are called clusters. Association is considered as the task of extracting associations of data items in the form of if-them. In Sect. 1.2, they will be described in detail.

In Table 1.1, the differences between the mining and the retrieval are presented. The output of data mining is the implicit knowledge which is necessary directly for making decisions, whereas that of retrieval is some of data items which are relevant to the given query. For example, in the domain of stock prices, the prediction of future stock prices is a typical task of data mining, whereas taking some of past and current stock prices is that of information retrieval. Note that the perfect certainty never exists in the data mining, compared to the retrieval. The more advanced computation for making knowledge from the raw data, which is called synthesis, is required for doing the data mining tasks.

1.2 Texts

This section is concerned with the concepts of texts before discussing the text mining. This section is divided into the two subsections: text components and formats. In Sect. 1.2.1, we explain the words, sentences, and paragraphs as the text components. In Sect. 1.2.2, we describe the three typical text formats: plain

Table 1.1 Mining vs retrieval

	Mining	Retrieval
Output	Knowledge	Relevant data items
Example	Predicted values	Past or current values
Certainty	Probabilistic	Crisp
Synthesis	Required	Optional

text, XML (Extensible Markup Language), and PDF (Portable Document Format). Therefore, this section is intended to describe texts as the source of text mining, with respect to their components and formats.

1.2.1   Text Components

In this study, the text is defined as the group of sentences or paragraphs which are written in a natural language. Even if what is written in an artificial language such as a source code belongs to the text in the broad view, it is excluded out from the scope of text, in this study. Individual words are basic text units, they are combined into a sentence, grammatically, and sentences are organized into a paragraph, logically. A full text consists of paragraphs with their variable lengths and its information such as data, author, title, and so on may be attached to the full text. Therefore, this subsection is intended to describe the text components: words, sentences, and paragraphs.

A word is considered as the basic text unit, here. Because a single word consists of several characters, a single character may be regarded as the further basic unit. The reason of setting a word as the basic unit is that a character has no meaning by itself, but a word has its own meaning. A word is mentioned as the meaningful basic unit, and it is distinguished from the strong which consists of arbitrary characters, ignoring the meaning. The grammatical words which are called stop words and used only for grammatical functions such as "a" or "the" have no meaning, so they are usually excluded in the text preprocessing.

Words are combined into a sentence by the rules which are called grammars. In English, each sentence is decomposed into words by white space in the process, called tokenization [65]. Each sentence starts with a capital letter and ends with the punctuation marks: period, question mark, or exclamation mark, especially in English. A sentence may have a single clause which basically consists of subjective and its verbs, or several clauses, depending on its types; a sentence may be decomposed into clauses. Since English is used most popularly in our world, texts will be characterized based on it, in this study.

In explaining a text, a paragraph is defined as an ordered set of sentences, keeping the consistence for a particular subtopic [49]. A paragraph is decomposed into sentences by their punctuation marks. A paragraph starts with an indented position and ends with its carriage return, as usual case in writing English texts. Each paragraph has its variable length in a natural language; it has a variable number of sentences with their variable lengths. Note that sentences are combined into a paragraph by the logical order or the semantic rules, rather than the grammar.

A text is viewed as an article such as a news article, a journal article, a patent, and a letter, and it consists of a single paragraph, a single group of paragraphs, and several groups of ones. An abstract of a technical article or a news article with its only text exists as a short text. A text which has a single group of paragraph which is called section may be considered as a medium one. A text with multiple groups of paragraphs as a multi-sectioned article is regarded as a long text.

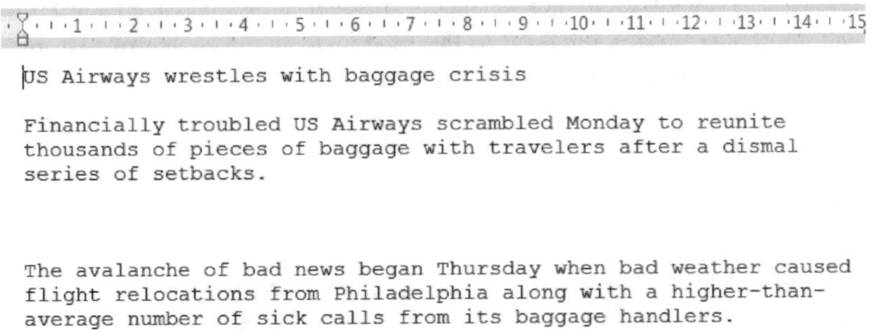

```
·⸮·ı·ı·1·ı·ı·2·ı·ı·3·ı·ı·4·ı·ı·5·ı·ı·6·ı·ı·7·ı·ı·8·ı·ı·9·ı·ı·10·ı·ı·11·ı·ı·12·ı·ı·13·ı·ı·14·ı·ı·15
```

US Airways wrestles with baggage crisis

Financially troubled US Airways scrambled Monday to reunite
thousands of pieces of baggage with travelers after a dismal
series of setbacks.

The avalanche of bad news began Thursday when bad weather caused
flight relocations from Philadelphia along with a higher-than-
average number of sick calls from its baggage handlers.

Fig. 1.1 Plain text file

1.2.2   Text Formats

Various text formats exist in processing texts. The popular text formats are those
which are made by the software MS Office, such as MS Word file whose extension
is "doc," MS PowerPoint file whose extension is "ppt," and MS Excel file whose
extension is "xls." Texts tend to be transferred most popularly in the PDF format
through the Internet between users by uploading and downloading them. An e-mail
message may be viewed as a kind of text; the spam mail filtering is a typical task of
text mining. Therefore, we mention some typical text formats in this subsection.

The plain text which is mentioned as the simplest format is the text which is made
by the text editor which is presented in Fig. 1.1. Each text usually corresponds to its
own file whose extension is "txt." The plain text collection is set as the course of text
mining in implementing the text categorization and clustering system in Chaps. 7
and 11. The collection is called corpus and given as a directory in implementing
them. Exceptionally, a file may correspond to a corpus rather than a text in the news
collection, "Reuter21578" [85].

The XML (Extensive Markup Language) may be considered as another text
format as shown in Fig. 1.2. The XML is the more flexible format of web document
which is designed based on the HTML as the standard format in web browsers. The
XML is currently used as the standard form for describing textual data items as their
relational forms. Each field is given as its start and end tag, and its value is given
between them. Because it is possible to convert the XML documents into records
[86], they are called semi-structured data [53].

Let us consider the PDF (Portable Document Format) as the file format which is
independent of system configurations for presenting the document. A plain text or a
text in any format in MS Office is able to be converted into a PDF file. The PDF file
may include images as well as texts. It is possible to convert a PDF file into a plain
text by an application program. The PDF file has been evolved into the multimedia
composite containing sounds, videos, text, and images.

In the information retrieval system, each text is viewed as a record. The fields of
each text are title, author, abstract, publication time, full text, and so on. Text with

```
<Document>
  <Title> Text Mining </Title>
  <Author> Taeho Jo</Author>
  <Abstract>Text mining refers to the process of
     extracting important knowledge from textual collection.
  </Abstract>
  <Publication>2001</Publication>
</Document>
```

Fig. 1.2 XML document

their variable length may be given as values of the fields: title, abstract, and full text. We need to define the similarity between texts for processing such kinds of records. In this study, we assume that each item is a record with its single field, "full text," whose value is absolutely a text, for simplicity.

1.3   Data Mining Tasks

Even if we provided the brief description of data tasks in Sect. 1.1, we need to explore individual tasks for specifying them further, and this section is divided into the three subsections. In Sect. 1.3.1, we cover the classification and regression. In Sect. 1.3.2, we describe the clustering as another task in detail. We mention also the association, in Sect. 1.3.3. Therefore, this section is intended to specify the concept of data mining by describing its individual tasks.

1.3.1   Classification

Classification is defined as the process of assigning a category or some categories among the predefined ones to each data item as shown in Fig. 1.3. As the preliminary tasks for the classification, a list of categories is predefined as the classification system and data items are allocated to each category as the sample data. For the task, we consider two kinds of approaches: the rule-based approaches where symbolic rules are defined manually and each data item is classified by the rules, and the machine learning-based approaches where the classification capacity is constructed by the sample data and each item is classified by it. In this study, the rule-based approaches are excluded because of their limits: the poor flexibility and the prior knowledge requirements. Therefore, this subsection is intended to describe the classification in its functional view, the process of applying machine learning algorithms, and the evaluation schemes.

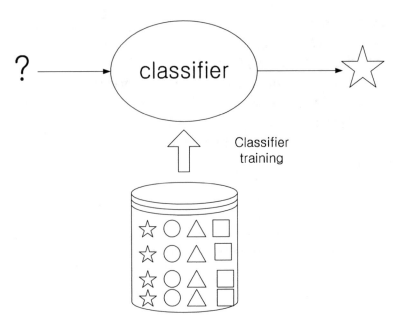

Fig. 1.3 Data classification

Classification is viewed as the black box whose input and output are a data item and its category, respectively, and let us explore types of classification. The hard classification refers to the classification where only one category is assigned to each data item, whereas the soft classification does to one where more than one category may be assigned to it [14]. In the flat classification, the categories are predefined as a single list, whereas in the hierarchical one, they are done as a hierarchical tree; nested categories exist in some categories [14]. The single viewed classification is regarded as one where only one classification system is predefined whether it is a flat or hierarchical one, whereas the multiple viewed classification is done as one where more than one classification system is predefined as same time. Before building the automatic classification system, it takes very much time for predefining categories and collecting sample data, depending on application areas [25].

Let us consider the steps of classifying data items by a machine learning algorithm. Categories are predefined as a list or a tree, and sample data is collected and allocated to each category, as mentioned above. By applying the machine learning algorithm to the sample data, the classification capacity is constructed which is given in the various forms: symbolic rules, mathematical equations, and probabilities [70]. By applying classification capacity, data items which are separated from the sample data are classified. Data items which are given subsequently as classification targets should be distinguished from the sample data items which are labeled manually and given in advance.

Let us consider the scheme of evaluating the results from classifying the data items. A test collection which consists of labeled data items is divided into the

Fig. 1.4 Data clustering

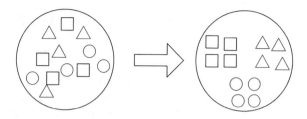

two sets: training set and test set. The training set is used as the sample data
for building the classification capacity using the machine learning algorithm. We
classify the data items in the test set and observe the differences between their
true and classified labels. The accuracy as the rate of consistently labeled items
to total and the F1 measure which will be described later in Chap. 8 are used as the
evaluation measures [95].

Regression is mentioned as another task which looks similar as the classification.
Regression refers to the process of estimating a continuous value or values by
analyzing the input data. Its difference from the classification is that the classification generates a discrete value or values as its output, while the regression does a
continuous one or ones. The nonlinear function approximation and the time series
prediction are typical examples of regression. There are two types of regression:
univariate regression where a single value is estimated and multivariate regression
where multiple values are done [29, 32].

1.3.2 Clustering

Clustering is defined as the process of segmenting a group of various items into
subgroups of similar ones as shown in Fig. 1.4. In the task, unlabeled data items
are given initially, and the similarity measures among them should be defined.
A group of items is segmented, depending on the similarities among them, into
subgroups. Let us assume that the supervised learning algorithms which are
described in Chap. 6 will be applied to the classification and the regression, whereas
the unsupervised ones which are described in Chap. 10 will be applied to it. In this
subsection, we will describe the clustering types, process, and evaluation.

Let us explore the types of clustering depending on views. The hard clustering
is the clustering where each item is arranged into only one cluster, whereas the
soft one is the task where each item may be arranged into more than one cluster
[14]. The flat clustering is the clustering where clusters are made as a single list,
whereas the hierarchical clustering is the task where they are made as a tree; nested
clusters in a cluster exist in the hierarchical one but do not in the flat one [14]. The
single viewed clustering means the task where only group of clusters is made as the
results, whereas the multiple viewed clustering does the task where several groups
of clusters may be constructed. The clustering is a very expensive computation; it
takes the quadratic complexity to the number of items [70].

Let us consider the process of clustering data items by the unsupervised machine learning algorithms. Initially, a group of unlabeled data items is given and the unsupervised learning algorithm as the clustering tool is decided. We define the scheme of computing the similarities among data items and parameter configurations depending on clustering algorithms. By running the unsupervised learning algorithm, the group of data items is segmented into subgroups of similar ones. The clustering may automate the preliminary tasks for the classification by predefining the categories as a list or a tree of clusters and arranging data items into them as the sample data [25].

The desired direction of clustering data items is to maximize the similarities among items within each cluster and to minimize the similarities among clusters [25, 44]. The value which averages the similarities among items within each cluster is called the intra-cluster similarity or the cohesion. The value which averages the similarities among clusters is called intercluster similarity, and reversed into the discrimination among clusters. Hence, the maximization of both the cohesion and the discrimination is the direction of clustering data items. The minimum requirement for implementing the clustering systems is that the cohesion and the discrimination are higher than results from clustering data items at random.

Let us consider the differences between the clustering and the classification. Classification requires the preliminary tasks, the category predefinition, and the sample data collection, whereas the clustering does not. Clustering is the task to which the unsupervised learning algorithms are applied as the approaches, whereas the classification is one to which the supervised ones are done. In the evaluation, classification requires the division of the test collection into the training set and the test set, whereas the clustering does not. The automatic data management system may be implemented by combining the clustering and the classification with each other, rather than leaving them separated tasks.

1.3.3   Association

Association is defined as the process of extracting the association rules in the form of if-then, as shown in Fig. 1.5. The association is initially intended to analyze the purchase trends of customers in big marts such Wall Mart; for example, it is intended to discover that if a customer buys a beers, he or she does also diapers. The groups of items called item sets are given as the input of this task and a list of association rules which are given as if-then ones is generated as the output. The data association should be distinguished from the clustering in that the similarity between two items is bidirectional in clustering, whereas if-then from item to another is unidirectional in the association. Hence, in this subsection, we describe the association with respect to its functional view, process, and evaluation.

The data association which was the initial data mining task was intended to get the purchase trends of customers. The collection of purchased items consists of item sets bought by customers; each item set is a list of ones which are purchased

Fig. 1.5 Data association

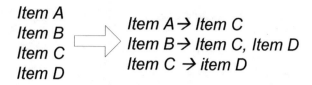

by the customer as a single transaction. For each item, the item sets including it are selected and association rules are extracted from the selected sets. The association rule is given as if-then rule symbolically; notate 'if A then B by $A \rightarrow B$. However, $A \rightarrow B$ is not always same to $B \rightarrow A$ in the association rules.

Let us describe briefly the process of extracting association rules from the item sets by Apriori algorithm. In the initial step, all possible items and item subsets which have a single item are generated as many as items from the list of item sets. For each item, the item subsets are expanded by adding itself, depending on their frequencies. The association rule is generated for each item subset, depending on its relative frequency [91]. The process of data association will be described in detail in Chap. 4.

Let us explain the important measures, support, and confidence, for doing the data association [91]. Support refers to the rate of item sets including the item subset to all; in $A \rightarrow B$, it is the ratio of the item sets including both A and B to all item sets [91]. Confidence refers to the rate of item sets including the item subset to the item sets including the conditional item; in $A \rightarrow B$, it is the ratio of item sets including A and B to the item sets including A [91]. In the process of data association, the support is used for expanding the item subsets, and the confidence is used for generating the association rule from each item subset. In comparing the support with the confidence, the numerator is the same in both, but the denominators are different.

Even if the association and the clustering look similar as each other, they should be distinguished from each other. In the clustering, a single group of unlabeled items is given as the input, whereas in the association, item sets are given as the input. The similarity between two items is the important measures for performing the clustering, whereas the support and the confidence are ones for doing the association. The clustering generates subgroups of similar items as results, whereas the association does a list of associations rules. The clustering may provide the subgroups of items which are able to become item sets as input to the data association in connecting both tasks with each other.

1.4   Data Mining Types

This section is concerned with the types of data mining, depending on the source, and it is divided into the three subsections. In Sect. 1.4.1, we mention relational data mining as the traditional type. In Sect. 1.4.2, we describe web mining as

ID	Authors	Title	Year
1	Taeho Jo	Text Mining	1999
2	Taeho Jo	Neural Networks	2000
3	Taeho Jo	Machine Learning	2001
4	Taeho Jo	Information Retrieval	2002

Fig. 1.6 Relational data

the expansion of text mining. In Sect. 1.4.3, we also consider big data mining as the recently challenging topic. Therefore, this section is intended to explore the three types of data mining for providing the background for understanding the text mining.

1.4.1   Relational Data Mining

This subsection is concerned with relational data mining as the traditional type. It is assumed that the collection of data items is given as a table or tables. The typical tasks of relational data mining are classification, clustering, and regression which are described in Sect. 1.2. The conversion of relational data items into numerical vectors as the preprocessing for using the machine learning algorithms is trivial in this type of data mining. Therefore, in this subsection, we describe the relational data mining tasks, and the preprocessing.

Relational data refers to the list of records in the tabular form shown in Fig. 1.6. Before making the relational data, the fields and their types are defined. Each field has its own value as the basic unit, and each record consists of a fixed number of field values. A collection of records is given as a table, and a table or tables are given depending on the given domain. In the relational data mining, it is assumed that a table or tables are given in the database independently of application programs [11].

The typical tasks for the relation data mining are classification, regression, and clustering, and it is assumed that a single table is given as the source. In the classification, the categories are defined in advance and records are classified into one or some among them. In the regression, an important value or values are estimated by analyzing attribute values of the given record. The records in the given table are organized into several tables of similar records in the clustering. Before doing the main data mining tasks, we need to clean data items by removing their noises, depending on the application area.

Because the different scales, string values, and noises always exist in raw data, we need to clean them for improving the performance [57]. All field values need to be normalized between zero and one using their maximum and minimum against the different scales over attributes. Each string value should correspond to its own numerical value; it is called data coding. Smoothing over attribute values and removal of very strange values need to be considered against their noises. Since a record looks similar as a numerical vector, the preprocessing which is required for the relational data mining tasks is very simple.

Let us consider the differences between relational data mining and text mining. The relational data items, a table or tables, are given as the source for relational data mining, whereas a text collection, which is called corpus, is given in text mining. The preprocessing, such as normalization, smoothing, and coding, in relational data mining is simpler than those in text mining such as text indexing and text encoding which will be covered in Chaps. 2 and 3. In relational data mining, the performance depends on only approaches, whereas in the text mining, it depends on both preprocessing schemes and approaches. The case where a list of XML documents is converted into that of records may be regarded as the mixture of both relational data mining and text mining.

1.4.2   Web Mining

Web mining refers to the special type of data mining where web data is given as the source. The web document is the data item which is expanded from a text by adding URL address, the hyperlinked words, and access logs. Web mining is divided into web content mining, web structured mining, and web usage mining [66]. The web document which is given as the source code in HTML as well as text, and the access logs to web documents are given as the source in this type of data mining. Therefore, in this subsection, we describe briefly the three types of web mining.

The web content mining refers to the type of web mining where contents of web documents are referred for the mining tasks [66]. Each web document is classified into its topic or topics as a task of web content mining. Web documents may be organized into subgroups of content-based similar ones as the clustering. The automatic summarization of web documents for the brief view needs to be considered as another task of web content mining. The web content mining looks almost the same to text mining, but fonts and colors of characters as well as words may be considered as its difference from the text mining.

The web structure mining refers to the type of web mining where structures or templates of web documents are considered for the mining tasks [66]. Unlike the text mining and the web contents mining, the categories are predefined by the templates or structures which are given as tag configurations, and web documents are collected as sample data, following them. We need the techniques of encoding tag configurations of web documents into numerical vectors, ignoring contents. Web documents are classified and organized by their templates and structures. Two web documents with their different contents but their similar structures are regarded as similar ones in this type of web mining.

The web usage mining refers to the type of web mining where the trends of accessing web documents are considered [66]. The trend of accessing web documents is represented into a directed graph where each node indicates its corresponding URL address and each edge does the access route from one URL address to another. The access behaviors of users within a fixed period may be given as source. We need the techniques of encoding graphs into numerical vectors or modifying existing machine learning algorithms into their graph-based versions for doing the tasks of web usage mining. The web documents are given as the source in the previous two types, whereas the access routes within a period are given as source in this type.

The text mining and the web usage mining are different clearly from each other, but it may look similar as the web contents mining and web structure mining. In the text mining, data items are classified and organized purely by their contents, whereas in web mining, they are done by templates and access trends as well as contents. In the web structure mining, items are classified by templates with regardless of contents as mentioned above. In the text mining, words and their posting information and grammatical ones are used as features or attributes, whereas in web mining, tags, hyperlinked, and other properties of characters in addition are used. However, the techniques of text mining are necessary for doing the web mining tasks as their core parts, especially in the web contents mining.

1.4.3   Big Data Mining

Big data mining is introduced recently as the challenging type of data mining. Big data is characterized as its variety, velocity, variability, and veracity, as well as its big volume [97]. Additional media such as mobile phones, sensors, and other ubiquitous equipment are to PCs exist for collecting data items; it causes to generate big data. The traditional algorithms of processing data items and PC-based software face the limits in processing big data. Therefore, this subsection is intended to characterize big data, present the three layers of big data mining, and mention the directions of developing the techniques.

As mentioned above, big data is characterized as follows:

- **Variety**: Big data consists of data items in their various formats, so the preprocessing is very complicated compared with relational data mining and web mining.
- **Velocity**: In big data, data items are updated very frequently; the very big volume of data items is added and deleted within a time.
- **Variability**: In big data, data items have very much inconsistency and noise, so it takes too much time for cleaning them.
- **Veracity**: In big data, data items are discriminated very much by their quality, and some data items are very reliable but others are very poor; the reliable data items are not known.

Fig. 1.7 Three tiers of big
data mining

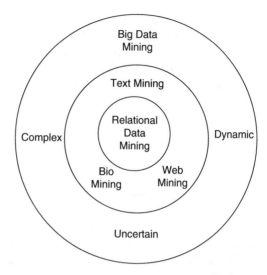

The three layers of big data mining may be considered as presented in Fig. 1.7. The techniques for relational data mining are shown in the inner circle of Fig. 1.7, as the basic ones. In the medium circle, the techniques for text mining, web mining, and bio mining, are shown. In the outer circle, we consider the techniques of big data mining where the data which has very complex structures, very frequent updates, and incomplete values is given as the source. The techniques for the relational data mining become the basis for developing ones for text mining, web mining, and bio mining; ones for the three types of data mining are used as the basis for developing the techniques of big data mining.

Because the individual techniques of encoding raw data into structured forms and classifying or organizing items which are used in the three kinds of data mining are not feasible to the big data mining tasks, we need to consider the directions for developing them. As the simplest one, we may mention the fusion of techniques which were developed for existing types of data mining into ones for processing data items in various formats. In applying the existing machine learning algorithms, we consider more cases such as the frequent updates of data items by the deletion and the insertion and discriminations among data items. We need to develop new machine learning algorithms, modify existing ones, or organize multiple machine learning algorithms for making them feasible to big data mining. The directions which are mentioned above become the challenges against big data mining.

Let us consider the relation of text mining with big data mining which is covered in this study. In text mining, textual data is given as the source in the uniform format, whereas in big data mining, various formats of data items including unfamiliar formats may be given. In text mining, it is assumed that the data items are fixed or updated not frequently, whereas in big data mining it is assumed that they are updated very frequently or constantly. In text mining, the data items are collected through the Internet, whereas in big data mining, they are collected by ubiquitous media such as sensors, RFID tags, and mobile phones. Even if the messages which are transferred among mobile phones belong to big data, they become the source of text mining, as well as texts provided by the Internet.

1.5   Summary

This section provides the overall summary of this chapter. Words, sentences, and paragraphs are regarded as the text components and the typical text formats are plain texts, XML documents, and PDF files. We studied classification, clustering, and association as the specific tasks of data mining. We mentioned the relational data mining as the basis for text mining, web mining as the subsequent area from text mining, and big data mining as the new challenging area. This section is intended to summarize the entire contents of this chapter.

Text mining refers to the process of extracting knowledge which is necessary directly for making the important decision from the textual data. Text mining is the special type of data mining and should be distinguished from the traditional data mining where knowledge is extracted from relational data items. The text classification, the text clustering, and the text summarization which will be described in the subsequent parts are the typical tasks of text mining. We adopted the machine learning algorithms as the approaches to the text mining tasks and describe them in Chaps. 6 and 10. Text mining should be also distinguished from the information retrieval where texts relevant to the query are retrieved.

The scope of text is restricted to what is written in a natural language; what is written in an artificial language such as source codes is excluded from the scope. The words are the basic semantic units of text and need to be distinguished from the string which is a simple combination of characters, ignoring its meaning. A text which refers to an article is decomposed into paragraphs, each paragraph is a semantically ordered set of sentences, and each sentence is a combination of words by the grammatical rules. The typical formats of text are the plain text which is made by a simple editor, the XML web document as the semi-structured data, and the PDF file as the standard format of documents. Text formats are plain texts in implementing the text mining systems in Chaps. 7 and 11, and the text collection is given as a directory.

We explored some data mining tasks: the classification, the clustering, and the association. The classification refers to the process of assigning a category or categories among the predefined ones to each data item, and the regression refers to that of estimating a continuous value or values by analyzing input values. The clustering refers to the process of segmenting a group of data items into subgroups of similar ones by computing their similarities. The association is the task of extracting the association rules which are given in the if-then form from a list of item sets. In this study, the supervised learning algorithms will be applied to the classification and the regression, and the unsupervised ones will be done to the clustering.

We explored other types of data mining: relational data mining, web mining, and big data mining. Relational data mining refers to the type of data mining where the relational data items are given as the source from which the implicit knowledge is extracted. Web mining is the type of data mining where the web documents and access logs are given as the source, and it is divided into the web contents mining, web structure mining, and web usage mining. Big data mining is the challenging area where we need to develop techniques by the three directions: fusion of existing approaches, consideration of dynamic status of data items such as frequent updates and various qualities, and proposal of approaches more suitable for big data mining through the modification and organizational combinations. This book on the text mining provides the techniques which may be advanced to ones for big data mining.

Chapter 2
Text Indexing

This chapter is concerned with text indexing which is the initial step of processing texts. In Sect. 2.1, we present the overview of text indexing, and explain it functionally step by step in Sect. 2.2. In Sect. 2.3, we present and explain the implementation of text indexing in Java. In Sect. 2.4, we cover the further steps for reinforcing text indexing with respect to its performance and efficiency. Therefore, this chapter is intended to study text indexing as the initial step.

2.1 Overview of Text Indexing

Text indexing is defined as the process of converting a text or texts into a list of words [54]. Since a text or texts are given as unstructured forms by itself or themselves essentially, it is almost impossible to process its raw form directly by using a computer program. In other words, the text indexing means the process of segmenting a text which consists of sentences into included words. A list of words is the result from indexing a text as the output of text indexing, and will become the input to the text representation which is covered in Chap. 3. Therefore, in this section, we provide the overview of text indexing before discussing it in detail.

Let us consider the necessities of text indexing [82]. Text is essentially the unstructured data unlikely the numerical one, so computer programs cannot process it in its raw form. It is impossible to apply numerical operations to texts and is not easy to encode a text into its own numerical value. A text is a too long string which is different from a short string, so it is very difficult to give text its own categorical value. Therefore, what mentioned above becomes the reason for the need to segment a text into words which are short strings as the process of text indexing.

The three basic steps of text indexing are illustrated in Fig. 2.1. The first step is tokenization which is the process of segmenting a text by white spaces or punctuation marks into tokens. The second step is stemming which is the process

© Springer International Publishing AG, part of Springer Nature 2019
T. Jo, *Text Mining*, Studies in Big Data 45,
https://doi.org/10.1007/978-3-319-91815-0_2

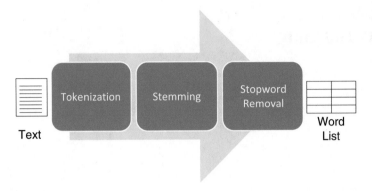

Fig. 2.1 The three steps of text indexing

of converting each token into its own root form using grammar rules. The last step is the stop-word removal which is the process of removing the grammatical words such as articles, conjunctions, and prepositions. Tokenization is the prerequisite for the next two steps, but the stemming and the stop-word removal may be swapped with each other, depending on the given situation.

The additional steps may be considered for improving the performance and the efficiency of the text mining and information retrieval tasks. We may attach the POS (Position Of Speech) tagging which is the process of classifying words into one of nouns, verbs, adjectives, and so on, grammatically to text indexing as an additional module. The weights of words may be computed as their importance degree, and words with relatively lower weights may be removed after the three steps, which is called index filtering. We take external words which are relevant to the strongly weighted words and, using search engines, add them to the results from indexing a text, which is called index expansion. We may consider the case of applying both of them, altogether, which is called index optimization.

The occurrences and weights of words may depend strongly on the text lengths; the overestimation and underestimation may be caused to long texts and short ones, respectively. We may reject too short text which consists of only their titles, and select only part in long texts as the text length normalization. We may use the relative frequencies or weights, instead of absolute ones, by dividing the absolute ones by their maximum. We consider applying the index filtering to long texts and index expansion to short ones. Without the above actions, long texts may retrieve more frequently to given query in the information retrieval tasks, and have strong influences in text mining tasks.

2.2 Steps of Text Indexing

This section is concerned with the three basic steps of text indexing. In Sect. 2.2.1, we explain tokenization as the first step, together with its examples. In Sects. 2.2.2 and 2.2.3, we study stemming and the stop-word removal, respectively. In Sect. 2.2.4, we cover schemes of computing weights of words. Therefore, this section is intended to study the process of indexing a text or texts with the three basic steps and computing the word weights.

2.2.1 Tokenization

Tokenization is defined as the process of segmenting a text or texts into tokens by the white space or punctuation marks. It is able to apply the tokenization to the source codes in C, C++, and Java [1], as well as the texts which are written in a natural language. However, the scope is restricted to only text in this study, in spite of the possibility. The morphological analysis is required for tokenizing texts which are written in oriental languages: Chinese, Japanese, and Korean. So, here, omitting the morphological analysis, we explain the process of tokenizing texts which are written in English.

The functional view of tokenization is illustrated in Fig. 2.2. A text is given as the input, and the list of tokens is generated as the output in the process. The text is segmented into tokens by the white space or punctuation marks. As the subsequent processing, the words which include special characters or numerical values are removed, and the tokens are changed into their lowercase characters. The list of tokens becomes the input of the next steps of text indexing: the stemming or the stop-word removal.

The process of tokenizing a text is shown in Fig. 2.3. The given text is partitioned into tokens by the white space, punctuation marks, and special characters. The words which include one or some of special characters, such as "16%," are removed. The first character of each sentence is given as the uppercase character, so it should be changed into the lowercase. Redundant words should be removed after the steps of text indexing.

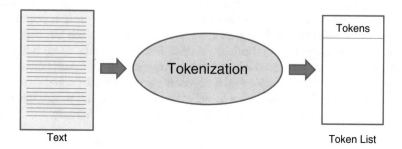

Fig. 2.2 Functional view of tokenization

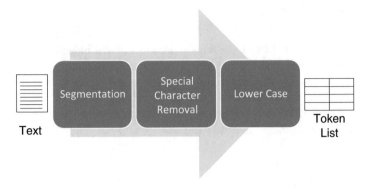

Fig. 2.3 The process of tokenizing text

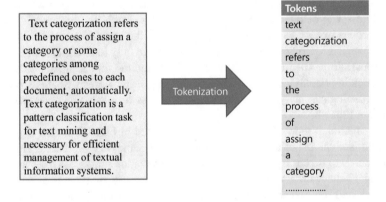

Fig. 2.4 The example of tokenizing text

The example of tokenizing a text is illustrated in Fig. 2.4. In the example, the text consists of the two sentences. Text is segmented into tokens by the white space, as illustrated in the right side in Fig. 2.4. The first word of the two sentences, "Text," is converted into "text," by the third tokenization step. In the example, all of words have no special character, so the second step is passed.

Let us consider the process of tokenizing the text written in one of the oriental languages, Chinese or Japanese. The two languages do not allow the white space in their sentences, so it is impossible to tokenize the text by it. It is required to develop and attach the morphological analyzer which segments the text depending on the grammatical rules for tokenizing the text. Even if the white space is allowed in the Korean texts, tokens which are results from segmenting each of them are not incomplete, so the morphological analyzer is also required for doing the further one. Since English is currently used as the global language in our world, the scope is restricted to only the texts which are written in English, in this study.

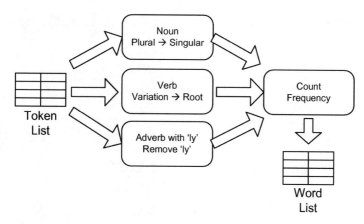

Fig. 2.5 Stemming process

2.2.2   Stemming

Stemming refers to the process of mapping each token which is generated from the previous step into its own root form [54]. The stemming rules which are the association rules of tokens with their own root form are required for implementing it. Stemming is usually applicable to nouns, verbs, and adjectives, as shown in Fig. 2.5. The list of root forms is generated as the output of this step. Therefore, in this subsection, we describe the stemming which is the second or third step of text indexing.

In the stemming, the nouns which are given in their plural form are converted into their singular form, as shown in Fig. 2.5. In order to convert it into its singular form, the character, "s," is removed from the noun in the regular cases. However, we need to consider some exceptional cases in the stemming process; for some nouns which end with the character, "s," such as "process," the post fix, "es," should be removed instead of "s," and the plural and singular forms are completely different from each other like the case of words, "child" and "children." Before stemming nouns, we need to classify words into nouns or not by the POS tagging. Here, we use the association rules of each noun with its own plural forms for implementing the stemming.

Let us consider the case of stemming verbs into their root forms. The verbs which are given in their third-person singular form need to be converted into their root forms by removing the postfix, "s" or "es." The verbs which are given as the noun or used as the present progressive form may be changed into their root forms by removing the post fix, "ing." If the verbs are used as the past tense or particle, they are converted into their root form by removing the postfix, "ed," in the regular cases. Note that the irregular verbs where their root and varied forms are different completely exist as exceptional cases.

Fig. 2.6 Hard vs soft
stemming

Varied Form	Root Form
better	good
best	good
simpler	simple
simplest	simple
assigning	assign
assigned	assign
assignment	assign
complexity	complex
analysis	analyze
categorization	categorize
categorizing	categorize
categorizes	categorize

Let us consider the case of stemming adjectives into their root forms. If the adjective is given as the adverb, it is changed into its root form by removing the postfix, "ly." If the adjectives are given as their comparative degree, the postfix, "er," is removed for taking its root form. If the adjectives are expressed as their superlative degree, the postfix, "est," is removed in each of them. However, the majority of adjectives are expressed as their comparative and superlative degree by adding the words, "more" or "most," before them, respectively.

The two versions of stemming rules are illustrated in Fig. 2.6. Nouns may be derived from verbs by adding the postfixes, such as "ation" and "ment." The two words, "assignment" and "categorization," are the examples of this case. The noun, "complexity," is derived from the adjective, "complex." In the soft stemming, the words are used as root forms by themselves, whereas in the hard stemming, they are converted into their verbs or adjectives.

2.2.3 Stop-Word Removal

Stop-word removal refers to the process of removing stop words from the list of tokens or stemmed words [54]. Stop words are the grammatical words which are irrelevant to text contents, so they need to be removed for more efficiency. The stop-word list is loaded from a file, and if they are registered in the list, they are removed. The stemming and the stop-word removal may be swapped; stop words are removed before stemming the tokens. Therefore, in this subsection, we provide the detailed description of stop words and the stop-word removal.

The stop word refers to the word which functions only grammatically and is irrelevant to the given text contents. Prepositions, such as "in," "on," "to"; and so on, typically belong to the stop-word group. Conjunctions such as "and," "or," "but," and "however" also belong to the group. The definite article, "the," and the infinite articles, "a" and "an," are also more frequent stop words. The stop words occur dominantly in all texts in the collection; removing them causes to improve very much efficiency in processing texts.

Let us explain the process of removing stop words in the ing. The stop-word list is prepared as a file, and is loaded from it. For each word, if it is registered in the list, it is removed. The remaining words after removing stop words are usually nouns, verbs, and adjectives. Instead of loading the stop-word list from a file, we may consider using the classifier which decides whether each word is a stop word or not.

Some nouns, verbs, and adjectives as the remaining words may occur in other texts as well as in the current one. The words are called common words, and they are not useful for identifying text contents, so they need to be removed like stop words. The TF-IDF (Term Frequency Inverse Term Frequency) weight is the criteria for deciding whether each word is a common word, or not, and will be explained in Sect. 2.2.4. The remaining words after removing stop words are useful in the news article collection, but not all of them are so in the technical one medical text collections. We may consider the further removal, depending on text lengths; it is applied to long texts.

Let us consider the sentimental analysis as the kind of text classification; it is the process of classifying texts into one of the three categories: positive, neural, and negative. We may need some of the stop words for doing the task, as exceptional cases. For example, the stop words, such as "but," "not," and "however," indicate the negative opinion. Not all of the stop words are useful for identifying the sentimental style of texts; we need to decide whether each stop word is useful, or not. If implementing the text processing system for doing the sentimental analysis, the process of text indexing needs to be modified.

2.2.4 Term Weighting

The term weighting refers to the process of calculating and assigning the weight of each word as its importance degree. We may need the process for removing words further as well as stop words for further efficiency. The term frequency and the TF-IDF weight are the popular schemes of weighting words, so they will be described formally, showing their mathematical equations. The term weights may be used as attribute values in encoding texts into numerical vectors, which will be covered in the next chapter. Therefore, in this subsection, we describe the schemes of weighing words: TF-IDF and its variants.

We may use the term frequency which is the occurrence of each word in the given text as the scheme of weighting words [63]. It is assumed that the stop words which occur most frequently in texts are completely removed in advance. The words are weighted by counting their occurrences in the given text in this scheme. There are two kinds of term frequency: the absolute term frequency as the occurrence of words and the relative term frequency as the ratio of their occurrences to the maximal ones. The relative term frequency is preferred to the absolute one, in order to avoid the overestimation and underestimation by the text lengths.

TF-IDF (Term Frequency-Inverse Document Frequency) is the most popular scheme of weighting words in the area of information retrieval but requires the references to the entire text collection which is called corpus. The word weights which are computed by the TF-IDF scheme are proportional to the occurrences in the given text, but reversely proportional to that in other texts. The TF-IDF weight, w_{ij}, of the word, t_i, in the text, d_i, is computed by Eq. (2.1) [83],

$$w_{ij} = log \frac{N}{DF_i}(1 + logTF_i) \tag{2.1}$$

where N is the total number of texts in the collection, DF_i is the number of texts which include the word, t_i, in the collection, and TF_i is the occurrence of the word, t_i, in the text, d_i. If the word occurs only one time in the text, its weight depends on the number of texts which include itself in the corpus, assuming that N is a constant. In encoding a text into a numerical vector, zero is assigned to the word which does not occur at all in the text.

We may derive some variants from the TF-IDF weight which is calculated by Eq. (2.1) [4]. By replacing the absolute term frequency by the relative one, Eq. (2.1) is modified into Eq. (2.2).

$$w_{ij} = log \frac{N}{DF_i}\left(1 + log \frac{TF_i}{TF_{max}}\right) \tag{2.2}$$

where TF_{max} is the maximal occurrences in the text, d_i. By modifying the inverse document frequency, $log \frac{N}{DF_i}$ into $1 + log \frac{N}{DF_i}$, the word weight is computed by Eq. (2.3),

$$w_{ij} = \left(1 + log \frac{N}{DF_i}\right)\left(1 + log \frac{TF_i}{TF_{max}}\right) \tag{2.3}$$

We may use $log \frac{N-DF_i}{DF_i}$ as the probabilistic inverse document frequency, instead of $log \frac{N}{DF_i}$ and $1 + log \frac{N}{DF_i}$, so the words are weighted by Eq. (2.4),

$$w_{ij} = \left(1 + log \frac{N - DF_i}{DF_i}\right)\left(1 + log \frac{TF_i}{TF_{max}}\right) \tag{2.4}$$

In implementing text indexing, we adopt Eq. (2.3) for computing the word weights.

In addition to text indexing and the term weighting, we may need the POS tagging. As mentioned in the previous section, the POS tagging refers to the process of classifying each word into one of the grammatical categories. Since the words around the given word are indicators, HMM (Hidden Markov Model) tends to be used as the approach [53]. We may apply the POS tagging to the stemming, the stop-word removal, and the term weighting as the optional step. However, the POS tagging is omitted in implementing the text indexing program, in Sect. 2.3.

2.3 Text Indexing: Implementation

This section is concerned with the process of implementing the text indexing program in Java. In Sect. 2.3.1, we define the classes which are involved in implementing the text indexing program. In Sect. 2.3.2, we mention the stemming rule for implementing the stemming process as the data structure. In Sect. 2.3.3, we implement the methods of the classes. Therefore, this section is intended to provide the guide for implementing text indexing.

2.3.1 Class Definition

In this subsection, we present the classes which are involved in implementing the text indexing program. Since the file access is the most frequent task in indexing a text, we need the class for doing it. Since a text is given as input and the word list is generated as the output, we need to define the classes for processing a text and words. We need to define the integrated class which is called API (Application Program Interface) class. In this subsection, we will explain the each class definition.

In Fig. 2.7, the class, "FileString," is defined. The property, "fileName," indicates the name of the file to which we try to access, and the property, "fileString," indicates the stream which is loaded and saved from and to the file; the object of the class "FileString" is created with the file name or with the file name and the file stream. The method, "loadFileString," loads the contents from the file as a string and assigns it to the property, "filesString." The method, "saveFileString" saves the property value, "fileString," into the file which is named by the property, "fileName." The method, "getFileString," gets the value which is assigned to the property, "fileString," and the method, "setFileString," mutates it.

The class, "Word," is defined in Fig. 2.8. The property, "wordName," indicates the word by itself, and the property, "wordWeight," indicates the importance degree, in the class, "Word." The object of class, "Word," is created by the value of the property, "wordName," and the property values are accessed and mutated by the methods whose prefix is "get" and "set," respectively. The method, "computeWordWeight," computes the word weight and assigns it to the property, "wordWeight." The weight is computed by Eq. (2.3), and its implementation is omitted in this book.

```
import java.io.*;
public class FileString {
    String fileName;
    String fileString;
    public FileString(String fileName){
        this.fileName = fileName;
        this.fileString = "";
    }
    public FileString(String fileName, String fileString){
        this.fileName = fileName;
        this.fileString = fileString;
    }
    public String getFileString(){
        return this.fileString;
    }
    public void setFileString(String fileString){
        this.fileString = fileString;
    }
    public void loadFileString(){}
    public void saveFileString(){}
}
```

Fig. 2.7   The definition of class: FileString

```
public class Word {
    String wordName;
    double wordWeight;

    public Word(String wordName){
        this.wordName = wordName;
        this.wordWeight = 0.0;
    }

    public String getWordName(){
        return this.wordName;
    }

    public double getWordWeight(){
        return this.wordWeight;
    }

    public void setWordWeight(double wordWeight){
        this.wordWeight = wordWeight;
    }

    public void computeWordWeight(String corpusPath,String fullText){}
}
```

Fig. 2.8   The definition of class: word

Fig. 2.9 The definition of
class: text

```
import java.util.*;
public class Text {
    String fileName;
    String textContent;
    Vector wordList;

    public Text(String fileName){
        this.fileName = fileName;
        this.textContent = "";
        this.wordList = new Vector();
    }

    public String getFileName(){
        return this.fileName;
    }

    public String getTextContent(){
        return this.textContent;
    }

    public Vector getWordList(){
        return this.wordList;
    }

    public void loadTextContent(){}
    private void tokenizeTextContent(){}
    private void stemWordList(){}
    private void removeStopWords(){}
    public void indexTextContent(){}

}
```

The class, "Text," is defined in Fig. 2.9. Each text is identified by its own file name by assuming that each text is given as a single file; the property, "fileName," becomes the text identifier. The property, "textContent," is the input of indexing the text, and the property, "wordList," is the output which consists of objects of the class, "Word." By the method, "loadTextContent," the text is loaded from the file, and it is indexed into the list of words by the method, "indexTextContent." The three methods whose access level is set "private" never were called externally but called in the method, "indexTextContent"; we will explain their implementations in Sect. 2.3.3.

The class, "TextIndexAPI," is defined in Fig. 2.10. The postfix, "API," which is attached to "TextIndex," indicates the interface between the main program and the involved classes. The object of the class is created with the file name and the directory path which identifies the corpus location, and the method, "indexText," is involved for indexing the text into the list of objects of the class, "Word," in the main program. The list of objects is converted into a stream for saving it into a file as the results. The final output of this program is given as a file where the list of words and their weights are saved.

```
import java.util.*;
public class TextIndexAPI {
    Text indivText;
    String corpusPath;

    public TextIndexAPI(String fileName,String corpusPath){
        this.indivText = new Text(fileName);
        this.corpusPath = corpusPath;
    }

    public Vector indexText(){
        return null;
    }
}
```

Fig. 2.10 The definition of class: TextIndexAPI

```
import java.util.*;

public class StemmingRule {
    String rootForm;
    Vector variedFormList;

    public StemmingRule(String line){
        String[] tokenList = line.split(" ");
        this.rootForm = tokenList[0];
        this.variedFormList = new Vector();
        for(int i = 1;i < tokenList.length;i++)
            this.variedFormList.addElement(tokenList[i]);
    }

    public boolean isRegistered(Word indivWord){
        return true;
    }
}
```

Fig. 2.11 The definition of class: StemmingRule

2.3.2   Stemming Rule

This subsection is concerned with the stemming rule as the data structure for implementing the stemming process. The stemming rule is given as the association of the root form with its derivations, for each word. We add one more class, "StemmingRule," to those which are mentioned in Sect. 2.3.1. We need to load the list of stemming rules from a file, initially. In this subsection, we define the class, "StemmingRule," and implement the involved methods.

The class, "StemmingRule," is defined in Fig. 2.11. The property, "rootForm," indicates the root of each word, and the property, "variedFormList," indicates the list of varied forms. It is assumed that in the file, the root form and its varied forms are

```
private Vector loadStemmingRuleList(String stemmingRuleFileName){
    Vector stemmingRuleList = new Vector();
    FileString fs = new FileString(stemmingRuleFileName);
    fs.loadFileString();
    String stemmingRuleStream = fs.getFileString();
    String[] lineList = stemmingRuleStream.split("\n");
    for(int i = 0;i < lineList.length ;i++){
        StemmingRule rule = new StemmingRule(lineList[i]);
        stemmingRuleList.addElement(rule);
    }
    return stemmingRuleList;
}

private String changeRootForm(Word indivWord,Vector stemmingRuleList){
    String wordName = indivWord.getWordName();
    int size = stemmingRuleList.size();
    for(int i = 0;i < size;i++){
        StemmingRule rule = (StemmingRule)stemmingRuleList.elementAt(i);
        if(rule.isRegistered(indivWord)){
            String rootForm = rule.getRootForm();
            return rootForm;
        }
    }
    return indivWord.getWordName();
}

private void stemWordList(String stemmingRuleFileName){
    Vector stemmingRuleList = this.loadStemmingRuleList(stemmingRuleFileNam
    int size = this.wordList.size();
    for(int i = 0;i < size;i++){
        Word indivWord = (Word)this.wordList.elementAt(i);
        indivWord.setWordName(this.changeRootForm(indivWord,stemmingRuleLis
        this.wordList.setElementAt(indivWord, i);
    }
}
```

Fig. 2.12   The implementation of method: stemWordList

given as a line: "root varied1 varied2 . . . ." An object is created line by line, reading the stream from the file, the line is segmented into tokens by the white space, and the first token is set as the root and the others are set as its varied forms. The method, "isRegistered," checks whether the word is registered in the list of varied forms, or not.

The method, "stemString," of the class, "Text," is implemented in Fig. 2.12. The list of stemming rules is loaded from the file, by calling the method, "loadStemmingRuleList." The objects of the class, "StemmingRule," are created line by line in the method, "loadStemmingRule." Each word is checked in each stemming rule whether it is registered as its varied form, or not, and if so, the word is changed into its root form, by getting it from the corresponding stemming rule. Redundant words are removed after stemming words.

```
public boolean isRegistered(Word indivWord){
    String wordName = indivWord.getWordName();
    int size = this.variedFormList.size();
    for(int i = 0;i < size;i++){
        String variedForm = (String)this.variedFormList.elementAt(i);
        if(wordName == variedForm)
            return true;
    }
    return false;
}
```

Fig. 2.13 The implementation of method: isRegistered

The method, "isRegistered," is implemented in Fig. 2.11. The method belongs to the predicate method which has the prefix "is," and checks the program status by returning true or false. If discovering the varied form which matches the value which is assigned to the property, "wordName," it returns true. The sequential search is adopted in the current implementation, but it needs to be replaced by the more advanced search strategies such as the binary search and the hashing, in the next version. When using the sequential search, it takes the quadratic complexity, $O(n^2)$, for stemming each token, assuming the n stemming rules and n varied forms per each stemming rule (Fig. 2.13).

Since it takes very high cost for stemming a word, as mentioned in Fig. 2.13, we need to improve the search for varied forms in each stemming rule and use one of more advanced search strategies. We need to sort the varied forms in each stemming rule, and replace the sequential search by the binary search. In each stemming rule, we may build a binary search tree or its variants such as RVL tree and Red-Black Tree. As an alternative strategy, we implement the search for the varied forms by the hashing which takes only constant complexity. We also need to improve the search for matching the stemming rule as well as the search for the varied forms.

2.3.3   Method Implementations

This section is concerned with implementing methods in the classes which were mentioned in Sect. 2.3.1. We already described the definitions and methods of the class, "StemmingRule," in Sect. 2.3.2. This subsection focuses on the two methods, "loadFileString" and "saveFileString" in the class, "FileString," and the three methods, "tokenizeString," "removeStopWords," and "indexTextContent" in the class, "Text." Since the class, "Word," has the methods with their trivial implementations, we omit the explanation about them. Therefore, in this subsection, we explain the method implementations of the classes, "FileString" and "Text."

We present implementing the methods, "loadFileString" and "saveFileString," in Figs. 2.14 and 2.15. The method, "loadFileString," takes what is stored in the given file with the exception handling for opening it. The file is opened, its content is read

```
public void loadFileString(){
    try{
        RandomAccessFile stream = new RandomAccessFile(this.fileName, "r");
        long length = stream.length();
        byte[] byArray = new byte[(int)length];
        stream.readFully(byArray);
        this.fileString = new String(byArray);
        stream.close();
    }catch (IOException e){
        System.out.println("There is the error in file processing!!" + e);
    }
}
```

Fig. 2.14   The implementation of method: loadFileString

```
public void saveFileString(){
    try{
        RandomAccessFile stream = new RandomAccessFile(this.fileName, "rw");
        stream.writeBytes(this.fileString);
        stream.close();
    }catch (IOException e){
        System.out.println("There is the error in file processing!!" + e);
    }
}
```

Fig. 2.15   The implementation of method: saveFileString

```
private void loadTextContent(){
    FileString fs = new FileString(this.fileName);
    fs.loadFileString();
    this.textContent = fs.getFileString();
}

private void tokenizeTextContent(){
    String[] tokenList = this.textContent.split(" .?!");
    int size = tokenList.length;
    for(int i = 0;i < size;i++){
        Word indivWord = new Word(tokenList[i].toLowerCase());
        this.wordList.addElement(indivWord);
    }
}
```

Fig. 2.16   The implementation of method: tokenizeTextContent

as a string by invoking "readFully," and it is assigned to the property, "fileString." The method, "saveFileString," saves the value of the property, "fileString," into the given file with the exception handling. In its implementation, the file is opened, the property value is written by invoking the method, "writeBytes."

The method, "tokenizeString," is implemented in Fig. 2.16. By invoking the method, "loadTextContent," the text which is the input of text indexing is loaded from the file. The text which is given as a string is segmented into tokens by invoking

```
private Vector loadStopWordList(String stopWordFileName){
    Vector stopWordList = new Vector();
    FileString fs = new FileString(stopWordFileName);
    fs.loadFileString();
    String stopWordStream = fs.getFileString();
    String[] lineList = stopWordStream.split("\n");
    for(int i = 0;i < lineList.length ;i++){
        stopWordList.addElement(lineList[i]);
    }
    return stopWordList;
}
private boolean isRegistered(Word indivWord,Vector stopWordList){
    int size = stopWordList.size();
    String wordName = indivWord.getWordName();
    for(int i = 0;i < size;i++){
        String stopWord = (String)stopWordList.elementAt(i);
        if(wordName == stopWord)
            return true;
    }
    return false;
}

private void removeStopWords(String stopWordFileName){
    Vector stopWordList = this.loadStopWordList(stopWordFileName);
    int size = this.wordList.size();
    for(int i = 0;i < size;i++){
        Word indivWord = (Word)this.wordList.elementAt(i);
        if(this.isRegistered(indivWord,stopWordList)){
            this.wordList.remove(i);
        }
    }
}
```

Fig. 2.17 The implementation of method: removeStopWords

the method, "split," and the string which is given as the argument in the method indicates the list of characters of partition boundaries. Each token is converted into its lowercases and added to the property, "wordList." The token list is generated as the results from the step.

The method, "removeStopWords," is implemented in Fig. 2.17. The list of stop words is loaded from the file, by invoking the method, "loadStopWordList." For each word in the property, "wordList," it is checked whether it is registered in the stop-word list, and if so, it is removed from the property. The method, "isRegistered," in the class, "StemmingRule," checks the registration in the varied form list, while the method, "isRegistered," in the class, "Text," does it in the stop-word list. By replacing the sequential search which is used in the current implementation by the more advanced search strategies, we improve the process of removing stop words.

```
public void indexTextContent(String stemmingRuleFileName,
        String stopWordFileName){
    this.loadTextContent();
    this.tokenizeTextContent();
    this.stemWordList(stemmingRuleFileName);
    this.removeStopWords(stopWordFileName)
}
```

Fig. 2.18 The implementation of method: indexTextContent

In Fig. 2.18, the method, "indexTextContent," is implemented. In its implementation, the methods whose access levels are set private are involved. The string which is loaded from the file by invoking the method, "loadTextContent," is tokenized into tokens by the method, "tokenizeTextContent," they are stemmed into their roots by invoking the method, "stemWordList," and among them, stop words are removed by invoking the method, "removeStopWords," following the steps of text indexing which were illustrated in Fig. 2.1. The additional steps which are mentioned in Sect. 2.4 may be implemented by adding more methods in the class. Therefore, the implementation is intended to provide the prototype version of indexing a text or texts rather than the real version.

2.4 Additional Steps

This section is concerned with the additional steps to ones which were covered in Sects. 2.2 and 2.3. In Sect. 2.4.1, we describe index filtering as the further removal of some words. In Sect. 2.4.2, we consider the index expansion which is the addition of more relevant words to the list of words, as the opposite one to index filtering. In Sect. 2.4.3, we cover index optimization as the composite of index filtering and expansion. In this section, we describe the three further steps to the basic ones which were mentioned in the previous sections.

2.4.1 Index Filtering

In Fig. 2.19, we illustrate index filtering which is the process of removing further words among remaining ones after doing stop words. The list of words which result from indexing a text with the basic three steps is given as the input, and the words are called actual words, here. The actual words are weighted by one of the equations which were presented in Sect. 2.2.4, and only words with their higher weights are selected among them. Index filtering generates a shorter list of words as its output and is usually applicable to long texts. Therefore, in this subsection, we describe the process of removing some words, further.

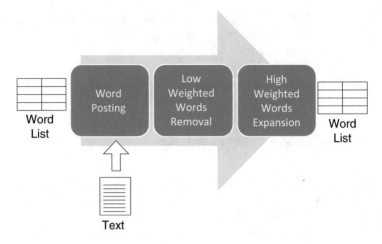

Fig. 2.19 Controlling word list from indexing text

The weights of words which are generated from indexing a text with the three basic steps are computed. As mentioned in the previous section, the weights are proportional to the absolute and relative frequencies of words in the given text, and reversely proportional to the number of other texts which include the words in the corpus. The words are sorted by their weights, and ones with their lower weights are cut off. Index filtering means the process of removing some of nouns, verbs, and adjectives in addition to the stop words for more efficiency. Therefore, in the process, the longer list of words is given as the input and the shorter list is generated as the output.

Let us consider the three schemes of selecting words by their weights. The first scheme is called rank-based selection where the words are sorted by their weights and the fixed number of words is selected among them. The second scheme is called score-based selection where the words with their higher weights than the given threshold are selected. The third scheme is the hybrid one which is the mixture of the first and second schemes. There is trade-off between the two schemes; the rank-based selection where the constant number of words is selected, but it is required to sort the words by their weights, and the score-based selection where it is not required to sort them, but the very variable number of words is selected.

We may consider the word positions in the given text for doing the index filtering, as well as their weights. Almost all of texts tend to be written, putting the essential part in the first and last paragraphs. Even if the words have their identical weights, they need to be discriminated among each other by their positions in the given text. More importance should be put to the words in the first and last paragraphs, than those in the medium ones. Grammatical information about words such as their grammatical categories and their roles in the sentences may be also considered, as well as the posting information.

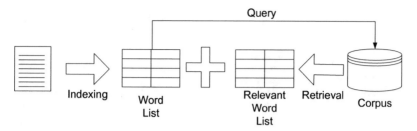

Fig. 2.20 Index expansion

Let us make remarks on index filtering. Before applying the index filtering as the additional step, we need to consider the text length. We expect more efficiency which results from applying the index filtering to long texts by getting more compact representations. When applying it to short texts, information loss is caused. We had better omit the index filtering to the short text, in order to avoid the information loss.

2.4.2   Index Expansion

Index expansion refers to the process of adding more relevant words to ones to which are indexed from the text, and it is illustrated in Fig. 2.20 [89]. The text is indexed into a list of words with the three steps. External words, called virtual words, which are relevant to the actual words are fetched through search engines. The external words are added to the list of actual words, and the longer list is generated as the output. Therefore, in this subsection, we describe the index expansion process.

The collocation of words is the essential criterion for fetching associated words from the external sources. Collocation refers to the occurrence of more than two words in the same text. The more collocations of words, the more semantic cohesion. The collocation of the two words, t_i and t_j, is computed by Eq. (2.5),

$$col(t_i, t_j) = \frac{2DF_{ij}}{DF_i + DF_j} \tag{2.5}$$

where DF_{ij} is the number of texts which include the two words, t_i and t_j in the corpus. The more the texts which include both the words, the higher the collocation of them, by Eq. (2.5).

In Fig. 2.20, the process of expanding the list of words is illustrated. The given text is indexed into the list of words with the three steps. For each word in the list, its collocations with other words in the external sources are computed by Eq. (2.5), and ones with higher collocations are fetched. The fetched words are added to the list of actual words, and the list is expanded with the virtual words. It is more desirable to fetch the external words which are relevant to only some actual words rather than all ones for more efficiency.

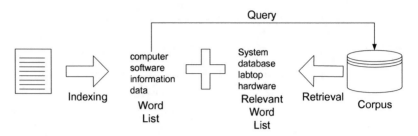

Fig. 2.21 Example of index expansion

The example of the index expansion is presented in Fig. 2.21. The text is indexed into words: "computer," "software," "information," and "data." The words are sent as the queries, and their associated words are retrieved as the virtual words. Both the actual and virtual words are given as the text surrogates, through the index expansion. However, note that the process of fetching the relevant words from the external sources is very expensive.

Let us consider the query expansion which is the process of adding more relevant words to the given query, as the similar task as the index expansion [78]. The query is given by a user, and he or she expresses his or her information needs, unclearly and abstractly. The initial query is given by the user, its collocations with other words are computed, and highly collocated words are fetched. The references to the user profiles may be used for doing the query expansion, as well as the collocations. Therefore, the query expansion is intended to make the query more clear and specific.

2.4.3 Index Optimization

Index optimization refers to the process of optimizing a list of words in order to maximize the information retrieval performance, and is illustrated in Fig. 2.22 [35]. It is viewed as the combination of the index filtering and expansion which were covered in Sects. 2.4.1 and 2.4.2. In other words, the index optimization contains the removal of unnecessary words for more efficiency and addition of more relevant words for more reliability. It is interpreted into an instance of word classification where each word is classified into one of the three categories: "expansion," "inclusion," and "removal." In this subsection, we provide the detailed description of the index optimization as the composite task of the index filtering and expansion.

In Sects. 2.4.1 and 2.4.2, we already studied the two tasks, and we need both for getting more efficiency and reliability. The index filtering is useful for long texts for more efficiency, and the index expansion is so for short texts for more reliability. The actual words in the given text are classified into the three groups: the

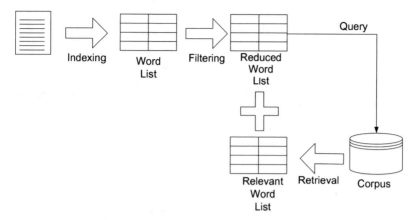

Fig. 2.22 Index optimization

words with their higher weights which should be reinforced by adding more relevant words to them, the words with their medium weights which should be remained, and the words with their lower weights which should be removed. The index expansion causes more unnecessary computations, and the index filtering causes the information loss. Therefore, we need the index optimization which is applicable to any text, regardless of its length for solving problems from the two tasks.

In Fig. 2.22, the index optimization process is presented. With the three basic steps, the text is indexed into a list of words. The shorter list of words is generated by the index filtering. The collocations of the words in the short list with other words are computed, and their associated words are fetched and added to the list. As illustrated in Fig. 2.22, the index optimization is viewed into the combination of the two tasks.

In 2015 and 2016, Jo interpreted index optimization into the classification task as another view, as illustrated in Fig. 2.23 [35, 41]. The three categories, "expansion," "inclusion," and "removal," are predefined, in advance. The words which are labeled with one of the three categories are prepared as the sample data, and the adopted machine learning is trained with them. The words which are indexed from a text are classified into one of the three categories. To the words which are labeled with "expansion," their associated words are added from the external sources; to the words which are labeled with "inclusion," they are remained; and to the words which are labeled with "removal," they are excluded.

We need to consider some issues in applying machine learning algorithms to index optimization which is viewed into the classification task. The first issue is how many sample words we need to prepare for the robustness, in advance. The second issue is to use which of the machine learning algorithms we use as the classifier. The third issue is how to define the domain scope of text as a source of sample words: the narrow scope where there is more reliability but less sufficiency of sample words, and the broad scope where there is more sufficiency but less reliability of them. The last issue is how to define the attributes for deciding one of the three categories.

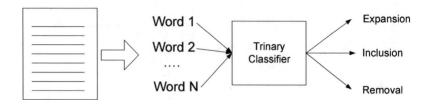

Fig. 2.23 View of index optimization into classification task

2.5   Summary

This chapter is concerned with the process of indexing a text or texts into a list of words. The three steps are basically needed for doing text indexing. Text indexing is implemented in Java for providing the guide for developing the text processing system. Index filtering, expansion, and optimization are considered as the further text indexing steps. In this section, we summarize the entire content of this chapter.

Text indexing refers to the process of mapping a text or texts into a list of words. It is needed for encoding texts into their structured forms. The list of nouns, verbs, and adjectives is usually the output of text indexing with the three basic steps. We consider the addition of index filtering, index expansion, and index optimization to the basic steps for improving the efficiency and the reliability. We consider other kinds of information about words such as the text length, the grammatical categories of words, and word positions, for doing the additional tasks.

A text is indexed into a list of words with the three basic steps: tokenization, stemming, and stop-word removal. Tokenization refers to the process of segmenting a text simply by white spaces and punctuation marks. Stemming refers to the process of changing each word into its own root form. Stop-word removal means the process of removing stop words which function only grammatically, regardless of the content. The words are usually weighted by the TF-IDF scheme and its variants.

We tried to implement the text indexing program in Java, in order to provide the development guides. We define the classes: "FileString," "StemmingRule," "Word," "Text," and "TextIndexAPI." The class, "StemmingRule," is for defining the data structure for associating a root form with its varied forms. We implemented the process of indexing a text by invoking the methods which correspond to the text indexing steps. We need to improve the search algorithm which is used in the implementation for more scalability.

Index filtering, expansion, and optimization are considered as the additional tasks for improving efficiency and reliability. Index filtering refers to the process of removing nouns, verbs, adjectives which are unimportant for identifying the content for more efficiency. Index expansion refers to the process of adding associated words which are fetched from the external sources to the original list of words. Index optimization, where important words are selected through the index filtering and their associated words are added through the index expansion, is viewed as the combination of both the tasks. Index optimization is regarded as a classification task, and we may apply a machine learning algorithm.

Chapter 3
Text Encoding

This chapter is concerned with the process of encoding texts into numerical vectors as their representations, and its overview will be presented in Sect. 3.1. In Sect. 3.2, we describe the schemes of selecting features from their candidates in generic cases, and point out its limits in applying them to the text mining cases. In Sect. 3.3, we cover the schemes of assigning values to the selected features as the main step of the text encoding. In Sect. 3.4, we point some issues in encoding texts into numerical vectors, and consider the solutions to them. Therefore, this chapter is intended to describe in detail the process of mapping a list of words, which is the output from the process which was described in Chap. 2, into a numerical vector.

3.1 Overview of Text Encoding

Text encoding refers to the process of mapping a text into the structured form as text preprocessing. The first half of text preprocessing is called text indexing which was covered in the previous chapter, and the output from the process becomes the input to the process which is covered in this chapter. In the process of text encoding, some among the words which were extracted from the text indexing are selected as features, and numerical values are assigned to them in encoding each text. The numerical vector which is generated from the process represents a text, but we need to consider some issues from the process. Hence, in this section, we explore the overview of text encoding before describing it in detail.

Let us review the results from indexing a text through the process which was covered in Chap. 2. A text is given as the input, and it is segmented into tokens in the tokenization process. Each token is mapped into its own root form by the stemming process. For more efficiency, stop words which carry out only grammatical functions irrelevantly to the given contents are removed. A list of words which result from indexing a text is given as the input to the text encoding process.

© Springer International Publishing AG, part of Springer Nature 2019
T. Jo, *Text Mining*, Studies in Big Data 45,
https://doi.org/10.1007/978-3-319-91815-0_3

Feature Candidates   Numerical Vector

Fig. 3.1 The steps of encoding text into numerical vector

The process of encoding a text into a numerical vector is illustrated in Fig. 3.1. A list of words which is generated from the process which was mentioned in Chap. 2 is given as feature candidates. Some among them are selected features, using one of the schemes which are described in Sect. 3.2. The numerical vector is constructed by assigning values to features with the reference to the text which is given as the initial input. Even if it is popular to encode texts into numerical vectors as text preprocessing, we need to consider some issues in doing so.

Let us present simple example for understanding the text encoding process, more easily. By indexing a corpus, ten words are generated as the feature candidates: "information," "data," "computer," "software," "hardware," "database," "structure," "programming," "text," and "string." By the process which is covered in Sect. 3.2, the four words, "information," "data," "computer," and "software," are selected as features. Their occurrences in the given text which we try to represent are assigned to the selected features as their values; the text is represented into a four-dimensional numerical vector. In real systems, several hundreds of features are selected among ten thousand feature candidates; a text is really represented into a vector whose dimension is several hundreds.

In 2006, Jo pointed out some issues in encoding texts into numerical vectors as their representations [25]. It requires keeping enough robustness to encode a text into a huge dimensional vector; three hundred features are usually selected for doing so. Another problem in encoding texts so is the sparse distribution in each numerical vector; zero values are dominant over nonzero ones in each numerical vector. The poor transparency usually underlies in numerical vectors which represent texts; it is very difficult to trace symbolically the results from classifying and clustering texts. However, there are some trials in previous works to solve the above issues by encoding texts into alternatives to numerical vectors.

3.2   Feature Selection

This section is concerned with the schemes of selecting some of words as features. In Sects. 3.2.1 and 3.2.2, we study the wrapper approach and the principal component analysis, respectively. In Sect. 3.2.3, we describe the independent component analysis as the popular approach to feature selection based on the signal processing theory. In Sect. 3.2.4, we explain the singular value decomposition as another feature selection scheme. Therefore, this section is intended to describe in detail the simple and advanced schemes of selecting features from candidates.

3.2.1   Wrapper Approach

The wrapper approach, which is covered in this section, refers to the feature selection scheme which depends on the results from classifying or clustering entities. The reason of calling this kind of approach wrapper approach is that the feature selection process is strongly coupled with the main task, classification or clustering. Increment and decrement of features depend on the results from doing a text mining task. In other words, the feature selection process is embedded in the given task, instead of an independent task which is separated from the task. Therefore, in this section, we present the two wrapper approaches, and analyze their complexities.

The forward selection which is a wrapper approach refers to one where the trial starts from one dimensional vector, and add more features, depending on the performance improvement. One feature is selected among candidates, at random, the classification or clustering is tried, and the performance is observed. One more feature is added, and the performances of the two cases before and after the addition are compared with each other. If the performance is improved, outstandingly, more features will be added, but the current dimension is selected, otherwise. The increment of features continues in the forward selection until the performance is improved little.

The backward selection which is opposite to the forward selection is the approach where classification or clustering is started with all candidates as features, and the number of features is decremented depending on the performance change. In the initial step, we use all feature candidates for carrying out classification or clustering. One of the features is deleted and the performances of cases before and after the deletion are compared with each other. If the two cases have very little performance difference, the deletion is continued, but it stops, otherwise. Therefore, in this approach, the feature is decremented continually until the outstanding degrade happens.

Let us consider the complexity of applying the wrapper approaches to the text classification or clustering. When denoting the final number of selected features and the total number of feature candidates by d and N, the dimension becomes from one to d in the forward selection, and from N to d in the backward selection. The summation over $1, 2, \ldots, d$ or $N, N - 1, \ldots, d$ takes the quadratic complexity, $O(d^2)$ to the number of selected features, d. In actual, the backward selection takes more computation cost than the forward selection. Hence, it takes very much cost for using the wrapper approach for selecting features [3].

Let us make some remarks on the wrapper approaches which are explained above. The wrapper approaches are characterized as very heuristic ones; additions and deletions of features are rules of thumbs. In this approach, the feature selection module should be embedded in the main module, rather than putting it as a separated one. Since it takes the quadratic cost to the dimension, it is not feasible to use it for text mining tasks in which texts are represented into huge dimensional vectors. We may consider the combination of the above two schemes where both increments and decrements of features are carried out.

3.2.2 Principal Component Analysis

The PCA (Principal Component Analysis) which is mentioned in this section refers to the scheme where features are selected based on their variances of input values. In the first step, the covariance matrix is constructed from the sample examples. Eigen vectors and values are computed from the covariance matrix. The features which correspond to maximal Eigen values are selected. Therefore, in this section, we describe the PCA as a feature selection scheme.

Let us explain the process of deriving the covariance matrix from the given examples. Initially, examples are given as input vectors which represent raw data with all of feature candidates,

$$\mathbf{X} = \{\mathbf{x}_1, \mathbf{x}_2, \ldots, \mathbf{x}_n\}$$

where n is the number of examples. Each input vector is given as a numerical vector,

$$\mathbf{x}_k = [x_{k1} \ x_{k2} \ \ldots \ x_{kN}] \, k = 1, 2, \ldots, n$$

where N is the total number of feature candidates, and the average for each element is computed by Eq. (3.1),

$$\bar{x}_i = \frac{1}{n} \sum_{m=1}^{n} x_{mi} \tag{3.1}$$

The covariance matrix is built as follows:

$$\Sigma = \begin{bmatrix} Cov(x_1, x_1) & Cov(x_1, x_2) & \dots & Cov(x_1, x_N) \\ Cov(x_2, x_1) & Cov(x_2, x_2) & \dots & Cov(x_2, x_N) \\ \vdots & \vdots & \ddots & \vdots \\ Cov(x_N, x_1) & Cov(x_N, x_2) & \dots & Cov(x_B, x_N) \end{bmatrix}$$

where $Cov(x_i, x_j) = \sum_{k=1}^{n}(x_{ik} - \bar{x}_i)^2(x_{jk} - \bar{x}_j)^2$. The covariance matrix is given as N by N matrix.

The Eigen vectors and values are computed from the above matrix. The relation of the covariance matrix, and its Eigen vectors and values are expressed as Eq. (3.2).

$$\Sigma v_i^T = \lambda_i v_i^T \tag{3.2}$$

We compute Eigen values, $\lambda_1, \lambda_2, \dots, \lambda_N$ by Eq. (3.2). The Eigen values are sorted in their descending order as $\lambda_{max_1}, \lambda_{max_2}, \dots, \lambda_{max_N}$ and selected d maximal Eigen values, $\lambda_{max_1}, \lambda_{max_2}, \dots, \lambda_{max_d}$. Therefore, in the PCA, the N dimensional input vector is projected into the d dimensional space based on the selected Eigen values.

Let us explain the process of mapping the N dimensional input vectors into the d dimensional ones through the projection as the next step. The mean vector is computed by averaging over input vectors in Eq. (3.3).

$$\bar{x} = \frac{1}{n}\sum_{i=1}^{n} x_i \tag{3.3}$$

The matrix W is built, using the Eigen vectors which are associated with the maximal Eigen values which are computed above, $v_{max_1}, v_{max_2}, \dots, v_{max_d}$ where $v_{maxi} = \begin{bmatrix} v_{max_i 1} & v_{max_i 2} & \dots & v_{max_i N} \end{bmatrix}$ as follows:

$$W = \begin{bmatrix} v_{max_1 1} & v_{max_1 2} & \dots & v_{max_1 N} \\ v_{max_2 1} & v_{max_2 2} & \dots & v_{max_2 N} \\ \vdots & \vdots & \ddots & \vdots \\ v_{max_d 1} & v_{max_d 2} & \dots & v_{max_d N} \end{bmatrix}$$

The d dimensional input vectors $Z = \{z_1, z_2, \dots, z_n\}$ are computed by Eq. (3.4).

$$z_i = W(x_i - \bar{x}) \tag{3.4}$$

Therefore, the N dimensional numerical vectors are reduced into d dimensional ones by the above process.

Let us make some remarks on the PCA in applying it to the text mining tasks. Using all the feature candidates as features, texts should be represented into N dimensional numerical vectors, in order to use the PCA. In this case, the huge covariance matrix which is a 10,000 by 10,000 one is built. It takes very much

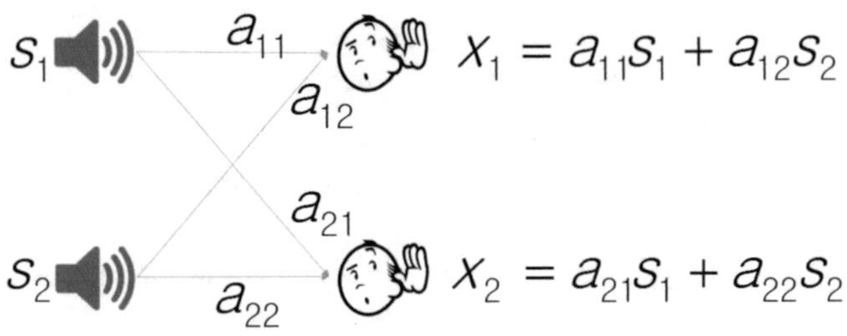

$$x_1 = a_{11}s_1 + a_{12}s_2$$

$$x_2 = a_{21}s_1 + a_{22}s_2$$

Fig. 3.2 Cocktail party: source and observed

time for computing ten thousands of Eigen values and vectors from the covariance matrix. Therefore, the PCA is not feasible as the feature selection scheme in text mining tasks.

3.2.3 Independent Component Analysis

This section is concerned with the ICA (Independent Component Analysis), and it refers to the process of mapping the input vectors whose features correlated with each other into ones whose features are independent of each other [22]. It is assumed that each element in an original input vector is a linear combination of independent components, as shown in Fig. 3.2. The original input vector is expressed as the multiplication of a matrix and a vector which consists of independent components, as shown in Eq. (3.5).

$$x = As \tag{3.5}$$

The independent components are computed by expressing the vector as the multiplication of its inverse matrix and the original input vector as shown in Eq. (3.6).

$$s = A^{-1}x \tag{3.6}$$

Therefore, in this section, we provide both detail description and mathematical expression of ICA as the dimension reduction scheme.

The two kinds of signals, source and observed signals, are illustrated in Fig. 3.3. In Fig. 3.3, the left two signals which are given as source signals correspond the sine function and the logarithmic one. The right two signals in Fig. 3.3 which are given as observed signals are linear combinations of the left two signals. It is assumed that the original data is given as the observed signals like the right side in Fig. 3.3, so search for the source signals is the goal of ICA. It is also assumed that the number of observed signals is greater than that of source signals.

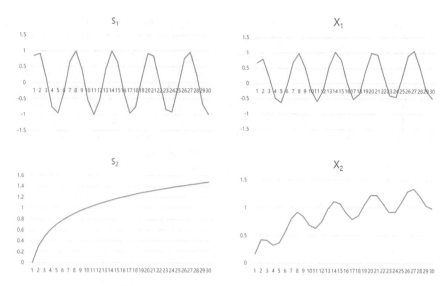

Fig. 3.3 Source signals: S_1 and S_2 vs observed signals: X_1 and X_2

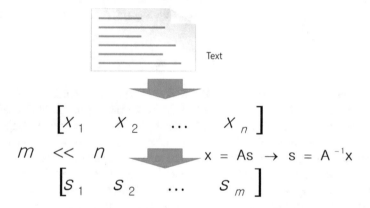

Fig. 3.4 Application of ICA to text encoding

Let us explain the process of reducing the dimension by ICA as shown in Fig. 3.4. The examples are given as the N dimensional input vectors as follows:

$$\mathbf{X} = \{\mathbf{x}_1, \mathbf{x}_2, \ldots, \mathbf{x}_n\}$$

$$\mathbf{x}_i = [x_{i1}\ x_{i2}\ \ldots\ x_{iN}]\, k = 1, 2, \ldots, n$$

Each input vector is expressed as the multiplication of a matrix and the dimensional vector which consists of independent components, as follows:

$$\mathbf{x}_i^T = \mathbf{A}\mathbf{s}_j^T$$

where $s_i = [s_{i1} \ s_{i2} \ \ldots \ s_{id}]$. If the matrix, A is decided initially, we find the vectors, $S = \{s_1, s_2, \ldots, s_n\}$ by Eq. (3.6), as the reduced dimensional vectors. If the matrix, A is known, we can find its inverse matrix, using the Gaussian Jordan method.

Let us mention the process of applying the ICA to the dimension reduction, in encoding texts into numerical vectors. The feature candidates are generated by indexing the corpus, and texts are represented into numerical vectors, using all candidates. The N dimensional input vectors are clustered into d clusters using a clustering algorithm, and we find prototypes of clusters which are given as N dimensional vectors, $c_i = [a_{i1} \ a_{i2} \ \ldots \ a_{iN}]$. The matrix, A is built by arranging the cluster prototypes as column vectors, as follows:

$$A = \begin{bmatrix} a_{11} & a_{12} & \ldots & a_{1d} \\ a_{21} & a_{22} & \ldots & a_{2d} \\ \vdots & \vdots & \ddots & \vdots \\ a_{N1} & a_{N2} & \ldots & a_{Nd} \end{bmatrix}$$

Afterward, its inverse matrix is computed using the Gaussian Jordan method, and vectors of independent components are computed by Eq. (3.6). Since $d << N$ is usual, we need to use the pseudo inverse of the matrix, A, instead of its inverse, as expressed in Eq. (3.7),

$$s_i T = A^+ x_i^T \tag{3.7}$$

We present the comparisons of the ICA with the PCA which was described in Sect. 3.2.2, in Table 3.1. The idea of PCA is the project of the N dimensional vector into the d dimensional space, whereas the idea of ICA is to find independent components from original components each which is the linear combination of independent ones. In the PCA, the covariance matrix is used as the mean of reducing the dimension, whereas in the ICA, the clustering algorithm is used. In the PCA, the matrix, W, is built by arranging Eigen vectors of covariance matrix as row vectors, whereas in the ICA, the matrix, A is built by doing cluster prototypes as column vectors. In the PCA, the d dimensional vectors are computed by multiplying the matrix, W by each original input vector, whereas in the ICA, they are computed by doing the inverse or pseudo inverse of the matrix, A by it.

Table 3.1 PCA vs ICA

	Principal component analysis	Independent component analysis
Idea	Projection into smaller dimensional space	Element as linear combination of independent smaller factors
Mean	Covariance matrix	Clustering algorithm
Matrix components	Eigen vectors	Cluster prototypes
Equation	$z_i = Wx_i$	$z_i = A^{-1}x_i$

3.2.4 Singular Value Decomposition

The SVD (Singular Value Decomposition) which is described in this section refers to the process of decomposing the $n \times N$ matrix into the three components: $n \times n$ orthogonal matrix, $n \times N$ diagonal matrix, and $N \times N$ orthogonal matrix [77]. If the matrix, $\mathbf{X}$ is the $n \times N$ matrix, its singular values which are denoted as $\sigma_1, \sigma_2, \ldots, \sigma_n$ are the square roots of Eigen values of the matrix, $\mathbf{X}^T\mathbf{X}$. The $n \times n$ orthogonal matrix as the first component consists of Eigen vectors of $\mathbf{XX}^T$, and the $N \times N$ orthogonal matrix as the last component consists of Eigen vectors of $\mathbf{X}^T\mathbf{X}$. The middle component in the SVD is the diagonal matrix which consists of the singular values. Therefore, in this section, we explain the process of applying it to the dimension reduction, together with its mathematical expression.

Let us mention the process of building the data matrix from the input vectors which are given as sample examples. The n input vectors are given as the N dimensional numerical vectors, $\mathbf{X} = \{\mathbf{x}_1, \mathbf{x}_2, \ldots, \mathbf{x}_n\}$. The data matrix is made as the $n \times N$ matrix as follows:

$$\mathbf{X} = \begin{bmatrix} x_{11} & x_{12} & \ldots & x_{1N} \\ x_{21} & x_{22} & \ldots & x_{2N} \\ \vdots & \vdots & \ddots & \vdots \\ x_{n1} & x_{n2} & \ldots & x_{nN} \end{bmatrix}$$

where input vectors are given as row vectors. The Eigen vectors are computed from the two square matrices: $\mathbf{X}^T\mathbf{X}$ and $\mathbf{XX}^T$. Since ten thousands of feature candidates are extracted from indexing a corpus, the number of input vectors tends to much less than the number of feature candidates in the text mining tasks as follows:

$$n << N$$

The above matrix, $\mathbf{X}$ is decomposed into the three components as expressed in Eq. (3.8),

$$\mathbf{X} = \mathbf{U\Sigma W} \tag{3.8}$$

We take the n Eigen vectors, $\mathbf{u}_1, \mathbf{u}_2, \ldots, \mathbf{u}_n$ from the matrix, $\mathbf{XX}^T$, and build the matrix, $\mathbf{U}$, by arranging them as its column vectors. Next, we take the n singular values, $\sigma_1, \sigma_2, \ldots, \sigma_n$ by getting square roots of Eigen values of $\mathbf{X}^T\mathbf{X}$, and build the matrix, $\mathbf{\Sigma}$, by arranging them as its diagonal elements. The third, we take the N Eigen vectors, $\mathbf{v}_{max_1}, \mathbf{v}_{max_2}, \ldots, \mathbf{v}_{max_N}$ from the matrix, $\mathbf{X}^T\mathbf{X}$, and build the matrix, $\mathbf{W}$, by arranging them as its row vectors. Some among the n singular values are given as nonzero values as follows:

$$\sigma_1 \geq \sigma_2 \geq \ldots \geq \sigma_r > 0$$

$$\sigma_r = \sigma_{r+1} = \ldots = \sigma_n = 0$$

As the next step, let us explain the process of deriving the reduced dimensional vectors by the singular value decomposition. We select the d maximal singular values denoted by $\sigma_1, \sigma_2, \ldots, \sigma d$ among them. We build the matrix, $\boldsymbol{\Sigma}'$ by filling values, $\Sigma_{11} = \sigma_1, \Sigma_{22} = \sigma_2, \ldots, \Sigma_{dd} = \sigma_d$ as diagonal elements. The matrix, $\mathbf{X}'$ is computed by Eq. (3.9),

$$\mathbf{X}' = \mathbf{U}\boldsymbol{\Sigma}'\mathbf{W} \tag{3.9}$$

We take nonzero d columns from the matrix, $\mathbf{X}'$, as the list of reduced dimensional vectors.

We mention the three schemes of reducing features, PCA, ICA, and SVS, as state-of-the-art schemes, in this section. They are feasible to case where, for example, fifty dimensions reduced into ten dimensions. In the text mining tasks, the number of feature candidates is at least several thousands, and the number of selected features is several hundreds; the dimension should be reduced from several thousands to several hundreds. It requires to represent each text into several thousands of dimensional vectors, initially, in order to apply one of the three schemes of reducing dimensions. Rather than using one of the three schemes in implementing text mining systems in the subsequent chapters, we take features by selecting some words depending on their coverages in the given corpus.

3.3   Feature Value Assignment

This section is concerned with the schemes of assigning values to the selected features. The task which is covered in this section is simpler than that covered in Sect. 3.2. In Sect. 3.3.1, we explain how to assign values to the selected features. In Sect. 3.3.2, we cover how to compute the similarity between texts once representing texts into numerical vectors. Hence, in this section, we describe the schemes of assigning values to features in encoding texts into numerical vectors, and those of computing similarities between texts or text groups.

3.3.1   Assignment Schemes

This section is concerned with the schemes of assigning values to features. Features are selected among candidates by one of the schemes which are mentioned in Sect. 3.2. The numerical values which are assigned to features in each text indicate their importance degrees in its contents. The numerical vectors as text representations are generated as the output from the process. In this section, we explore the schemes of setting values to features.

Let us mention the simplest scheme of assigning values to features in encoding texts into numerical vectors. Once some words are selected as features, binary values may be assigned as feature ones in this scheme. The text and a list of words is given as input, and if each word occurs at least one time in the text, one is assigned to the word as the feature. For example, the words, "computer," "information," "data," and "retrieval," are selected as the features, and if the given text includes the two words, "computer," and "information," the text is represented into the four-dimensional numerical vector, [0 0 1 1]. The demerit of this scheme is that there is no discrimination among available words by their importance degree.

The occurrences of words corresponding to features in the given text are assigned as feature values in another scheme. The features are denoted by $f_1, f_2, \ldots, f_d$, and the values are assigned to them by Eq. (3.10),

$$f_i = tf_{ij} \tag{3.10}$$

where tf_{ij} is the frequency of the word, t_i, and in the text, d_j. The overestimation and underestimation by the text length may be caused by using the frequency computed by Eq. (3.10) directly, so it is recommended to use the relative frequency which is expressed by Eq. (3.11),

$$f_i = rf_{ij} = \frac{tf_{ij}}{tf_{\max j}} \tag{3.11}$$

where $tf_{\max j}$ is the maximum frequency in the text, d_j. For example, we select the four words, as above, "computer," "information," "data," and "retrieval," and if the maximum frequency is 10 in the given text and their occurrences are 4, 5, 0, and 0, respectively, the text is represented into the numerical vector, [0.4 0.5 0 0]. Even if the discrimination among words in the given text is available, unimportance from occurrences in other texts is not considered in this scheme.

Instead of frequencies or relative frequencies, TF-IDF weights may be used as features in encoding texts into numerical vectors as well as in indexing a text or a corpus. The TF-IDF weight is computed to each word by Eq. (2.1) in Sect. 2.2.4. Their derived equations, Eqs. (2.2)–(2.4), may be used for computing TF-IDF weights. In the equations, two is used as the base of logarithmic. The higher document frequencies, the lower TF-IDF weight, as shown in Eq. (2.1)–(2.4).

Let us consider other kinds of factors for computing word weights as feature values. We may consider the grammatical properties such as subjective, objective, verb, or nouns which are obtained by POS tagging, for calculating the weights. As another factor, we may consider the posing information which indicates where a word positions in the given text; higher weights are assigned to features which position in the first or last paragraph. Feature values are adjusted by considering contexts given as words around the given feature; the feature may have its different meaning depending on context. We may consider the writing style and kinds of articles such as technical texts, novels, and essays, for computing the feature values.

3.3.2   Similarity Computation

This subsection is concerned with the process of computing the similarity between texts. This computation is essentially necessary for doing the text classification and the text clustering which are covered in the subsequent parts. Once texts are represented into numerical vectors, we compute cosine similarity or other similarities between vectors as the similarity between texts. Once we define the scheme of computing the similarity between texts, we may consider how to compute the similarity between text groups. Therefore, in this section, we describe the scheme of computing the similarity between individual texts or between text groups.

The two texts are encoded into the numerical vectors denoted by $\mathbf{d}_1 = [f_{11} \; f_{12} \; \ldots \; f_{1d}]$ and $\mathbf{d}_2 = [f_{21} \; f_{22} \; \ldots \; f_{2d}]$. The cosine similarity between two vectors which is the most popular measure is computed by Eq. (3.12).

$$sim(\mathbf{d}_1, \mathbf{d}_2) = \frac{2\mathbf{d}_1 \cdot \mathbf{d}_2}{\|\mathbf{d}_1\|^2 + \|\mathbf{d}_2\|^2} \tag{3.12}$$

We may consider the Euclidean distance which is computed by Eq. (3.13) as the inverse similarity.

$$\text{dist}(\mathbf{d}_1, \mathbf{d}_2) = \sqrt{\sum_{k=1}^{d} (f_{ik} - f_{2k})^2} \tag{3.13}$$

The Jaccard similarity exists as a variant of the cosine similarity and is computed by Eq. (3.14),

$$sim(\mathbf{d}_1, \mathbf{d}_2) = \frac{2\mathbf{d}_1 \cdot \mathbf{d}_2}{\|\mathbf{d}_1\|^2 + \|\mathbf{d}_2\|^2 - \mathbf{d}_1 \cdot \mathbf{d}_2} \tag{3.14}$$

The value which is computed by the Euclidean distance is converted into similarity, by Eq. (3.15) or (3.16).

$$sim(\mathbf{d}_1, \mathbf{d}_2) = \frac{1}{\sqrt{\sum_{k=1}^{d} (f_{ik} - f_{2k})^2}} \tag{3.15}$$

$$sim(\mathbf{d}_1, \mathbf{d}_2) = C - \sqrt{\sum_{k=1}^{d} (f_{ik} - f_{2k})^2} \tag{3.16}$$

where C is an arbitrary constant.

Let us explain the process of computing the similarity between texts using one of the above equations. The two texts are given as the input, and they are encoded into numerical vectors as their representations. The similarity between the numerical vectors is computed using one of the above equations. The similarity is given as the output. The value which is computed by Eq. (3.12) or (3.14) is always given as a normalized value between zero and one.

Once we get the scheme of computing the similarity between individual texts, we may consider the similarity between text groups. Mean vectors of groups are computed, and the similarity between the two mean vectors is set as the similarity between groups. In another scheme, we compute the similarities of all possible pairs of texts between two groups, and average over them as the similarity between the groups. We may set the maximum or the minimum among similarities among all possible pairs as the similarity between groups as an alternative scheme. The first scheme is used most popularly among them, so it is adopted for implementing text mining systems in the subsequent parts.

The similarities among texts or text groups are very important measures for doing text clustering. Text clustering refers to the process of segmenting a group of texts into subgroups by their similarities. Clustering texts into subgroups of similar ones depends strongly on their similarities. We also need the process of computing similarities among texts for doing the text classification as well as the text clustering. The word and text association will be covered in Chap. 4, and need the process of computing similarities among texts.

3.4 Issues of Text Encoding

In this section, let us consider some issues in encoding texts into numerical vectors and the solutions to them. In Sect. 3.4.1, we discuss the huge dimensionality of numerical vectors representing texts as the first issue. In Sect. 3.4.2, we consider the sparse distribution in each numerical vector as another issue. In Sect. 3.4.3, we regard numerical vectors as non-symbolic representations of texts and point out the difficulties in tracing the results from classifying and clustering texts. Therefore, this section is intended to present some issues in encoding texts into numerical vectors and discuss the solutions to them.

3.4.1 Huge Dimensionality

This section is concerned with the huge dimensionality as the issue of encoding texts into numerical vectors. This issue means that too many features, at least several hundreds, are required for maintaining the system robustness. More than ten thousand feature candidates are generated from the given corpus through text indexing, and several hundreds of ones among them are selected as features. Any of schemes which are described in Sect. 3.2 is not feasible for reducing the feature candidates, because of too much time for doing the computations involved in doing so. In this section, we consider the huge dimensionality as one of the important issues of text encoding, and present some solutions to it.

A corpus is given as the source from which feature candidates are generated. In previous literatures, Reuter21578, 20NewsGroups, and OSHUMED were used as the typical corpus [36]. Each of them usually consists of 20,000 texts, and more than 20,000 vocabularies are extracted as feature candidates. Even if the corpus, NewsPage.com, was used as the smallest one which consists only one thousand texts, for evaluating text categorization schemes, at least almost five thousands of words are extracted [31]. If we use one of the schemes mentioned in Sect. 3.2 as the state-of-the-art one, we must build a 20,000 by 20,000 matrix.

Let us explore the dimensions of numerical vectors which represent texts after the feature selection, in previous literatures. In 2002, by Sebastiani, it was mentioned that each text is represented into three hundred dimensional numerical vectors in his survey article on text classification [85]. In implementing the text clustering system which was called WEBSOM by Kaski et al., texts are encoded into about five hundred dimensional vectors [50]. Jo et al. applied the KNN (K Nearest Neighbor) and the SVM (Support Vector Machine) which are very popular classification approaches to text classification tasks, but it resulted in very poor performance [26, 27, 46]. Even if the text classification system where texts are encoded into one hundred dimensional vectors works well to some domain, it works very poorly to other domains; it was discovered that the system is very unstable [36].

Let's see demerits which are caused by the huge dimensionality in encoding texts into numerical vectors. Basically, it costs too much time for processing vectors. It requires more sample examples for maintaining enough robustness of learning, proportionally to the dimension. The huge dimension is also the cause of overfitting in learning sample examples for doing the text classification. Hence, by the huge dimensionality, it is not feasible to apply one of the three schemes to this case.

We need to consider the schemes of solving the problem in order to avoid the above results. We use groups of semantically similar words as features instead of individual words, but it requires much computation cost for clustering words semantically, as its payment. In 2015, Jo proposed to use the semantic similarities among features as well as those among feature values, but it costs very high complexity for doing them [35, 37–39]. Texts were encoded into tables as more compact representations than numerical vectors; it resulted into better performance with the smaller sizes in both text classification and text clustering. It was proposed to use the string kernel in applying the SVM to text classification by Lodhi; it was successful in doing classifications in bioinformatics, but not in text classifications [58, 61].

3.4.2 Sparse Distribution

In this section, let us point out another issue, sparse distribution, in encoding texts into numerical vectors. Each word which is selected as a feature has very small coverage; it is included in only couple of texts except stop words. So, it is not able to avoid zero values which are dominant over nonzero ones in each numerical vector;

zero values occupy more than 95% in it. Discrimination among numerical vectors is lost as results of the sparse distribution. Therefore, in this section, we visualize the effect from the sparse distribution, and present the solutions to it.

Let us consider the specific cases of sparse distribution in numerical vectors. Zero vector is the typical example of sparse distribution where all elements are given as zero values. It is most popular that only less 5% of elements are nonzero values and the others are zero values in a numerical. Even if more than 5% of elements are nonzero values, they are very tiny ones which are close to zeros. In order to avoid the problem, increasing the number of features causes the huge dimensionality which was mentioned in Sect. 3.4.1.

The sparse distribution in each numerical vector is very fragile to the process of computing similarities by one of the equations which was described in Sect. 3.3.2. When two vectors are zero ones, zero value is absolutely generated as their similarity by any of cosine similarity, Euclidean distance, and Jaccard similarity. If two vectors have nonzero values in their different features, the similarity between texts or text groups is computed as a zero value by the cosine similarity or the Jaccard one. In using one of them, the similarity between texts which are represented into sparse vectors may be underestimated. Especially, short texts tend to be encoded into sparse vectors; index expansion is needed before encoding them.

We need to consider results which are caused by zero values as results from computing text similarities. Very few association rules may be generated in the case of word and text association which will be covered in Chap. 4. Among sample texts, contradiction of labels may happen; for example, some zero vectors belong to the positive class and the others belong to the negative one, in a binary classification. The sparse distribution of numerical vectors results into many small sized clusters, almost like the singletons as many as individual items; it is very few chance that very few couple of texts into cluster. Therefore, poor performance of any tasks of text mining is caused by the sparse distribution.

Let us discuss some solutions to the sparse distribution. When zero vectors are included in sample examples after encoding texts into numerical vectors, we need to remove them in order avoid the contradiction of labels. When nonzero values are given in different features of two vectors, we may use similarity between features which was proposed by Jo in 2015, in order to avoid that similarity between texts becomes zero [35, 37–39]. Word clusters, instead of individual words, may be used as features for solving both the huge dimensionality and the spare distribution, at the same time. In previous works, it was proposed that texts should be represented into symbolic forms, in order to avoid the sparse distribution, completely, but it requires to define similarity measures, correspondingly [36].

3.4.3 Poor Transparency

This subsection is concerned with the last issue which we need to consider in encoding texts. Numerical vectors which are regarded as the popular representations

	word 1	word 2	word 3	word 4	word 5	word 6
text 1	0.5026	0.6500	0.5255	0.9990	0.3132	0.7651
text 2	0.9065	0.2402	0.1554	0.1021	0.0434	0.6344
text 3	0.8878	0.0109	0.0646	0.2007	0.1207	0.0424
text 4	0.9985	0.6824	0.2072	0.3365	0.3893	0.3957
text 5	0.0957	0.6635	0.5941	0.0317	0.9340	0.0744
text 6	0.6487	0.8335	0.3195	0.5797	0.4163	0.7906
text 7	0.7124	0.1916	0.4836	0.0982	0.0154	0.5528
text 8	0.7173	0.8387	0.6763	0.5768	0.6810	0.3140
text 9	0.5083	0.0228	0.6717	0.7805	0.3816	0.1034
text 10	0.2714	0.2993	0.1868	0.8266	0.9127	0.2448

Fig. 3.5 Poor transparency in encoding texts into numerical vectors

of texts are essentially characterized far from symbolic ones. Although texts are regarded as symbolic data by themselves, numerical vectors which are encoded from them characterized as numerical values. In other words, there is no way of guessing contents only by their representations. Therefore, in this section, we point out the poor transparency from encoding texts into numerical vectors, and mention the symbolic representations of texts.

In Fig. 3.5, we illustrate numerical vectors which represent texts. In this example, it is assumed that a text is encoded into a numerical vector where six words are used as features. Once texts are mapped into their own numerical vectors, it is almost impossible to guess their contents by their representations, hiding the features. Users tend to want evidences for classifying or clustering texts by tracing them, as well as results from doing so. Hence, it is unnatural to encode texts which are essentially symbolic into numerical vectors.

We try to trace contents from numerical vectors, providing features. Features are selected from candidates and a numerical vector is generated from text by assigning values to the features. The elements of numerical vector are scanned, associating their features. Nonzero elements are selected from the numerical vector, the list of their corresponding features is made, and the content is guessed by them. The words which correspond to nonzero features are discriminated by their importance degree in case of assigning TF-IDF weights.

In previous works, it was proposed that texts should be encoded into symbolic forms as shown in Fig. 3.6, in order to solve the problem. In 2008 and 2015, texts were represented into tables each of which consists of entries of words and their weights by Jo [36, 42]. Since 2005, Jo proposed that texts should be encoded into the string vector which is a finite ordered set of words in text mining tasks [28, 30, 43]. By doing so, it is expected to be easier to trace results from doing both the tasks for getting the evidences.

We need to modify correspondingly machine learning algorithms for processing the symbolic structured forms. In 2015, the KNN was modified into its version

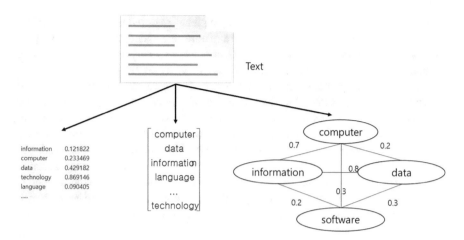

Fig. 3.6 Symbolic representations of text

where similarities among features are considered as well as feature values. In 2015, a table matching algorithm was proposed as the approach to the text classification and in 2016, the KNN and the AHC (Agglomerative Hierarchical Clustering) algorithms into their table-based version to both the tasks: text classification and clustering. The KNN and the SVM were modified into their string vector-based version, and semantic operations on words are defined as the foundation for modifying the subsequent machine learning algorithms, in 2015. We need more research on operations on symbolic structured forms, in order to modify more advanced machine learning algorithms.

3.5   Summary

In this chapter, we describe the process of encoding texts into numerical vectors, and point out the issues of doing so. We mention the four schemes of reducing dimensions in encoding raw data into numerical vectors. We present the schemes of assigning values to features, once features are selected. We point out the huge dimensionality, the sparse distribution, and the poor transparency, as the main issues of encoding texts into numerical vectors. In this section, we summarized what is covered in this chapter.

Text encoding refers to the process of mapping texts into numerical vectors as its representation, in context of text mining. A list of words which is generated from indexing a corpus is given as a collection of feature candidates. Some of them are selected as features. Values are assigned to the selected features, and a numerical vector is constructed as the representation of text. Therefore, from this process, a numerical vector is generated as the representation of text.

Let us mention the four schemes of selecting features or reducing the dimensions. The wrapper approach is the scheme where features are added or deleted, observing the performance change. The principal component analysis is the dimension reduction scheme where the covariance matrix is made from the input vectors where all feature candidates are used, some of Eigen values and their associated Eigen vectors are selected, and the reduced dimensional vectors are generated by product of the matrix which consists of the Eigen vectors and the input vector. The independent component analysis is one where cluster vectors are made by clustering input vectors, the matrix is made by them as column vector, and reduced ones are generated by product of its inverse and input vector. The singular value decomposition is one where data matrix is decomposed with the three components, some singular values are selected, the middle component is replaced by one with only selected singular values, and the data matrix for reduced vectors is computed with the replaced one.

Once features are selected as the attributes of numerical vectors, we consider some strategies of assigning values to them. Binary values are assigned to features, depending on their occurrences in the given text. In another strategy, frequencies or relative frequencies are assigned to features. We may use the TF-IDF weights which are computed by one of the equations which were described in Chap. 2 as feature values. We also mention the process of computing the similarity between texts and text groups.

We point out some issues in encoding texts into numerical vectors. The huge dimensionality is the issue where many features are required for encoding texts into numerical vectors. The sparse distribution is the dominance of zero values more than 95% in each numerical vector which represents a text. The poor transparency is the difficulty in guessing contents by the numerical vector, hiding its features, and tracing symbolically results from classifying or clustering texts. The solutions to the issues were proposed in previous research.

Chapter 4
Text Association

This chapter is concerned with text association as the fundamental task of text mining. In Sect. 4.1, we present the overview of text association, and in Sect. 4.2, we describe the generic data association and the Apriori algorithm as the approach. In Sect. 4.3, we mention the process of constructing the word text matrix from a corpus and explain the word association using the Apriori algorithm, together with its example. In Sect. 4.4, we explain the text association which is specific to texts together with its example. Therefore, this chapter is intended to provide the detailed description of text association as the elementary text mining task.

4.1 Overview of Text Association

Text association refers to the process of extracting association rules as the if-then template from word sets or text sets. Text association is derived from the association of data items from item sets as the fundamental data mining task. In this chapter, the Apriori algorithm is decided as the approach to the text association, assuming that word or text sets are given as the input. In this chapter, we restrict the scope of text association to the two types: word association and document association, even if other types are available. Therefore, in this section, we explore the overview of text association, together with its approach, Apriori algorithm.

We mention the generic data association from which the text association is derived, before discussing it. The data association is originally intended to discover customer's purchase patterns, such as, "if customers buy A, they tend to buy B." From each customer, his or her purchased items are collected through their payment process; a set of items is obtained continually from each customer. Association rules are extracted from item sets, using the association algorithm such as Apriori

© Springer International Publishing AG, part of Springer Nature 2019
T. Jo, *Text Mining*, Studies in Big Data 45,
https://doi.org/10.1007/978-3-319-91815-0_4

algorithm as the form "$A \rightarrow B$" which means the above sentence. Hence, results from doing the data association become the important references for arranging items in stores.

The Apriori algorithm is mentioned as the important tool of data association. Item sets are given as the input, and the support and confidence thresholds are given as the configurations for doing the tool. The support threshold is used for generating frequent sets and the confidence threshold is used for generating association rules from each frequent set. Each association rule consists of its conditional item and its causal sets; it is interpreted into the statement, "if the conditional item is given, the causal sets happen or are given." The Apriori algorithm consists of the three steps: frequent set generation, rule candidate generation, and rule selection.

Let us consider several types of text association, depending on the kinds of textual units. Word association which is the typical type of text association is one where association rules of words are generated from word sets. In the narrow view, text association which is called document or article association is one where association rules of articles or documents are generated from document sets.[1] Sentence or paragraph association which is called subtext association is one where association rules of sentences or paragraphs are generated from sets. The scope of type is set to word association and text association; they are covered in Sects. 4.3 and 4.4.

The association should be distinguished from the clustering, even if they look similar as each other, so their differences are presented in Table 4.1. In the association, item sets are given as input, whereas in the clustering, a group of items is given as input. In the association, association rules each of which consists of the conditional item and its causal sets are generated as the output, whereas in the clustering, subgroups of similar items are generated as the output. The association is intended to discover the causal relations among items, whereas the clustering is intended to discover similarities among items. Each causal relation is characterized as the unidirectional one, whereas the similarity between two items is characterized as the bidirectional one.

Table 4.1 Association vs clustering

	Association	Clustering
Input	Multiple item sets	Single item set
Output	If-then rules	Clusters
Relation	Influence	Similarity
Direction	One direction	Two directions

---

[1]Text association which is mentioned in Sect. 4.4 is restricted to the association of documents or articles.

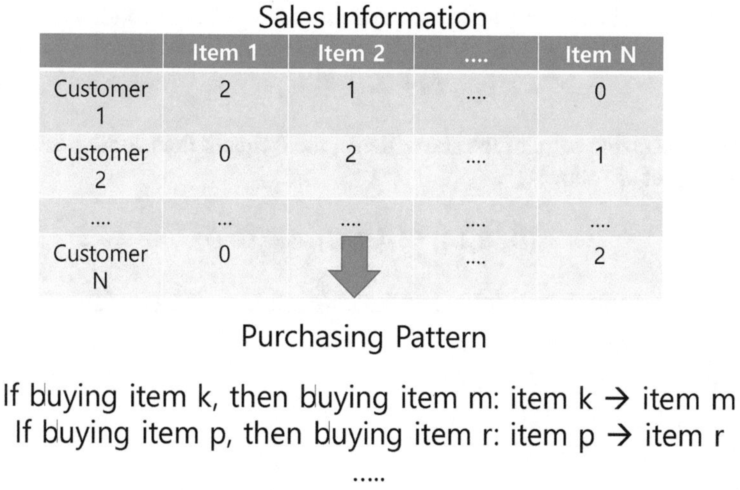

Sales Information

	Item 1	Item 2	....	Item N
Customer 1	2	1	....	0
Customer 2	0	2	....	1
....	...	....	....	....
Customer N	0		....	2

Purchasing Pattern

If buying item k, then buying item m: item k → item m
If buying item p, then buying item r: item p → item r
.....

Fig. 4.1 Data association

4.2 Data Association

This section is concerned with the generic data association and it consists of the three subsections. In Sect. 4.3.2, we explain the data association in the functional view. In Sect. 4.2.2, we cover the support and the confidence as the measures for performing the data association. In Sect. 4.2.3, we describe the Apriori algorithm which is the approach to both word and text association, presenting its pseudocode. Therefore, in this section, we cover the functional and conceptual view of data association, the measures, support and confidence, and the Apriori algorithm.

4.2.1 Functional View

Data association is defined as the process of extracting the association rule as if-then form from item sets as shown in Fig. 4.1. The data association is initially intended to discover the purchasing patterns of customers in big shopping malls [91]. The item sets which are collected from customers during their payments are given as the input. From the data association, the association rule is extracted; it is interpreted into the statement: if a customer buys a particular item, he or she tend to buy other items. In this section, we explore the data association, conceptually and functionally.

Let us consider the process of collecting item sets which are given as the input to the data association. We may define the matrix where each column corresponds to an item and each row to a customer, as the frame. During payments, items which are purchased by a customer are collected, and they are filled with the row which corresponds to the customer. An item set is generated from each row in the matrix. Depending on the application, it is possible to select items whose number is greater than threshold for generating item sets.

For example, items are initially selected as frequent ones as follows:

$$A, B, C, \text{and } D$$

Among all possible pairs of the above items, the frequent item subsets are selected by the support as follows:

$$\{A, B\}\{A, C\}, \{B, C\}, \text{and} \{C, D\}$$

For each subset, the association rule candidates are extracted as follows:

- $\{A, B\} : A \rightarrow B \text{ and } B \rightarrow A$
- $\{A, C\} : A \rightarrow C \text{ and } C \rightarrow A$
- $\{B, C\} : B \rightarrow C \text{ and } C \rightarrow B$
- $\{C, D\} : C \rightarrow D \text{ and } D \rightarrow C$

Only some among the eight rule candidates are selected by the confidence. The support is used for finding the frequent item subsets, and the confidence is used for selecting association rules among candidates; they will be covered in Sect. 4.2.2.

Unique three item subsets are generated by attaching each frequent item which is not included in the original subset as follows:

$$\{A, B, C\}, \{A, B, D\}, \{A, C, D\}, \text{ and } \{B, C, D\}$$

Two of the above subsets are selected as the frequent subsets as follows:

$$\{A, B, C\} \text{ and } \{B, C, D\}$$

The six rule candidates are generated from the above frequent subsets as follows:

- $\{A, B, C\} : A \rightarrow \{B, C\}, B \rightarrow \{A, C\} \text{ and } C \rightarrow \{A, B\}$
- $\{B, C, D\} : B \rightarrow \{C, D\}, C \rightarrow \{B, D\} \text{ and } D \rightarrow \{B, C\}$

Some of them are also selected as the final association rules by the confidence. In the association rule, $A \rightarrow \{B, C\}$, A is the conditional item and $\{B, C\}$ is the causal set; it is interpreted as the statement, if A is given, then B and C are also given.

4.2.2 Support and Confidence

We need the two measures for proceeding the data association which was described in Sect. 4.3.2. One measure is the support, and the other is the confidence. The support and the confidence are used for selecting frequent item subsets and association rules, respectively, as mentioned previously. Both measures are the configurations for carrying out the data association. Therefore, in this section, we describe the support and the confidence, with respect to their computations.

The support refers to the ratio of item sets including all items of the given subset to all item sets in the collection. For example, the support of item subset, $\{A\}$, is the ratio of item sets including A to all item sets shown in Eq. (4.1).

$$sup(A) = \frac{\#item\_set\_with\ A}{\#all\_item\_set} \tag{4.1}$$

As another example, the support of item subset, $\{A, B\}$, the ratio of item sets including A and B to all item sets, is shown in Eq. (4.2),

$$sup(A, B) = \frac{\#item\_set\_with\ A\ and\ B}{\#all\_item\_set} \tag{4.2}$$

As the general form, the support of the subset, $\{I_1, I_2, \ldots, I_n\}$ is expressed in Eq. (4.3)

$$sup(I_1, I_2, \ldots, I_n) = \frac{\#item\_set\_with I_1, I_2, \ldots, I_n}{\#all\_item\_set} \tag{4.3}$$

The support of item subset is usually decreased, as the subset size increases as follows:

$$sup(I_1) \leq sup(I_1, I_2) \leq \cdots \leq sup(I_1, I_2, \ldots, I_n)$$

The confidence refers to the ratio of item sets including all items in the frequent subset to the item set which includes the item in the conditional part. The confidence of the association rule, $A \rightarrow B$, is expressed in Eq. (4.4).

$$con(A \rightarrow B) = \frac{\#item\_set\_with\ A\ and\ B}{\#item\_set\_with\ A} \tag{4.4}$$

As another example, the confidence of the rule, $A \rightarrow \{B, C\}$, is given as Eq. (4.5),

$$con(A \rightarrow \{B, C\}) = \frac{\#item\_set\_with\ A, B\ and\ C}{\#item\_set\_with\ A} \tag{4.5}$$

As the general form, the confidence of rule, $C \rightarrow \{I_1, I_2, \ldots I_n\}$, is expressed as Eq. (4.6),

$$con(C \rightarrow \{I_1, I_2, \ldots I_n\}) = \frac{\#item\_set\_with\ C, I_1, I_2, \ldots, I_n}{\#item\_set\_with\ C} \tag{4.6}$$

The confidences of two rules, $A \rightarrow B$ and $B \rightarrow A$, are expressed differently as follows:

$$con(A \rightarrow B) = \frac{\#item\_set\_with\ A\ and\ B}{\#item\_set\_with\ A}$$

$$con(B \rightarrow A) = \frac{\#item\_set\_with\ A\ and\ B}{\#item\_set\_with\ B}$$

In other words, the numerator is shared by the two rules, but the denominator is different from each other.

For example, the six item sets are given as follows:

$$\{A, B, C, D\}, \{A, B, C\}, \{A, C, D\}, \{A, D\}, \{B, C, D\}, \text{ and } \{C, D\}$$

In computing, the support of subset $\{A, B\}$, since the two sets $\{A, B, C, D\}$ and $\{A, B, C\}$ have A and B among six sets, the support becomes 0.3333. In computing the confidence of the rule, $A \rightarrow B$, since the two sets have A and B, and the four sets have A, the confidence becomes 0.5. In computing the confidence of rule $B \rightarrow A$, since the three sets have B, the confidence is 0.6777. From this example, we know that the confidences of $A \rightarrow B$ and $B \rightarrow A$ are different from each other.

The supporting and confidence thresholds are given as the configurations of the data association system. The subsets whose supports are greater than or equal to the supporting threshold are as the frequent subsets. If the support threshold is set closely to zero, it is more probable to select many frequent item subsets. From each frequent subset, the association rule candidates whose confidences are greater than or equal to the confidence threshold are selected as the final rules. If the confidence is set closely to zero, it is more probable that almost all of candidates becomes association rules.

4.2.3 Apriori Algorithm

This section is concerned with the Apriori algorithm which is the approach to the data association. Apriori algorithm illustrated in its pseudocode in Fig. 4.2, and the support and confidence thresholds are given as the external parameters. As shown in Fig. 4.2, the Apriori algorithm consists of the three parts: extractAssociation-RuleList, generateApriori, and generateAssociationRuleList. Each association rule is given as an object whose properties are the conditional item and the causal item set. In this section, we describe the Apriori algorithm based on the pseudocode which is presented in Fig. 4.2.

Let us explain the main procedure of Apriori Algorithm, extractAssociation-RuleList. In the procedure, it takes the two arguments: itemSetList which is the collection of item sets as input and itemList which is the list of all items. In the for loop, the frequent items are generated and the frequent item subsets are initialized based on their supports. In the second loop of repeat, by calling the procedure, generateApriori, the frequent item subsets are expanded and by calling the procedure, generateAssociationRuleList, association rules are generated. The main procedure, extractAssociationRuleList, returns the association rule list, and the two procedures which are called inside this procedure will be explained in the subsequent paragraphs.

```
List extractAssociationRuleList(List itemSetList, List itemList)
    //Extract Frequent 1-item Sets
    for each item in itemList
        if(support(item) > supportThreshold)
            frequentItemSubsetList.add(itemSet.add(item))
            frequentItemList.add(item)
    Repeat until No FrequentSubset
        frequentItemSubsetList = generateAprori(frequentItemSubsetList, frequentItemList)
        for each subset in frequentItemSubsetList
            associationRuleList.concat(generateAssociationRuleList(subset))
    return associationRuleList
```

```
List generateAprori(List itemSetList, List itemList)
    for each set in itemSetList
        for each item in itemList
            if(set.notContain(item) and (set.support(item) > supportThreshold))
                set.add(item)
                if(newItemSetList.notContain(set))
                    newItemSetList.add(set)
    return newItemSetList
```

```
List generateAssociationRuleList(Set itemSet)
    for each set in itemSet
        if(set.cardinality > 1)
            for each item in set
                ruleCandidate ← set.MakeRule(item, the others)
                if(confidence(ruleCandidate) > confidenceThreshold)
                    associationRuleList.add(ruleCandidate)
    return associationRuleList
```

Fig. 4.2 Pseudocode of Apriori algorithm

The procedure, generateApriori, is for expanding the frequent item subsets by checking their supports. As its arguments, the procedure takes iemSetList which is the frequent item subsets which is named and itemList which is the list of items. In the nested loop, for each of the existing frequent item subset and each item, if the item is new to the frequent item subset and the support of all elements of the subset and the item is higher than the threshold, the item is added to the given frequent item subset. If the frequent item subset which is expanded by the above process is unique, it is added to the frequent item subset list. Therefore, the new frequent item subset is generated by adding an item to the existing frequent one, based on its support, and it is added to the frequent item subset list, depending on its uniqueness.

For each frequent item subset, association rules are generated by calling the procedure, generateAssociationRuleList. The condition for generating association rules is that the cardinality of item subset should be greater than one. For each item in the item subset, one item is set as the conditional part, and the others are set as the casual set, in making each rule candidate. The confidence is computed for each rule candidate, and if the confidence is higher than the threshold, it is added to the association rule list. The association rules are generated for each frequent item subset and they are integrated over all subsets by concatenating in the procedure, extractAssociationRuleList.

Let us consider the complexity of Apriori algorithm to the number of all item sets and items, and they are denoted by N and n, respectively. It takes $O(nN)$ for selecting frequent items from item sets and $O(N)$ for computing the support of each item; it repeats this process n times. In the worst case, it takes $O(nN)$ for invoking the procedure, generateApriori, and it is invoked at most n time; it takes $O(n^2N)$ for executing the statements in the second loop of the procedure, extractAssociationRuleList. We need $O(N^2)$ for executing statements of the for-loop which is nested in the second loop, so we need $O(n^2N) + O(nN^2)$ as the total computation complexity. If N is equal to n, it takes the cubic complexity, $O(n^3)$ for executing the Apriori algorithm.

4.3   Word Association

In this section, we specialize the data association into one of text mining tasks. Word association is covered in this section, and the word-document matrix is mentioned as the preprocessing step before executing the word association, in Sect. 4.3.1. In Sect. 4.3.2, we describe the word association in its functional view. In Sect. 4.3.3, we present the process of executing the word association with its simple example. Therefore, this section is intended to describe in detail the word association as an instance of data association.

4.3.1   Word Text Matrix

This section is concerned with the word-document matrix as the source from which word sets are generated. Each text is associated with its included words by the text indexing process which was described in Chap. 2, as shown in Fig. 4.3. From the results from indexing texts in the collection, the word-document matrix is generated. In indexing texts in real versions of systems, statistical, posting, and grammatical information as well as words themselves may be included. In this section, we explain the concept of word-document matrix, the process of building it, and the process of generating word sets from the matrix.

Let us mention the text collection which was used for evaluating text mining performances in previous works [85]. NewsPage.com was mentioned as the relative small corpus which consists of 2000 texts [36]. The text collection, 20NewsGroups, consists of 20 categories and 20,000 texts with their balanced distributions over categories [79]. The collection, Reuter21578, consists of more than 100 categories and approximately 20,000 texts with their unbalanced distribution over categories [85]. In this book, it is assumed that the text collection is given as a directory where individual texts are given as text files whose extension is "txt."

The process of generating the word-document matrix is illustrated in Fig. 4.4. An individual text is indexed by the process which was described in Chap. 2, into a list

Fig. 4.3 Frame of
word-document matrix

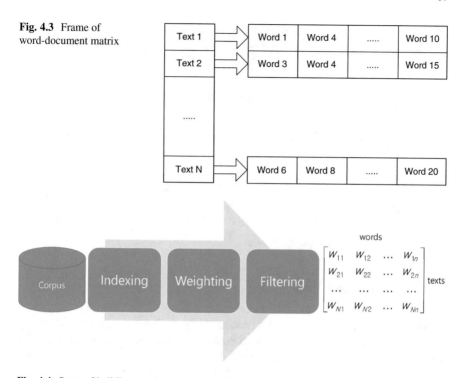

Fig. 4.4 Steps of building word-document matrix in corpus

of words. The weights of words are computed in each text by the weighting process
which was mentioned in Chap. 2. In the filtering process, the words with their
relative lower weights may be removed. As the output of the process, the word text
matrix is generated where each row corresponds to a text, each column corresponds
to a word, and each entry is given as the word weight in the corresponding text.

The simple word text matrix is illustrated in Fig. 4.5. In the given matrix, the rows
correspond to the eight texts which are identified by their file names and the columns
correspond to the four words. The weights may be given as binary values which
indicate the presence or absence of corresponding words; zero means that the word
is not available in the text. For example, text 1 is linked to the words: "business,"
"computer," and "information," and text 2 is linked to the words: "company" and
"computer." Text by text, the word set is extracted by selecting words whose weights
are 1.0.

The item sets are extracted from the word text matrix by a simple process. Each
row in the word-document matrix is mapped into a set by selecting the columns
with nonzero values. The words and texts correspond to items and customers when
compared with the case in Sect. 4.2, respectively. Each text is represented into a
word for preparing word association. The word association will be described in
detail in Sect. 4.3.2.

	business	company	computer	information
Text 1	1	0	1	1
Text 2	0	1	1	0
Text 3	0	1	1	1
Text 4	1	0	1	0
Text 5	1	0	0	1
Text 6	1	0	0	0
Text 7	0	1	0	1
Text 8	0	1	1	0

Fig. 4.5 Example of word-document matrix

4.3.2 Functional View

Word association is defined as the process of extracting the associations among words as if-then rules. Word sets are extracted from the word-document matrix by the process which was mentioned in Sect. 4.3.1. The association rules of words are generated by the Apriori algorithm which was described in Sect. 4.2.3. The word association rules may be utilized for generating associated queries automatically in using search engines. In this section, we explain the task, word association, in the functional view, assuming that the Apriori algorithm is applied to it.

We need to collect word sets which are given as item sets before applying the Apriori algorithm. The word text matrix is constructed from a corpus by the process which was mentioned in Sect. 4.3.1. The word sets are built row by row by taking words whose weights are nonzeros. A collection of word sets and a list of words which correspond to columns in the matrix are given as arguments for the Apriori algorithm. The word sets may be adjusted depending on text lengths by index expansion, index filtering, or index optimization which were mentioned in Chap. 2.

We prepare the collection of word sets and the word list as the input of Apriori algorithm and configure the external parameters, support and confidence threshold. Some among words are selected by their supports as the initial frequent item subsets and the frequent item list. Each of the frequent item subsets is expanded by calling the procedure, generateApriori, which is shown in Fig. 4.2. For each frequent item subset, we generate association rules as the form if a word is given, its causal list of words is reminded by it. In the word association, each association rule consists of the conditional word and its causal set of words.

Let us consider the association rule which is extracted in the word association. For example, the association rule, "computer" → "software," is interpreted into the statement that the word, "software," is reminded from the word, "computer." There is no guarantee of rule, "software" → "computer," since the association is a unidirectional relation; we need to distinguish the association from the similarity between words. If the rule, "computer" $\rightarrow$ "software," is extracted, and the rule "software" $\rightarrow$ "computer" is not extracted by their confidences, the word, "computer," becomes the sufficient condition to the word, "software," and the latter becomes the necessary condition to the former. When both rules are extracted, both words become the necessary and sufficient conditions to each other.

Let us consider the issues in executing word association which are distinguished from that of the generic data association. Lexically, different words which have different spellings may have the same meaning as the case of words, "vehicle" and "automobile"; the association rule, "vehicle" $\rightarrow$ "truck," is available but the rule "automobile" $\rightarrow$ "truck" is not. The identical association rule has its different confidence depending on the given corpus; for example, the rule, "vehicle" $\rightarrow$ "truck" may be extracted in a particular corpus, but it is not in another corpus, because different confidence values are computed in the two corpora. We need to discriminate association rules depending on their confidences: association rules whose confidences are close to one and ones whose confidences are slightly higher than threshold. Because of very high computation cost of the Apriori algorithm, we also need to consider the computation feasibility to a very large number of word sets.

4.3.3 Simple Example

This section is concerned with the demonstration of doing the word association using the Apriori algorithm. The word text matrix is constructed from the process which was mentioned in Sect. 4.3.1, and the word sets are extracted from the matrix. The word association rules are extracted from the word sets by the Apriori algorithm. Even if the example is far from the real case, it is used for understanding the word association easily. Therefore, in this section, we present the demonstration through the simple example.

The word text matrix is constructed from a corpus as illustrated in Fig. 4.5. The word text matrix in the real case is a very big matrix; it is usually a ten thousands and ten thousands matrix. In Fig. 4.5, the eight sets of words are extracted; each row corresponds to a set. In each text, the words whose weights are 1.0s are selected as elements of sets. The cardinality of each word set is in the range of two or three.

In Fig. 4.6, we illustrate the process of extracting the frequent word sets and frequent words. The left table in Fig. 4.6 is the collection of word sets which are extracted from the word text matrix which was given in Fig. 4.5, and the support and confidence thresholds are set to 0.25s. In the right table, all possible pairs of the four words as two item subsets and their supports are computed in the next column. The two item subsets whose supports are higher than the threshold are marked with circles in the right column in the table. Hence, the two item subsets marked with circles are selected as the frequent ones, in this example.

The process of generating the association rules and expanding the frequent item sets is illustrated in Fig. 4.7. The rule candidates are generated from each frequent

Text	Word List
Text 1	computer business information
Text 2	company computer
Text 3	company computer information
Text 4	business computer
Text 5	business information
Text 6	business
Text 7	company information
Text 8	company computer

Item Set	Support	Selection
{company, information}	0.25	O
{computer, information}	0.125	X
{business, information}	0.25	O
{company, computer}	0.375	O
{business, company}	0.0	X
{business, computer}	0.25	O

Support Threshold = 0.25   Confidence Threshold = 0.25

FrequentWordList = {business company computer information}
FrequentWordSetList = {{business}, {company}, {computer}, {information}}

Fig. 4.6   Two item sets

Apriori

Item Set	Support	Selection
{company, information}	0.25	O
{computer, information}	0.125	X
{business, information}	0.25	O
{company, computer}	0.375	O
{business, company}	0.0	X
{business, computer}	0.25	O

Association Rules

Association Rules	Confidence
company → information	0.5
information → company	0.5
business → information	0.5
information → business	0.5
company → computer	0.75
computer → company	0.6
business → computer	0.5
Computer → business	0.4

FrequentItemList = {business company computer information}
FrequentItemSetList = { {company, information}, {business, information}, {company, computer}, {business, computer}}

Item Set	Support	Selection
{company, computer, information}	0.125	X
{business, company, information}	0.0	X
{business, computer, information}	0.125	X
{business, company, computer}	0	X

Fig. 4.7   Multiple item sets

word subset which is marked with a circle in the top left table. The confidence is computed and filled in each rule candidate, and all candidates are selected as rules because their confidences are greater than or equal to 0.25. The three item subsets are made in the bottom table, by adding the frequent item to the existing frequent item subsets. Because supports of all subsets are lower than 0.25, nothing is selected and the word association proceeds no more.

Finally, the word association rules in the top right table of Fig. 4.6 are generated as the output of this example, but we need to consider the adjustment of the support and confidence thresholds. The current results are characterized as the two facts; one is that the three frequent item subsets are not available, and the other is that all candidates are extracted as the association rules from the frequent two item subsets. As the adjustment, we try to decrease the support threshold from 0.25 to 0.1 and increase the confidence threshold from 0.25 to 0.55. In the adjusted configurations, the three item subsets in the bottom table whose support is 0.125 are selected as the frequent item subsets, and only two candidates, "computer" $\rightarrow$ "company" and "company" $\rightarrow$ "computer," are selected as association rules. From this example, we know that the support and confidence thresholds are proportional inversely to the number of extracted association rules.

4.4 Text Association

This section is concerned with text association which is another instance of data association. Text association is also based on the word-document matrix which was described in Sect. 4.3.1, like the case of word association. In Sect. 4.4.1, we explain the text association in its functional view. In Sect. 4.4.1, we demonstrate that the text association is executed using the Apriori algorithm through a simple example. In this section, we explain and demonstrate the text association.

4.4.1 Functional View

Text association refers to the process of extracting the text association rules from a corpus or text sets. The text association which is covered in this section is called document association or article association, and is specialized for articles each of which consists of more than one paragraph, in the specific view; it should be distinguished from text association which was mentioned in Sect. 4.1 in the general view. In the text association, a word is represented into a text set, and the association rule is extracted as the form: text $A \rightarrow$ text B. The results from executing the text association are used for displaying directional relations among texts in the given corpus. Therefore, in this section, we provide the detailed description of text association as one which is specific to articles.

The text association is a different kind of text mining task; it is different from the word association. In word association, a text is represented into a word set, whereas in the text association, a word is represented into a text set. In word association, the text index where each text is linked with its included words is the basis, whereas in the text association, the inverted index where each word is linked with texts which include it is the basis. The results from executing word association are used for getting the associated queries to the given query in implementing search engines, whereas those from executing the text association are used for representing texts into directed graph where vertices are texts and edges are their associations. As alternative types, we may consider the subtext association which extracts association rules of sentences or paragraphs with others.

Once a collection of text sets is prepared, the Apriori algorithm is applied for executing the text association, as well as the word association. Among texts, only some are selected by their supports as the initial frequent item subsets and the frequent items. The frequent item subsets are expanded by invoking the procedure, generateApriori, where more frequent item subsets are generated by adding a text. For each frequent item subset, association rules are generated as the statement, if a text is given, other texts are reminded. Each association rule is composed with the conditional text and its causal text set.

Let us analyze the association rules which are results from executing the text association with the Apriori algorithm. In the case of rule, text $A \rightarrow$ text B, since text A is less frequent than text B, it is more probable that text A is short and text B is long. If text B includes the entire content of text A as its part, it is certain to generate text $A \rightarrow$ text B as the association rule, since its confidence reaches 1.0. If the association rule, text $A \rightarrow$ text B, text C, and text D, text A may be a summarized version to the others. The text association rule may be used for deciding the representative text in a text group, based on the fact.

We need to consider some issues of text association as well as those of word association. The number of association rules is very dependent on whether a corpus is broad or specific; more association rules are usually extracted from a specific corpus than from a broad one, since item sets are more similar as each other in the specific one. Text length may become a bias for carrying out the text association; as mentioned above, short texts become the conditional part, and long texts locate in the causal set. Results from performing the text association depend on the scheme of indexing texts; text sets are built differently depending on with or without the index optimization. Like the case of word association, we need to consider the discriminations among association rules by their confidences.

4.4.2 Simple Example

This section is concerned with the demonstration of executing the text association using the Apriori algorithm. The word text matrix is constructed from a corpus by the process which was mentioned in Sect. 4.3.1, and text sets are extracted from the matrix. The text association rules are extracted from text sets by the Apriori algorithm. Even if the word text matrix which is presented in Fig. 4.8 is far from the real case, it needs to be used for understanding the process of extracting text association rules easily. Therefore, in this section, we present the demonstration of proceeding the text association by the Apriori algorithm, using the simple example.

As illustrated in Fig. 4.8, the transpose of word-document matrix is constructed. In the matrix, the columns correspond to the four texts and the rows correspond to the eight words. The matrix entries are given as binary values which indicate the presence or the absence of words in the given text. We select texts whose weights are 1.0s in each word as set elements. The cardinality of each text set is between one and three in this example.

	Text1	Text2	Text3	Text4
Agent	0	0	0	1
business	1	1	0	1
company	0	1	1	0
computation	1	0	0	1
computer	0	1	0	1
information	1	1	1	0
intelligence	1	0	0	1
technology	0	1	0	1

Fig. 4.8 Example of document-word matrix

Word	Texts including Word
information	text1, text2, text3
computer	text2, text4
business	text1, text2, text4
company	text2, text3
intelligence	text1, text4
agent	text4
computation	text1, text4
technology	text2, text4

Item Set	Support	Selection
{text1, text2}	0.25	O
{text1, text3}	0.125	X
{text1, text4}	0.25	O
{text2, text3}	0.375	O
{text2, text4}	0.0	X
{text3, text4}	0.25	O

Support Threshold = 0.25   Confidence Threshold = 0.25

FrequentWordList = {text1, text2, text3, text4}
FrequentWordSetList = {{text1}, {text2}, {text3}, {text4}}

Fig. 4.9 Two item sets

The process of extracting the frequent two item subsets and association rules is illustrated in Fig. 4.9. The left table in Fig. 4.9 is the collection of text sets which are extracted from the matrix which is given in Fig. 4.8, and the support and confidence thresholds are set to 0.25s. In the right table, all possible pairs of four texts as two item sets and their supports which are computed from the left table are given. The two item sets whose supports are greater than or equal to 0.25 are marked with circles as shown in the right table. The two item sets which are marked with circles are selected as the frequent item subsets.

The process of generating the association rules and expanding the frequent item subsets is illustrated in Fig. 4.10. The rule candidates are generated from each frequent word subset which are selected. In each rule candidate, its own confidence is computed and given; all candidates are selected as association rules, because their confidences are greater than or equal to 0.25. By expanding two item subsets

Apriori

Item Set	Support	Selection
{text1, text2}	0.25	O
{text1, text3}	0.125	X
{text1, text4}	0.25	O
{text2, text3}	0.375	O
{text2, text4}	0.0	X
{text3, text4}	0.25	O

Association Rules

Association Rules	Confidence
text1 → text2	0.5
text2 → text1	0.5
text1 → text4	0.75
text4 → text1	0.5
text2 → text3	0.5
text3 → text2	1.0
text2 → text4	0.75
text4 → text2	0.4

FrequentItemList = {text1, text2, text3, text4}
FrequentItemSetList = { {text1, text2}, {text1, text4}, {text2, text3} , {text3, text4}}

Item Set	Support	Selection
{text1, text2, text3}	0.125	X
{text1, text2, text4}	0.125	X
{text1, text3, text4}	0.0	X
{text2, text3, text4}	0.0	X

Fig. 4.10 Multiple item sets

with each of frequent items, the three word subsets are made in the bottom table. However, nothing is selected because all of three item subsets have their supports which are less than 0.25, and the process is terminated.

In this example, results are similar as those in the example in Sect. 4.3.3, as shown in Fig. 4.10. The current results shown in Fig. 4.10 are characterized identically to those in Fig. 4.7. Let u try to decrease the support threshold from 0.25 to 0.1 and increase the confidence threshold from 0.25 to 0.55. By adjusting the parameter, two more three item subsets whose support is 0.125 in the bottom table may be selected as the frequent item subsets and the three association rules from each of frequent two item subsets, text 1 → text 4, text 3 → text 2, and text 2 → text 4, are selected rather than all. We try to execute text associations with their different configurations several times, and select the best one by subjectivity, in using the text association systems.

4.5 Overall Summary

This section is concerned with the brief summary of the entire content of this chapter. As the fundamental task of text mining, the text association is defined as the process of getting association rules of textual units in the broad view. However, it should be distinguished from one which is covered in Sect. 4.4 in its specific view. In this chapter, the Apriori algorithm is described as the approach to both word association and text association as well as genetic data association. In this section, we summarize what is studied in this chapter.

Depending on what kind of text unit is associated with others, the text association is divided into the three types: word association, article association, and sentence association. The article association among them is mentioned as text association in its specific meaning in Sect. 4.4. Word association refers to the process of extracting association rules of words and the article association refers to that of extracting rules of texts. It is possible to generate association rules of texts with words and vice versa. We may consider the subtext association where association rules of sentences or paragraphs are extracted or the hybrid association where association rules of entity with another type of entities are extracted, even if they are out of the scope of this chapter.

The data association which is the fundamental task of the traditional data mining is derived from the idea of discovering the purchase patterns of customers in big stores. From each customer, a set of his or her purchased items is collected. From each frequent item set, using the Apriori algorithm the association rule, item $A \rightarrow$ item B, is extracted and interpreted into the statement: if a customer buys item A, he or she tends to buy item B. The support and confidence are used for selecting frequent item subsets and association rules, respectively, as the numeric metrics. It takes the cubic complexity for using the Apriori algorithm for doing the data association to the size of item sets and items.

We studied the word association which is specialized for words as items. The word text matrix is constructed from a given corpus and word sets are extracted, text by text. Association rules are generated from word sets using the Apriori algorithm. Each word association consists of its conditional word and its causal word set. Results from doing the word association are used for extracting the associated queries in implementing search engines.

We studied the text association which is another task of text mining. Text sets are extracted text by text from the transpose of word text matrix. Using the Apriori algorithm, the association rules are generated from text sets; the Apriori algorithm is applicable to both word association and text association. If no additional process is applied, it is more probable that short texts are given as the conditional parts. The results from executing the text association are used for displaying the directional relationships among texts in the corpus.

Part II
Text Categorization

Part II is concerned with the four aspects of text categorization: concepts, approaches, implementations, and evaluations. This part begins with the specification of text categorization with respect to its functions. We mention the supervised machine learning algorithms as the approaches to text categorization, and present its implementation in Java. We consider the schemes of evaluating the text categorization system, statistically. Therefore, Part II is intended to cover the text categorization with the four aspects.

Chapter 5 is concerned with the functional view of text categorization. We present the process of decomposing the text categorization into binary classifications and provide the guide of applying the machine learning algorithms which will be mentioned in Chap. 6 to the given task. The text categorization is classified into the four types: hard text categorization, soft text categorization, hierarchical text categorization, and multiple viewed text categorization, and each of them and their differences are explained. We will explore the special instances of text categorization, such as spam mail filtering, information personalization, and opinion classification. Therefore, this chapter is intended for describing the tasks of text categorization in the functional view.

Chapter 6 is concerned with the representative machine learning approaches to the tasks of text categorization. The chapter begins by mentioning the k Nearest Neighbor as the typical machine learning algorithm which is simple and practical. Next, we explain the Naive Bayes as another popular approach to the tasks. The SVM (Support Vector Machine) will be also described as the most recent approach which has the best performance among the approaches. Therefore, Chap. 6 is intended to describe briefly the three representative machine learning algorithms as the text categorization tools.

Chapter 7 is concerned with implementing the text categorization system in Java. We present the text categorization system specification and its class list, together with the system roles. We will explain the Java source codes for implementing the system, in detail. As an independent prototype system, we will demonstrate the text categorization system in its execution phase. The K Nearest Neighbor (KNN) is adopted as the approach.

Chapter 8 is concerned with the schemes of evaluating the text categorization systems. We describe the F1 measure which is the most popular evaluation metric and is introduced from the information retrieval area. The statistical test is applied to comparing the two text categorization approaches with each other. We also cover the advanced statistical method for comparing more than two approaches, such as ANOVA. Therefore, this chapter provides the statistical scheme for comparing text categorization system with each other, based on the F1 measure.

Chapter 5
Text Categorization: Conceptual View

This chapter is concerned with the conceptual view of text categorization. We cover the overview of text categorization in Sect. 5.1, and explain the classification in its conceptual view in Sect. 5.2. In Sect. 5.3, we will explore the types of text categorization by the different dichotomizations. We mention the typical real tasks which are derived from the text categorization, in Sect. 5.4, and make the summarization and further discussions on this chapter in Sect. 5.5. In this chapter, we describe the text categorization tasks with respect to their types and real examples.

5.1 Definition of Text Categorization

Text categorization is defined as the process of assigning one or some of the predefined categories to each text, basically, through the three steps which are shown in Fig. 5.1. The text categorization which is called text classification is an instance of classification task, where a text is given as the classification target. Texts are encoded into numerical vectors by the processes which were described in Chaps. 2 and 3. We use mainly the machine learning algorithms which are described in the next chapter, as the approaches to the text categorization. In this section, we explore the overview of text categorization, before covering it in detail.

The preliminary tasks are required for executing the text categorization system, even in the simplest version which will be mentioned in Chap. 7. It is required to pre-define a list or a tree of categories as the frame of classifying data items. Texts should be allocated to each category, as sample ones. All sample texts are indexed into a list of words which are called feature candidates, and some of them are selected. As the additional preliminary task, we may decide the classification algorithm and the type of classification such as exclusive or overlapping classification and flat or hierarchical one.

© Springer International Publishing AG, part of Springer Nature 2019
T. Jo, *Text Mining*, Studies in Big Data 45,
https://doi.org/10.1007/978-3-319-91815-0_5

Fig. 5.1 Text categorization steps

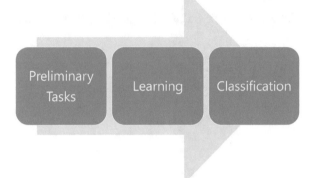

Once accomplishing the above preliminary tasks including the decision of which classification algorithm is adopted, the classification capacity should be constructed using sample texts. The sample texts which are allocated to categories in the preliminary task are encoded into numerical vectors whose attributes are the selected features. Using the training examples which are encoded from the sample texts, the classification capacity which is given as one of various forms such as equations, symbolic rules, or optimized parameters, depending on the classification algorithm, is constructed; this is called learning process. Optionally, the results from the learning process may be validated using a set of examples, called the validation set which is separated from training examples. In this case, the set of sample texts may be divided into two sets: the pure training set and the validation set; the validation set is not involved in the learning process.

After the learning process, the texts which are given, separated from the sample texts, are classified. The classification capacity is constructed through the learning process which is mentioned above, including the validation process. The classification performance is evaluated, using the test set which is initially left separated from the training set. Depending on the classification performance, we decide whether we adopt or not the classification algorithm. The set of texts which is given in a real field after adopting the classification is called the real set.

Let us consider some issues from implementing the text categorization systems. A list or tree of topics, what is called a classification frame, is predefined, depending on subjectivity, before gathering sample texts. It is not easy to gather labeled sample texts; it is very tedious to decide manually target categories of texts. Independencies among categories are not guaranteed; some categories are correlated with others. The classification frame for maintaining texts continually is usually fixed; it takes very much cost for changing it in maintaining text categorization systems.

5.2   Data Classification

This section is concerned with the data classification in its conceptual view and consists of the four sections. In Sect. 5.2.1, we describe binary classification conceptually as the simplest classification task. In Sect. 5.2.2, we cover the multiple classification which is expanded from the binary classification. In Sect. 5.2.3, we explain the process of decomposing a single multiary-classification task into binary classification tasks as many as the predefined categories. In Sect. 5.2.4, we describe the regression to which the supervised learning algorithms which we mention in Chap. 6.

5.2.1   Binary Classification

Binary classification is referred to the simplest classification task where each item is classified into one of the two categories, as illustrated in Fig. 5.2. The assumption underlying in the binary classification is that each item belongs to one of the two classes, exclusively. We need to define criteria, in order to decide each item belongs to which of the two classes. Some items may belong to both categories in real classification tasks; such a classification task is called overlapping or fuzzy binary classification. Therefore, in this section, we describe the binary classification task as the entrance for understanding the classification task.

Let us present a simple example of binary classification. Points may be plotted in the two-dimensional space; each point expressed as (x_1, x_2). The points belonging to the positive class are plotted in the area where $x_1 \geq 0$ and $x_2 \geq 0$. The points belonging to the negative class are plotted in the area where $x_1 < 0$ and $x_2 < 0$. The points of the two classes are plotted in the two-dimensional space, and let us consider the dichotomy which separates the two classes from each other.

Let us define the symbolic classification rules for classifying the points into one of the two classes. The rule for classifying a point into the positive class is defined as if $x_1 \geq 0$ and $x_2 \geq 0$ then the point belongs to the positive class. The rule for classifying a point into the negative class is defined as if $x_1 < 0$ and $x_2 < 0$ then the

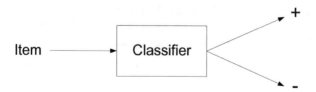

Fig. 5.2 Binary classification

point belongs to the negative class. According to the above rules, the points, where $x_1 \geq 0$ and $x_2 < 0$, or $x_1 < 0$ and $x_2 \geq 0$, are rejected; they are out of the above rules. We may consider the alternative dichotomy to the above rules for classifying a point into one of the two classes.

Machine learning algorithms are considered as the alternative kinds of classification tools to the rule-based approaches. Instead of above rules, we gather examples which are labeled with positive or negative class; they are called training examples. By analyzing the training examples, the classification capacity is constructed and given as various forms such as equations, symbolic rules, or neural network models. The examples in the separated set, which is called the test set, are classified by the classification capacity. In Chap. 6, we will describe the classification capacity which is generated from the training examples, in detail.

Even if the binary classification looks a very simple toy task, it may exist as a real task. Spam mail filtering where junk emails are automatically filtered from incoming ones is the typical instance of binary classification. Detecting whether a report about a traffic accident is true or false is a real example which is used in insurance companies. The information retrieval task is viewed as a binary classification which decides whether each text is relevant or irrelevant to a given query. The keyword extraction and the text summarization are instances of binary classification.

5.2.2   Multiple Classification

Multiple classification is illustrated as a block diagram in Fig. 5.3. Each item is classified into one of the two classes in binary classification, as illustrated in Fig. 5.2. Multiple classification is referred to the classification task, where each item is classified into one of at least three categories. The multiple classification may be decomposed into binary classification tasks as many as categories; it will be described in Sect. 5.2.3. Therefore, in this section, we describe the multiple classification, in its conceptual view.

In order to understand it easily, we present a simple example of multiple classification in the two-dimensional space. It is assumed that the four classes are given, class 1, class 2, class 3, and class 4. The constraints for class 1 are given as $x_1 \geq 0$ and $x_2 \geq 0$ and those of class 2 are given as $x_1 < 0$ and $x_2 \geq 0$. The constraints, $x_1 \geq 0$ and $x_2 < 0$, are considered for class 3, and $x_1 < 0$ and $x_2 < 0$ are considered for class 4. A point, (x_1, x_2) in the two dimensional space is classified into one of the four classes by the above constraints.

Fig. 5.3 Multiple classification

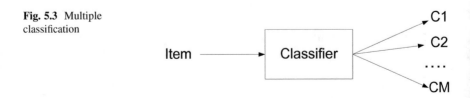

The if-then rules for classifying points in the two-dimensional space into one of the four classes are defined as follows:

$$\text{if } x_1 \geq 0 \text{ and } x_2 \geq 0 \text{ then class 1}$$

$$\text{if } x_1 < 0 \text{ and } x_2 \geq 0 \text{ then class 2}$$

$$\text{if } x_1 \geq 0 \text{ and } x_2 < 0 \text{ then class 3}$$

$$\text{if } x_1 < 0 \text{ and } x_2 < 0 \text{ then class 4}$$

For example, the point, (5, 4), is classified into class 1 by the first rule, because both x_1 and x_2 are greater than or equal to zero. As another example, the point, $(-7, 2)$, is classified into class 2, applying the second rule to it. The point, $(6, -3)$, is classified into class 3, according to the third rule. It is possible that no rule is applicable to an input in real tasks which are complicated much more than the current multiple classification.

Let us consider the soft classification where an item is allowed to be classified into more than one class. The classification which is mentioned above belongs to the hard classification where no overlapping among the four classes is allowed from the rules. The first rule is modified as follows: if $x_1 \geq -2$ and $x_2 \geq -2$ then class 1. For example, the point, $(-1, -1.5)$, may be classified into class 1 and class 4, because the first and the second are applied to the input. Therefore, the area, $-2 \leq x_1 < 0$ and $-2 \leq x_2 < 0$, is overlapping between class 1 and class 4.

Let us present some real multiple classification instances. The case of classifying news articles into one or some of predefined sections may be mentioned as a typical case. The optical character recognition which is referred to the process of classifying a character image into one of ASCII codes is mentioned as another case. The POS (Part of Speech) tagging which is the process of classifying a word by its grammatical function is a multiple classification instance in the area of natural language processing. Sentimental analysis which is a special type of text classification is the process of classifying a text into one of the three categories: positive, neutral, and negative.

5.2.3 Classification Decomposition

This section is concerned with the process of decomposing the multiple classification task into binary classification ones, as illustrated in Fig. 5.4. It is very risky of misclassifying items to apply a single classifier directly to the multiple classification task where each item is classified into one of multiple categories. The multiple classification is divided into binary classifications as many as categories; each item is classified into positive as the corresponding category or negative; it is intended to reduce the misclassification risk. The category or categories which correspond to what the binary classifiers produce positives are assigned to each item. In this section, we describe the process of decomposing the multiple classification into binary classifications.

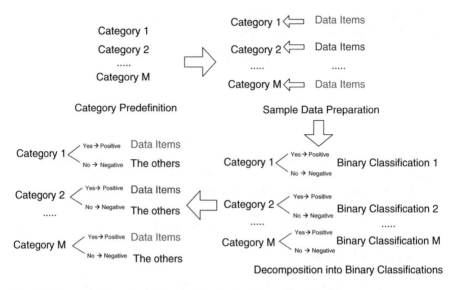

Fig. 5.4 Decomposition of multiple classification into binary classifications

Let us mention the binary classification task which is derived from a multiple classification task by the decomposition. The task which is initially defined is the multiple classification where each item is classified into one of M categories. The M binary classification tasks which correspond to the M categories are derived by decomposing the initial task, and two labels are defined in each binary classification as positive and negative. The positive class stands for the fact that the item belongs to the corresponding category, while the negative class stands for the fact that it does not. The multiple classification task is interpreted into a group of binary classification tasks through the decomposition.

Let us mention the process of decomposing the multiple classification into the M binary classification tasks. The M classifiers are allocated to the M categories as the binary classifiers. The training examples which belong to the corresponding category are labeled with the positive class, and some of the training examples which do not belong to it are selected as many as the positive examples, and labeled with the negative class. The classifiers which correspond to the category learn their own positive examples and negative examples. If all of training examples which do not belong to the corresponding category are used as the negative examples, the classifier tend strongly to classify novice examples into negative class by the unbalanced distribution which is biased toward it.

Let us explain the process of classifying a novice item after learning the sample examples. A novice item is submitted to the classifiers which correspond to the predefined categories as the classification target. It is classified by the classifiers into a list of positive and negative classes. The categories which correspond to the classifiers which classify the novice example into the positive class are assigned to it. If the given classification is exclusive, the category with its maximum categorical score among them is selected.

Note that much overhead is caused by decomposing the multiple classification into binary classifications. The training examples which are initially labeled with one or some of the predefined categories should be relabeled with the positive class or the negative class, and they should be reassigned to each classifier as the payment for doing the decomposition. The classifiers as many as the categories should learn training examples, in parallel. We must wait until all of the classifiers make their decisions, instead of taking an output of single classifier. Because the classifiers are independent of each other with respect to their learning and classification, it is possible to implement them as the parallel and distributed computing.

5.2.4 Regression

Regression is referred to the process of estimating an output value or output values as continuous ones by analyzing input values. The supervised learning algorithm is also applied to the task like the classification. In the classification, one of output values is given as a discrete value, whereas, in the regression, it is given as a continuous one. The regression may be mapped into the classification in some areas by discretizing each continuous value into a finite number of intervals. In this section, we describe the regression in its functional view and map it into the classification.

As mentioned above, it is possible to map the regression into the classification. An output value is discretized into a finite number of intervals; the intervals are given as the predefined categories. The target output value is replaced by its corresponding interval for each training example; they are learned by the given supervised machine learning algorithm. A novice item is classified into one of the predefined intervals; the classified one is the label which indicates the interval within which its continuous value exists. The regression is mapped into a binary classification by discretizing the output value into only two intervals; if the number of intervals is more than two, it is mapped into a multiple classification.

Figure 5.5 illustrates the process of decomposing the regression into the binary classification tasks. Regression is mapped into the multiple classification by discretizing the output continuous value into several intervals. The multiple classification is decomposed into the binary classifications by the process which was described in Sect. 5.2.3. The classifiers are allocated to the corresponding intervals

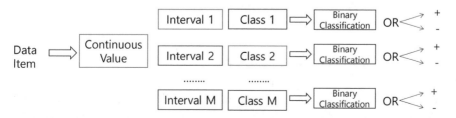

Fig. 5.5 Decomposition of regression into binary classifications

and learn their corresponding training examples, but the process of preparing the training examples will be mentioned in the next paragraph. The classification mapped from the regression belongs to the exclusive one where each item is classified into only one category; only one classifier is allowed to classify the item into the positive class.

The training examples which are prepared for the regression task are labeled with their own continuous value. The label which is given as a continuous value is changed into its corresponding internal in mapping so. By changing one more time the discrete label into the positive or the negative, the training examples are prepared for each classifier in mapping the multiple classification into the binary classifications. The classifiers are allocated to the intervals and trained with the prepared examples. In Sect. 5.2.3, we already explained the meaning of the positive and negative to the binary classifiers.

The classification and the regression are compared with each other in terms of their differences and sharing points. The supervised learning algorithms are applied to both kinds of tasks as their sharing point. The difference between them is that the classification generates a discrete value or values as its output whereas the regression does a continuous value or values, as mentioned above. The classification instances are spam mail filtering, text categorization, and image classification, and the regression instances are nonlinear function approximation and time series prediction. The error rate which is called risk function is defined as the rate of misclassified items to the total items in the classification and average over differences between desired and computed values in the regression.

5.3   Classification Types

This section is concerned with the types of text categorization, depending on the dichotomy criterion. In Sect. 5.3.1, we will explain the hard classification and the soft classification depending on whether each item is allowed to belong to more than one category, or not. In Sect. 5.3.2, we mention the flat classification and the hierarchical classification, depending on whether any nested category in a particular one is allowed or not. In Sect. 5.3.3, we describe the single viewed classification and the multiple viewed one, depending on whether multiple classification frames are accommodated or not. In Sect. 5.3.4, we consider the independent classification and the dependent one, depending on whether a current classification is influenced by the results from classifying items previously.

5.3.1   Hard vs Soft Classification

The hard classification and the soft one are illustrated in Fig. 5.6. The hard classification is one where no overlapping exists among categories; all items belong

Fig. 5.6 Hard vs soft
classification

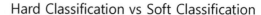

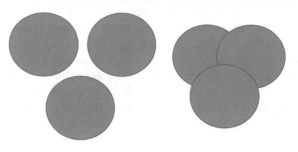

Hard Classification					Soft Classification			

Hard Classification

	Category 1	Category2	Category3	Category4
Item 1	O	X	X	X
Item 2	X	O	X	X
Item 3	X	X	X	O
Item 4	X	O	X	X
Item 5	X	O	X	X
Item 6	O	X	X	X
Item 7	X	X	O	X
Item 8	X	O	X	X

Soft Classification

	Category 1	Category2	Category3	Category4
Item 1	O	X	O	X
Item 2	X	O	O	O
Item 3	O	X	X	O
Item 4	X	O	X	X
Item 5	O	O	X	X
Item 6	O	X	O	X
Item 7	X	X	O	X
Item 8	X	O	O	X

Fig. 5.7 Example of hard and soft classification

to only one category. The soft one is one where any overlapping exists; one item
belongs to two categories, at least. The criterion of deciding the hard classification
and the soft one is whether the overlapping is allowed or not. In this section, we
describe the two classification types in the conceptual view.

The hard classification is defined as the classification where every item is
classified into only one category, as shown in the left part of Fig. 5.6. In this
classification type, training examples which are labeled with only one category are
prepared. When a small number of categories are predefined, we may use a single
classification without decomposing the task into binary classifications. For each test
item, one of a fixed number of predefined categories is decided by the classifier. The
optical digit recognition and the spam mail filtering belong to the hard classification.

The soft classification is referred to one where, at least, one item is classified into
more than one category. Some or almost all of training examples are initially labeled
with more than one category. This type of classification should be decomposed into
binary classifications as the requirement for applying machine learning algorithms.
The F1 measure is used as the evaluation metric, rather than accuracy in this case.
The news article classification and the image classification become instances of this
classification type.

Both the classification types are demonstrated by a simple example in Fig. 5.7.
The eight items and the four categories are prepared in the example. In the hard
classification, the eight items are labeled on only one of the four categories, as shown
in the left part of Fig. 5.7. In the soft classification, the six items are labeled with

more than one category, as shown in the right part of Fig. 5.7. It is possible to assign the category membership values to each item, instead of some categories.

Text categorization belongs to the soft classification, more frequently than to the hard classification. The spam mail filtering in the email system is the typical case of hard classification. Because each text tends to cover more than one topic, the text categorization belongs to the soft classification. The text collection, Reuter21578, where almost all of texts are labeled with more than one topic is the most standard one which is used for evaluating text categorization systems. So we consider segmenting a text into subtexts based on its topics which will be mentioned in Chap. 14.

5.3.2   Flat vs Hierarchical Classification

Figure 5.8 illustrates the two kinds of classification: flat classification and hierarchical classification. The dichotomy criterion for dividing the classification into the two types is whether any nested category is allowed in a particular one, or not. Flat classification is one where the predefined categories are given as a list and no nested category is allowed, whereas the hierarchical category is one where the predefined categories are given as a tree and any nested category is allowed. In hierarchical classification, we need to consider the case where a data item is classified correctly at the abstract level, but incorrectly at the specific level. In this section, we describe the two types of classification in detail.

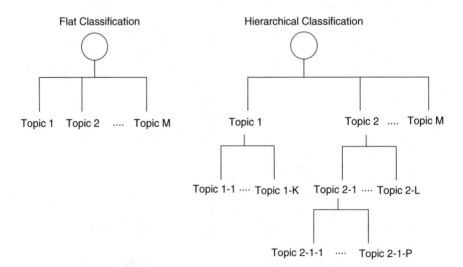

Fig. 5.8 Flat vs hierarchical classification

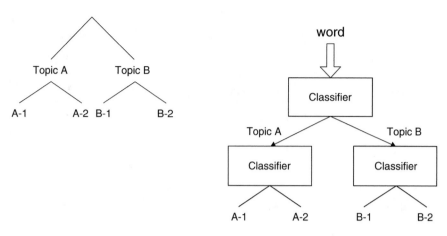

Fig. 5.9 Process of applying classifiers to hierarchical Classification

Flat classification is illustrated in the left part of Fig. 5.8. It is one where no nested category is available in any category. In this classification type, the categories are predefined as a list. Among the predefined categories, one or some are assigned to each item. The digit recognition, the optical character recognition, and spam mail filtering are instances of flat classification.

Hierarchical classification is shown in the right part of Fig. 5.8. It is one where any nested category is allowed in a particular category. In this classification type, categories are predefined as a tree. An item is classified in the top-down direction along the classification tree. The scheme of evaluating the performance is more complicated in the hierarchical classification than in the flat one.

Figure 5.9 illustrates the process of applying the classifiers to the hierarchical classification task. In the first level, there are two categories, A and B, category A has the nested categories, A-1 and A-2, and category B has B-1 and B-2, as shown in the left part of Fig. 5.9. The classifier is prepared in the first level, for classifying an item into A or B. In the second level, two classifiers are needed: one that classifies the items which are classified into A into A-1 or A-2 and the other that does ones which are classified into B into B-1 or B-2. In the hierarchical classification, classifiers are allocated in the root node and the interior nodes as the basic scheme of applying them.

In implementing the classification systems, there is the trend of expanding the flat classification into the hierarchical one. The optimal character recognition and the spam mail filtering belong to the flat classification, typically. The classification module in the digital library system and patent classification system tends to be implemented as the complicated hierarchical classification system in the early 2000s [15, 17]. Because we need to consider the case of classifying items correctly at the higher level, but incorrectly at the lower level, the process of evaluating the results becomes much more complicated. Various schemes of applying classifiers to the hierarchical classification exist other than what is mentioned above.

Classification Categories 1	Classification Categories 2	Classification Categories 3
• Topic 1-1	• Topic 2-1	• Topic 3-1
• Topic 1-2	• Topic 2-2	• Topic 3-2
•	•	• ....
• Topic 1-K	• Topic 2-L	• Topic 3-M

Fig. 5.10 Different classification categories

5.3.3 Single vs Multiple Viewed Classification

Figure 5.10 illustrates several different classification categories which are defined differently even within the same domain by subjective. A fixed single classification frame is required for implementing classification systems; it takes too much cost for deciding a single classification frame in the process of doing so. It very tedious to update and maintain continually the current classification frame, dynamically. Even if the classification frame is updated and changed, it does not guarantee that the new one is more suitable to the current collection of texts than the previous one. In this section, we propose the multiple viewed classification and compare it with the traditional classification type.

The single viewed classification is referred to the classification type where only one group of categories is predefined as a list or a tree. Until now, it has been the classification paradigm which underlies inherently in existing classification programs. Only one group of categories is predefined in the standard text collections which have been used for evaluating text categorization systems, such as Reuter21578, 20NewsGroups, and OSUMED. In the hard and single viewed classification, only one category is assigned to each item. Coordinating different opinions about the category predefinition is required for keeping this classification type.

The multiple viewed classification is one where at least two groups of classification categories, each of which is called view, are allowed as trees or lists. The training examples even in the hard classification are labeled with multiple categories corresponding to groups of predefined categories. The classifiers are allocated to the groups of predefined categories, assuming that the machine learning algorithms applied to the classification tasks without any decomposition. Novice items are classified independently of views. The multiple viewed classification is interpreted as independent classifications as many as groups of predefined categories.

A simple example of the multiple viewed classification is illustrated in Fig. 5.11. The two views are given in Fig. 5.11 as the two groups of predefined categories:

View 1   View 2

Class A	Class B	Class C	Class D
Text 1	Text 5	Text 1	Text 2
Text 2	Text 6	Text 3	Text 4
Text 3	Text 7	Text 5	Text 6
Text 4	Text 8	Text 7	Text 8

	Class A	Class B	Class C	Class D
Text 1	O	X	O	X
Text 2	O	X	X	O
Text 3	O	X	O	X
Text 4	O	X	X	O
Text 5	X	O	O	X
Text 6	X	O	X	O
Text 7	X	O	O	X
Text 8	X	O	X	O

Fig. 5.11 Example of multiple viewed classification

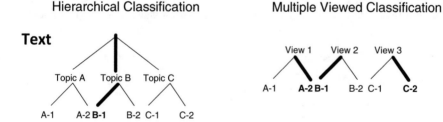

Fig. 5.12 Hierarchical vs multiple viewed classification

view 1 where classes A and B are predefined, and view 2 where classes C and D are predefined. Each of eight texts is labeled with exactly two categories; one is from view 1 and the other is from view 2. We need the two binary classifiers: one is for classifying an item into class A or B, and the other is for classifying it into class C or D. The two binary classifiers correspond to the two views under a single set of training examples.

The differences between the hierarchical classification and the multiple viewed one are illustrated in Fig. 5.12. The hierarchical classification is shown in the left part of Fig. 5.12; there are three categories in the first level and each of them has its own two nested categories. The multiple viewed classification is presented in the right part of Fig. 5.12; there are three different independent binary classifications. In the hierarchical classification, for example, text is classified into the category, B-1, by means of the category B, whereas in the multiple viewed category, it is classified into the three independent categories, A-2, B-1, and C-2. The predefined category structure is given as a tree in the hierarchical classification, whereas the structure is given as a forest which consists of more than independent trees or lists, in the multiple viewed classification.

5.3.4 Independent vs Dependent Classification

This section is concerned with the two types of classification, depending on whether the results from classifying items before have influence on the current classification, or not. The two types, the independent classification and the dependent one, are illustrated in Fig. 5.13. The independent classification is one where the results from previous and current classifications are independent of each other, whereas the dependent classification is one where the current classification results depend on the results of classifying data items previously. The three independent binary classifications are presented in the left part of Fig. 5.13, whereas the two binary classifications which depend on a binary classification are shown on the right side of Fig. 5.13. In this section, we describe the two kinds of classification in detail and compare them with each other.

The independent classification is one where there is no influence of classifying an item into a particular category on decision into another category. The flat and

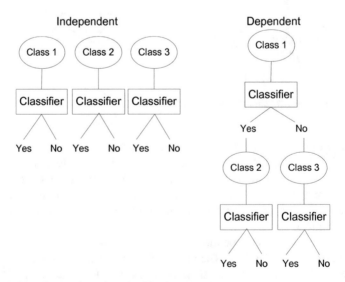

Fig. 5.13 Independent vs dependent classification

exclusive classification belongs strictly to the independent classification, whereas the hierarchical classification belongs to the independent classification, in that the scope of specific categories is dependent on their higher category. There are two cases which are required for the independent classification: the prior classification is not required for classifying the item, currently, and the category into which the item was classified before does not influence the current classification. The spam mail filtering, the optical character recognition, and the digital recognition belong to the independent classification which are not influenced by the results from classifying items, previously. The flat classification may belong to the independent classification and the dependent one, by correlation among categories.

The dependent is one where a data item is classified into a category, influenced by a prior classification or prior classifications. The hierarchical classification belongs typically to the independent classification, as mentioned above. The dependent classification may be applicable to the flat overlapping classification where if an item is classified into a particular category; it may be classified into its related ones with higher probabilities. For example, if a news article is classified into business, it may be classified more easily into the topic, IT, compared with the topic, sports. The text classification where texts tend to be classified into related categories becomes a typical example of the independent classification.

Figure 5.14 presents the specific examples of two kinds of classification. The example of independent classification where the four classifiers classify the given text independently into the positive class or the negative class is presented in the left part of Fig. 5.14. The right part of Fig. 5.14 presents the example of dependent

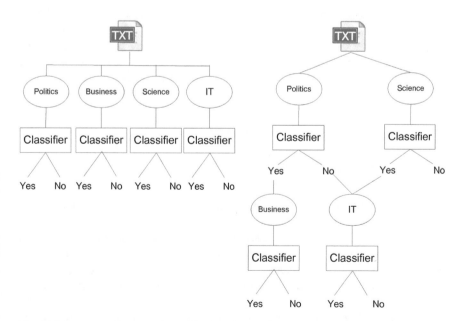

Fig. 5.14 Examples of independent and dependent classification

classification where the classifier which corresponds to the category, business, is dependent on the classifier corresponding to the category, politics, and the classifier corresponding to the category, IT, is dependent on the two classifiers which correspond to the categories, politics and science, respectively. When the classifier corresponding to the category, politics, classifies a text into the positive class, the classifier corresponding to the category, business, is able to classify a text into one of the two classes. When the classifier corresponding to the category, politics, classifies it into the negative class and the classifier corresponding to the category, science, does it into the positive, the classifier corresponding to the category, IT, is able to classify it.

We need to consider the relations among categories in doing classification tasks. The independent relations among classes have been assumed in the traditional classification tasks, such as digit recognition, optical character recognition, spam mail filtering, and other single binary classification tasks. More than ten categories exist in the standard text collections such as Reuter21578 and 20NewsGroups which are mentioned in Chap. 8. The structure of categories is hierarchical in the collection, 20NewsGroups, and more than 100 categories in the collection Reuter21578 are related semantically to each other. So, the text classification usually belongs to the dependent classification.

5.4   Variants of Text Categorization

This section is concerned with the tasks which are derived from the text categorization, in their conceptual views. In Sect. 5.4.1, we cover the spam mail filtering which decides whether each email is junk or sound. In Sect. 5.4.2, we study the sentimental analysis which classifies an opinion into one of the three categories, negative, positive, and neutral. In Sect. 5.4.3, we describe information filtering which decides whether a text is interesting or not. In Sect. 5.4.4, we mention the topic routing which is the reverse task to the text categorization.

5.4.1   Spam Mail Filtering

The spam mail filtering is an instance of binary classification where each email is classified into ham (sound) or spam (junk), as shown in Fig. 5.15. Users of email account tend to take very much time for removing their unnecessary emails. The task may be automated by the spam mail filtering which is a special instance of text categorization. An email is assumed to be a text and the categories, spam and ham, are predefined. In this section, we describe the spam mail filtering as an instance of text categorization, in detail.

It is necessary to clear junk emails in managing email accounts. Users usually have more than one email account. Too many junk emails tend to arrive every day

Fig. 5.15 Spam mail filtering

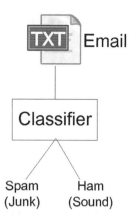

and it takes too much time for removing junk emails, scanning them individually. They need to remove junk emails automatically before they arrive at users. So, the spam mail filtering system should be installed in almost all of email accounts.

It is considered to apply the machine learning algorithms to the spam mail filtering, viewing it into a binary classification task. The sample emails which are labeled with spam or ham are prepared as training examples. The classifier learns the sample labeled emails. An email which arrives subsequently is classified into ham or spam. In the real version, it does by symbolic rules, rather than machine learning algorithms.

Let us review some representative cases of applying the machine learning algorithms to the spam mail filtering. In 2003, the Naive Bayes was applied to the spam mail filtering by Schneider [84]. In 2003, the neural networks were proposed as the approach to the email classification by Clark et al. [10]. In 2007, the four main machine learning algorithms, the Naive Bayes, the K Nearest Neighbor, the decision tree, and the support vector machine, are compared with each other by Youn and McLeod [100]. In 2010, using the resampling, the KNN-based spam mail filtering system was improved by Loredana et al. [62].

Let us consider some issues in implementing the spam mail filtering systems. We need to discriminate the two types of misclassifications: misclassification of spam into ham and vice versa. Emails are classified into spam or ham, depending on sample emails which are previously gathered. Spam mails tend to be different from previous ones; alien mails tend to be misclassified. An email is given as very short and colloquial text, so it is not easy to analyze it.

5.4.2   Sentimental Analysis

Sentimental analysis is referred to the process of classifying an opinion into the positive, the neutral, or negative, as shown in Fig. 5.16. An opinion is given a textual input data, and one of the three attitudes is generated as the output. The sentimental

Fig. 5.16 Sentimental
analysis

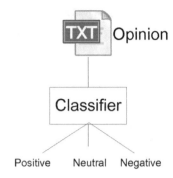

analysis is used for classifying automatically opinions to commercial products or
political issues based on the attitudes. It should be distinguished from the topic
spotting where a text is classified by one or some of topics. In this section, we
describe the sentimental analysis as a specialized instance of text categorization.

Let us mention the three categories in the sentimental analysis. The positive means the opinion which described something or somebody with positive
expressions such as good, excellent, and great. The neutral means one which
describes something objectively without positive nor negative, or with the mixture
of both of them. The negative means one which describes something with negative
expressions, such as bad, terrible, and poor. The neutral may be divided into the two
types: one with no sentimental expression and one with a mixture of positive and
negative.

The machine learning algorithm could be applied to the sentimental analysis by
viewing the task into a classification task. The texts which are labeled with one of
the three categories are collected and encoded into numerical vectors. The machine
learning algorithm learns the numerical vectors which are encoded from the sample
labeled texts. If a novice text is given, it is encoded into a numerical vector, and
classified into one of the three classes. Even if the sentimental analysis is an instance
of text classification, its classification criteria differ from that of the topic-based one.

Let us introduce some previous approaches to the sentimental analysis. In 2004,
the support vector machine was applied to the sentimental analysis based on diverse
information sources by Mullen and Colier [71]. In 2008, Pang and Lee explained the
sentimental analysis and the feature extraction in detail [75]. In 2009, by Boiy and
Moens, both learning algorithms, the support vector machine and the Naive Bayes,
are applied to the sentimental analysis of web texts which are written in English,
Dutch, and French [6]. In 2011, words are defined as features and the support vector
machine was applied to the task by Maas et al. [64].

We need to consider some issues in the sentimental analysis which is distinguished from the topic-based text categorization. The sentimental analysis tends to
depend strongly on the positive and negative terms. Because an opinion is given
as a very short text, information is not sufficient for distinguishing positive and
negative opinions from each other. An opinion or a thread tends to include colloquial
expressions and slangs. Negative opinions which are expressed softly in neutral
words may be misclassified into the neutral one.

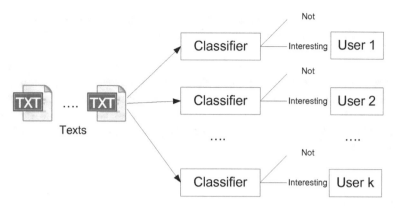

Fig. 5.17 Information filtering

5.4.3   *Information Filtering*

Information filtering is referred to the process of providing interesting texts for users among incoming ones, as shown in Fig. 5.17. It assumed that texts are incoming continually to users. A classifier is allocated to each user, and learns sample texts which are labeled interesting or not. Each incoming text is classified into one of the two labels and the texts which are classified as interesting are transferred to users. In this section, we describe the information filtering which is derived from text categorization.

Let us mention the web crawling before describing the information filtering. Web crawling means the process of collecting web documents which are interesting to users based on their profiles, and the user profile means the historical records of making queries and accessing web documents. The user profiles are gathered and relevant documents are retrieved through the Internet. Information retrieval is the process of retrieving relevant documents only in the given query, whereas web crawling is the process of retrieving ones continually by collecting the user profiles. The important issue is how to make the schedule of performing the web crawling.

The task of information filtering is viewed as a binary classification task. For each user, sample texts which are labeled interesting or not are collected. The sample texts are encoded into numerical vectors and the allocated machine learning algorithm learns them. After learning, incoming web documents are classified as interesting or not. The web documents which are labeled uninteresting are discarded and only ones which are labeled interesting are delivered to the user.

Let us explore previous works on the information filtering system. In 1995, some heuristic algorithms of information filtering were proposed by Shardanand and Maes [87]. In 1996, the Rocchio algorithm which is an instance of the machine learning algorithm was applied to the information filtering by Kroon and Kerckhoffs [56]. In 2001, schemes of extracting features and various kinds of approaches about the information filtering were mentioned by Hanani et al. [20]. In 2010, the bag-of-words matching algorithm was applied to the information filtering by Sriram et al. [88].

Let us mention some issues in implementing the information filtering system. We need to consider the scheme of gathering sample texts from each user before deciding the approach. Depending on the user, even the same text should be labeled differently. If a user changes his/her interest, some of sample texts should be labeled differently. As well as texts, images may be considered for performing the information filtering.

5.4.4   Topic Routing

This section is concerned with the topic routing which is derived from the text categorization. The text classification which allows the soft classification is called topic spotting, whereas the topic routing is the inverse task. Topic routing is defined as the process of retrieving texts which belong to the given topic; a topic or topics are given as the input and texts which belong to it or them are retrieved as the output. Because a topic or topics are given as the query in the topic routing, it looks similar as the information retrieval. So, in this section, we describe the topic routing in the functional view.

The overlapping text categorization was called topic spotting as its nickname where more than one category are allowed to be assigned to each text [95]. The process of assigning more than one category is viewed as the process of spotting topics on a text. In [95], the Perceptron and MLP (Multiple Layer Perceptron) were proposed as the approach and the task was decomposed into binary classifications. A single text is given as the input and its relevant topics are retrieved as the output, in the topic spotting. The topic spotting has been mentioned in subsequent literatures [45, 60, 72].

Let us consider the process of doing the topic routing. A list of categories is predefined and texts should be tagged with their topic or topics. A topic is given as a query and texts whose tags match with it are retrieved. If a topic is given out of the predefined ones, texts whose contents match it are retrieved. The difference from the information retrieval is to predefine the categories in advance and label texts with one or some of them.

The topic routing may be evaluated likely as the information retrieval task. The categories are predefined as mentioned above and the labeled texts are prepared. The recall and the precision are computed from texts which are retrieved from the topic. The two metrics are combined into the F1 measure. The process of computing the F1 measure is described in detail, in Chap. 8.

We need to point out some issues in implementing the topic routing system. If a topic is given out of the predefined ones as a query, the system ignores it or converts into the information retrieval system. When the topics are predefined as a hierarchical structure, we need to refine the abstract topic which is given as a query into its specific ones. We need to consider the data structures for string topic associations to their relevant texts. If a text consists of paragraphs each of which covers its own different topic, heterogeneously, it is considered to retrieve subtexts, instead of full texts.

5.5 Summary and Further Discussions

This chapter is characterized as the functional description of text categorization. The class task is defined as a text mining task, and the multiple classification is decomposed into binary classifications. Depending on the dichotomy views, we surveyed the various kinds of classification. We covered spam mail filtering, sentimental analysis, information filtering, and the topic routing, as the variant tasks which are derived from the text categorization. In this section, we make further discussion about what we study in this chapter.

Clustering which is covered from Chap. 9 plays the role of automating the preliminary tasks for the text categorization. The preliminary tasks are to predefine categories as a list or a tree and to allocate sample texts. The results from doing the preliminary tasks are accomplished by the clustering task. In 2006, Jo proposed the combination of the text categorization and the text clustering with each other for managing texts automatically, based on the idea [25]. It will be mentioned in detail in Chap. 16.

Let us characterize the text categorization in the functional view. The manual preliminary tasks which were mentioned above are required for executing the main task. The results from doing the text categorization are evaluated by the accuracy or the F1 measure. The supervised learning algorithms which are described in Chap. 6 are applied to the task. In the text categorization, it is assumed that the frame of organizing texts is given as a list of a tree of predefined categories.

Text categorization is actually the semiautomated management of texts by itself. It requires the preliminary manual tasks, category predefinition, and sample text preparation. Learning of sample texts and classification of novice texts belong to the automatic portion in the text categorization. As texts are added and deleted, we need to update categories. The full automation of text management is implemented by combining the text categorization with the text clustering, and it will be described in detail in Chap. 16.

Machine learning algorithms are adopted as approaches to text categorization, in this study. The K Nearest Neighbor and its variants are adopted as the simple and practical approaches to the task. The probabilistic learning such as Bayes classifier, Naive Bayes, and Bayesian learning are considered as typical approaches. The support vector machine is the most popular tool of any kind of classification task, as well as the text categorization. The machine learning algorithms will be described in Chap. 6.

Chapter 6
Text Categorization: Approaches

This chapter is concerned with some machine learning algorithms which are used as the typical approaches to text categorization. We explain the machine learning algorithms, conceptually, in Sect. 6.1, and deal with the KNN (K Nearest Neighbor) algorithm and its variants in Sect. 6.2. In Sect. 6.3, we cover another type of machine learning algorithm which is called probabilistic learning. We describe the SVM (Support Vector Machine), which is the most popular approach to classification tasks, in Sect. 6.4, and make the summarization and further discussions on this chapter, in Sect. 6.5. In this chapter, we describe the machine learning algorithms which are the approaches to text categorization with respect to their learning and classification process.

6.1   Machine Learning

Machine learning refers to the computation paradigm where the ability of classification, regression, or clustering is obtained by previous examples, as shown in Fig. 6.1. The first step of executing the machine learning algorithms is to prepare examples which are called training examples or sample data, depending on literatures. The capacity of classification, regression, or clustering is given as various forms depending on types of machine learning algorithms, and is constructed by prepared examples; this is called the learning process. Novice examples which are called test examples are classified in the classification task, and output values are estimated in the regression task; this is called generalization. In this section, we describe the machine learning algorithms in their general view.

The machine learning algorithms are mainly applied to the tasks: classification, regression, and clustering. The classification refers to the task where one or some of predefined categories are assigned to each object. The regression is defined as the process of estimating an output value or output values which are related linear or

© Springer International Publishing AG, part of Springer Nature 2019
T. Jo, *Text Mining*, Studies in Big Data 45,
https://doi.org/10.1007/978-3-319-91815-0_6

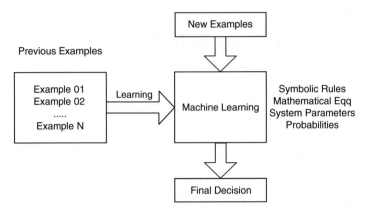

Fig. 6.1 Overview of machine learning

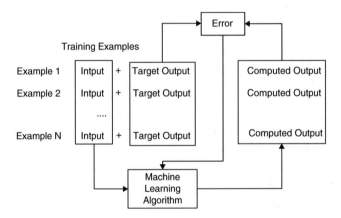

Fig. 6.2 Supervised learning

nonlinear to input values. The clustering means the process of partitioning a group of objects into subgroups of similar ones. There are two kinds of machine learning algorithms, supervised learning and unsupervised learning; the former is used for the classification and the regression, and the latter is used for the clustering.

The supervised learning algorithm is illustrated in Fig. 6.2 as a diagram. It is assumed that all of training examples are labeled with their own output values which are called target labels. The parameters of the machine learning algorithm are initialized at random, and the output values of training examples which are called computed outputs are computed with the parameters. The difference between the two kinds of outputs, target outputs and computed ones, is called error and becomes the basis for correcting the parameters. The supervised learning refers to the process of optimizing the machine learning parameters to minimize the error.

The unsupervised learning algorithm is represented in Fig. 6.3, as a diagram. Training examples have no labels, different from the ones in the supervised learning,

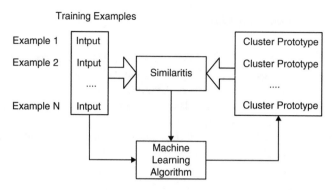

Fig. 6.3 Unsupervised learning

in comparing Fig. 6.2 with Fig. 6.3. Cluster prototypes are initialized at random, and their similarities with training examples are computed. Cluster prototypes are optimized to maximize their likelihoods to training examples; this is the unsupervised learning process. The reinforced learning is one where input is given as an action, and decision of penalty or reward is made to each input; this type of learning algorithm is different from both supervised and unsupervised ones.

The machine learning approaches are preferred to the rule-based ones where classification rules are given manually in the form of if-then. Preparing training examples is less tedious than manually classifying them for building the classification system. It is more flexible to classify items by machine learning algorithms than to do them by the classification rules; it is able to classify slightly noisy and alien items by the machine learning algorithms. We avoid completely the contradiction among classification rules, if we use the machine learning algorithms for the classification tasks, rather than the rule-based systems. It does not require the knowledge about the given application domain, in using machine learning algorithms.

6.2 Lazy Learning

This section is concerned with the K Nearest Neighbor and its variants which belong to lazy learning and consists of the four sections. In Sect. 6.2.1, we describe the initial version of the KNN as one of the main approaches to text categorization. In Sect. 6.2.2, we mention the radius nearest neighbor which is an alternative lazy learning to KNN. In Sect. 6.2.3, we cover the advanced version of KNN which discriminates the nearest neighbors by their distances. In Sect. 6.2.4, we explain another advanced version which computes distances with discriminations among attributes.

6.2.1   K Nearest Neighbor

This section is concerned with the initial version of KNN, as illustrated in Fig. 6.4.
The assumption in using the KNN for classification tasks is that data items
are represented into numerical vectors. Similarities of each novice example with
training examples are computed and the most similar training examples are selected
as the nearest neighbors. The label of the novice example is decided by voting the
labels of the nearest neighbors. Therefore, in this section, we describe the original
version of KNN as a classification algorithm, in detail.

Computing a similarity between two examples is the core operation for executing
the KNN. If two vectors are expressed into $\mathbf{x_1}$ and $\mathbf{x_2}$, the cosine similarity between
them is expressed as Eq. (6.1)

$$sim(\mathbf{x_1}, \mathbf{x_2}) = \frac{\mathbf{x_1} \cdot \mathbf{x_2}}{||\mathbf{x_1}|| \cdot ||\mathbf{x_2}||} \tag{6.1}$$

The alternative similarity metrics between the two vectors are expressed by
Eqs. (6.2) and (6.3).

$$sim(\mathbf{x_1}, \mathbf{x_2}) = \frac{\mathbf{x_1} \cdot \mathbf{x_2}}{||\mathbf{x_1}|| \cdot ||\mathbf{x_2}|| - \mathbf{x_1} \cdot \mathbf{x_2}} \tag{6.2}$$

$$sim(\mathbf{x_1}, \mathbf{x_2}) = \frac{2\mathbf{x_1} \cdot \mathbf{x_2}}{||\mathbf{x_1}|| \cdot ||\mathbf{x_2}||} \tag{6.3}$$

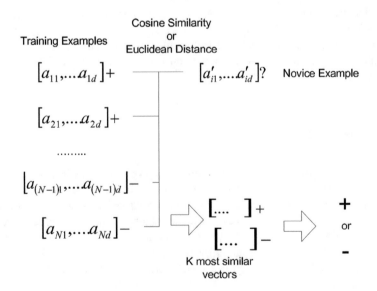

Fig. 6.4 The K Nearest Neighbor

Input: Labeled training examples, an unlabeled novice example
        , and the number of nearest neighbors →K
    for each training example
        compute the similarity between the training example the novice example
        (similarity: Euclidean distance or cosine similarity)
    Sort the training examples in descending order of the similarity
    (in ascending order of Euclidean distance or in descending order of cosine similarity)
    select the K training examples with highest similarity as neighbors
    return the majority of labels of neighbors

Fig. 6.5 K nearest neighbor algorithm

The inverse Euclidean distance which is a similarity metric is expressed by Eq. (6.4).

$$sim(\mathbf{x_1}, \mathbf{x_2}) = \frac{1}{\text{dist}(\mathbf{x_1}, \mathbf{x_2})} = \frac{1}{\sqrt{\sum_{i=1}^{d}(x_{1i}^2 - x_{2i}^2)}} \tag{6.4}$$

We may use one of the four Eqs. (6.1)–(6.4) as a similarity metric for classifying items by the KNN.

The process of classifying the novice which is represented into a numerical vector is illustrated in Fig. 6.5. The training examples, one test example, and the number of nearest neighbor, K, are given as input to the KNN algorithm. The similarities of the test example with the training examples are computed and the training examples are sorted by their similarities with the test example. The most K similar training examples are selected as the nearest neighbors, and the label of the text example is decided by voting their labels. The number of the nearest neighbors, K, is usually set as an odd number such as 1, 3, and 5, in order to avoid the balanced distribution over labels in a binary classification.

Let us mention the scheme of applying the K Nearest Neighbor to the text categorization as a tool. The sample texts are encoded into numerical vectors by the process which was described in Chaps. 2 and 3. The novice text which we try to classify is encoded into a numerical vector, and its similarities with the sample ones are computed. Several most similar sample ones are selected as its nearest neighbors and its label is decided by voting their labels. Encoding texts into numerical vectors and defining a similarity metric are requirements for applying the KNN to the text categorization.

The KNN algorithm which is described in this section belongs to lazy learning. Lazy learning means the machine learning paradigm where the learning starts when a test example is given. No learning of training examples happens in advance; before a novice example is given, any training example is touched. The KNN algorithm never touched any training example in advance, as an instance of lazy learning. Lazy learning will be covered entirely in Sect. 6.2, and other instances will be studies in subsequent subsections.

6.2.2 Radius Nearest Neighbor

The Radius Nearest Neighbor is mentioned as an alternative lazy learning algorithm to the KNN in this section. In the KNN, the k training examples which are most similar as the given test example are selected as its nearest neighbor. Instead of the number of nearest neighbors, k, the radius is used as the external parameter in the Radius Nearest Neighbor. As its nearest neighbors, the k training examples whose similarities with the test example are greater than or equal to the critical similarity which is called radius. In this section, we describe the RNN (Radius Nearest Neighbor) as one of approaches to text categorization.

We point out some demerits in the KNN algorithm as the motivation for switching it to the RNN, even if the KNN is a simple and practical approach. In the KNN, the selected nearest neighbors are treated with their identical importance for voting the labels of the selected ones, even if they are discriminated by their distances. The number of neighbors should be variable to distribution over distances; a variable number of neighbors is selected for voting their labels, in order to be more reliable. It may be fragile to the small number of training examples with the large value of k; it is more probable to misclassify items by the many nearest neighbors. Sorting training examples by their similarities with the test example is required for executing the KNN.

The RNN algorithm is expressed into the pseudo code which is illustrated in Fig. 6.6. The training examples with their labels, a test example with no label, and the threshold similarity are given as the input. For each test example, its similarities with training examples are computed and ones with higher similarity than the threshold are selected as the nearest neighbors. The label of the test example is decided by voting the tables of the nearest neighbors. Therefore, the RNN algorithm is the same as the KNN algorithm except its scheme of selecting the nearest neighbors.

The demerit of the RNN is the possibility that no nearest neighbor or too many nearest neighbors may be selected, so we mention its variants in order to solve it. We mention the dynamic RNN where increasing or decreasing the radius is allowed, depending on the number of selected nearest neighbors. The RNN where multiple radii are given as its external parameters and the similarities with the

Input: Labeled training examples, an unlabeled novice example
         , and the number of nearest neighbors →K
      for each training example
          compute the similarity between the training example the novice example
          (similarity: Euclidean distance or cosine similarity)
      Sort the training examples in descending order of the similarity
      (in ascending order of Euclidean distance or in descending order of cosine similarity)
      select the K training examples with highest similarity as neighbors
      return the majority of labels of neighbors

Fig. 6.6 Radius nearest neighbor algorithm

Table 6.1 KNN vs RNN

	KNN	RNN
Selection	Based on rank	Based on threshold
Parameter	# nearest neighbors	Distance or similarity threshold
Adv	Always nearest neighbor	Outlier detection
Dis	Outlier classification	No nearest neighbor

training examples are computed again by weighting differently to the radii may exist as another version. Depending on the attribute importance, different lengths of radius may be assigned. However, its merit over the KNN algorithm is the faster processing, because it does not require sorting training examples by their similarities.

In Table 6.1, we illustrate the comparison of the two lazy learning algorithms: KNN and RNN. From the training examples, in the KNN, the nearest neighbors are selected by ranking the training examples by their similarities, whereas in the RNN, they are selected by comparing the similarities with the threshold. In the KNN, the number of nearest neighbors is given as its parameter, whereas in the RNN, the similarity or distance threshold is given. A fixed number of nearest neighbors is guaranteed for any test example in the KNN, whereas any test example without its nearest neighbor may be detected as an outlier in the RNN. In the KNN, since a fixed number of nearest neighbors is taken absolutely, there is no way of detecting outliers, whereas in the RNN, it not guaranteed to take nearest neighbors for classifying novice examples.

6.2.3 Distance-Based Nearest Neighbor

This section is concerned with a KNN variant which provides discriminations among nearest neighbors by their similarities or distances. The nearest neighbors are treated as identical ones in voting their labels in the initial version of KNN. We will mention another version of KNN where selected neighbors are discriminated by their distances or similarities. The weights are assigned to nearest neighbors proportionally to their similarities or distances in voting their labels. In this section, we describe the schemes of assigning weights to the nearest neighbors as their discriminations.

Let us mention some schemes of weighting the nearest neighbors proportionally to their similarities for voting their labels. K notates the number of nearest neighbors, and i does the index of i most similar training examples among nearest neighbors with $1 \leq i \leq K$. The ith nearest neighbor is weighted by Eq. (6.5),

$$w_i = \frac{2(K - i + 1)}{K(K - 1)}, \tag{6.5}$$

and the sum of the nearest neighbors is 1.0 as expressed by Eq. (6.6),

$$\sum_{i=1}^{K} w_i = \sum_{i=1}^{K} \frac{2(K - i + 1)}{K(K - 1)} = 1 \qquad (6.6)$$

The additional schemes of weighting the nearest neighbors for voting their labels are expressed by Eqs. (6.7) and (6.8),

$$w_i = \frac{e^{c(K-i)}}{e^{cK}} = \exp(-ci) \qquad (6.7)$$

$$w_i = \frac{2(K - i + 1) + c}{K(K - 1) + c} \qquad (6.8)$$

where c in both equations, (6.7) and (6.8), is a constant real value. This KNN version is close to the initial version of the KNN or the 1 Nearest Neighbor, depending on the value of c in both equations, (6.7) and (6.8).

Let us mention the process of classifying novice items by weighting labels of nearest neighbors. The K nearest neighbors are selected based on the similarities of novice items with the training examples. The selected nearest neighbors are computed by one of equations, (6.5), (6.7), and (6.8). It is assumed that the given task is a binary classification where the categories are given as positive and negative. The categorical score is computed by Eqs. (6.9) and (6.10),

$$CS_+ = \sum_{i=1|+}^{K} w_i \qquad (6.9)$$

where $\sum_{i=1|+}^{K}(\cdot)$ means the summation of K items which belong to the positive class,

$$CS_- = \sum_{i=1|-}^{K} w_i \qquad (6.10)$$

where $\sum_{i=1|-}^{K}(\cdot)$ means the summation of K items which belong to the negative class. The weighted voting is applicable to the RNN which was described in Sect. 6.2.2, as well as the KNN which was covered in Sect. 6.2.1.

It is possible to derive the simplest version Bayes Classifier from this kind of the KNN. A novice example is classified depending on its distance or similarity with the mean vectors corresponding to the categories, in the Bayes classifier. The labeled training examples are prepared and the mean vector of identically labeled examples is computed category by category, as the learning process. The similarities with the mean vectors are computed and the category where its corresponding mean is most similar is decided as label of the given novice item. Therefore, the Bayes classifier is viewed as the 1 Nearest Neighbor to the mean vectors rather than individual ones.

In this section, the local learning is introduced as a specialized learning paradigm rather than a specific machine learning algorithm. The learning paradigm is applicable to any type of machine learning algorithms. The local learning is referred to the learning paradigm where the machine learning algorithm learns only neighbors rather than learning all training examples in advance, whenever a novice example is given. To the next novice example, the learning algorithm is initialized again and learns its neighbors, again. It is expected to show better classification performance than the traditional one, but because much time is required for learning neighbors to each novice example, it is not suitable for implementing a real-time classification system.

6.2.4 Attribute Discriminated Nearest Neighbor

Let us mention another KNN variant which discriminates the attributes which are involved in classifying data items. Various correlations among attributes with output values motivate for deriving the KNN variant. In computing a similarity between a test example and a training example, different weights are assigned to the attributes, depending on the correlations with the output value. In advance, we need to do the attribute analysis in the training examples for computing attribute correlations. In this section, we will describe the KNN variant which considers the correlations between the attribute values and the output value.

We need to assign weights different to attributes, $a_1, a_2, \ldots, a_d$ before using the KNN. We compute the variances and the covariance from the attribute values and the output values from the training examples; $S_{a_i a_i}$, S_{yy}, and $S_{a_i y}$ stand for the attribute variance, the output variance, and the covariance of them, respectively. We compute the correlation coefficient between the attribute and the output, by Eq. (6.11),

$$r_{a_i} = \frac{S_{a_i y}}{\sqrt{S_{a_i a_i} S_{yy}}} \tag{6.11}$$

If the value of r_{a_i} is close to -1.0 or 1.0, it means strong correlation, and if it is around zero, it means week correlation. The weights which are assigned to the attributes are computed by normalizing the correlation coefficients by Eq. (6.12),

$$w_{a_k} = \frac{r_{a_k}}{\sum_{i=1}^{d} r_{a_i}} \tag{6.12}$$

The assumption underlying in the above weighting scheme is that output values are given as continuous ones.

The similarity between numerical vectors is computed after defining the attribute weights by the above process. The normalized attribute weights which are computed by Eq. (6.12) are notated by $w_{a_1}, w_{a_2}, \ldots, w_{a_d}$ and are given as the dimensional vector, $\mathbf{w}$, where $||\mathbf{w}|| = 1$. The similarity between two vectors which considering the attributes is defined by modifying Eq. (6.1), as Eq. (6.13)

$$sim(\mathbf{x_1}, \mathbf{x_2}) = \frac{\mathbf{w} \cdot \mathbf{x_1} \cdot \mathbf{x_2}}{||\mathbf{x_1}|| \cdot ||\mathbf{x_2}||} \tag{6.13}$$

The modified inverse Euclidean distance is also defined by modifying Eq. (6.4) as Eq. (6.14),

$$sim(\mathbf{x_1}, \mathbf{x_2}) = \frac{1}{\text{dist}(\mathbf{x_1}, \mathbf{x_2})} = \frac{1}{\sqrt{\sum_{i=1}^{d} w_{a_i} (x_{1i}^2 - x_{2i}^2)}} \tag{6.14}$$

It is possible to modify Eqs. (6.2) and (6.3) for computing the attribute weighted similarity.

Let us consider the similarities among attributes, especially in the word categorization. The attributes of numerical vectors which represent words are given as text identifiers. The attributes in the task never be independent of each other; similarities among attributes exist. The similarity between attributes is called feature similarity and the similarity between vectors such as the cosine similarity is called feature value similarity; another version of K Nearest Neighbor which considers both was proposed by Jo, in 2015 [35, 37]. By Jo, the proposed version was applied to the tasks: word categorization, word clustering, keyword extraction, and index optimization [38, 39].

Let us consider some points in adopting this version of KNN algorithm as the approach to classification tasks. It is not ignorable to weights attributes by the statistical analysis. The complexity for doing so is $O(Nd)$ where N stands for the number of training examples and stands for the dimension of a numerical vector which represents an example. This version of KNN is not feasible to the text categorization where each text is represented into a several hundred-dimensional numerical vector at least. The correlation coefficients may replace for weighting the attributes by the information gain based on the entropy.

6.3   Probabilistic Learning

This section is concerned with another type of supervised learning algorithm which is called probabilistic learning. In Sect. 6.3.1, we describe mathematically the Bayes rule in order to provide the basis for deriving the probabilistic learning algorithms. In Sect. 6.3.2, we explain the Bayes classifier as the simplest probabilistic learning algorithm. In Sect. 6.3.3, we describe the Naive Bayes as one of the popular classification algorithms. In Sect. 6.3.4, the Naive Bayes will be expanded into the Bayesian learning.

6.3.1 Bayes Rule

The Bayes rule is defined as the theoretical rule which relates the two conditional probabilities, $P(A|B)$ and $P(B|A)$, with each other. The probability of the category, C, and the probability of the input vector, $\mathbf{x}$ are denoted by $P(C)$ and $P(\mathbf{x})$, respectively. The probability of the category, C given the input vector, $\mathbf{x}$ is notated by $P(C|\mathbf{x})$, which is called posteriori probability, and the input vector, $\mathbf{x}$, is classified based on the posteriori probability, $P(C|\mathbf{x})$. In the learning process, the probability, $P(\mathbf{x}|C)$, which is called likelihood of the input vector, $\mathbf{x}$ to the category, C, is computed from the training examples. In this section, we study the Bayes rule which represents the relation the probabilities, $P(C|\mathbf{x})$ and $P(\mathbf{x}|C)$, for the classification task.

The conditional probability is defined as the probability of event under the assumption that another event happens. We notate the probability of event A and the probability of event B as $P(A)$ and $P(B)$, respectively. The probability of event A under the assumption that the event B happens is denoted by $P(A|B)$, and the probability of event B under the assumption that event A happens is denoted by $P(B|A)$. The two conditional probabilities, $P(A|B)$ and $P(B|A)$, are expressed as Eqs. (6.15) and (6.16).

$$P(A|B) = \frac{P(AB)}{P(B)} \tag{6.15}$$

$$P(B|A) = \frac{P(AB)}{P(A)} \tag{6.16}$$

When the two events, A and B, are independent of each other, the probabilities of the two events are expressed by Eq. (6.17), by $P(A) = P(A|B)$ and $P(B) = P(B|A)$.

$$P(AB) = P(A)P(B) \tag{6.17}$$

The relation between the two conditional probabilities, $P(A|B)$ and $P(B|A)$, is expressed as the Bayes rule which is given as Eq. (6.18),

$$P(A|B) = \frac{P(B|A)P(A))}{P(B)} \tag{6.18}$$

The probabilities, $P(A|B)$ and $P(B|A)$, were already expressed into Eqs. (6.15) and (6.16), respectively. $P(AB)$ is generally expressed into Eq. (6.19),

$$P(AB) = P(B|A)P(A)) \tag{6.19}$$

Equation (6.18) is derived by substituting Eq. (6.19) for Eq. (6.15). Equation (6.17) which expresses the Bayes rule consists of the posterior, $P(A|B)$, the likelihood, $P(B|A)$, the prior, $P(A)$, and the evidence, $P(B)$.

Equation (6.18) is expanded into the general form where events are given as $A_1, A_2, \ldots, A_n$ where $B = \cup_{i=1}^n A_i B$. The probability, $P(B)$, is computed by Eq. (6.20),

$$P(B) = \sum_{i=1}^n P(A_i B) = \sum_{i=1}^n P(B|A_i)P(A_i) \tag{6.20}$$

The conditional probability, $P(A_i|B)$, is computed by Eq. (6.21),

$$P(A_k|B) = \frac{P(B|A_k)P(A_k)}{P(B)} = \frac{P(B|A_k)P(A_k)}{\sum_{i=1}^n P(B|A_i)P(A_i)} \tag{6.21}$$

The evidence is given as the summation over the products of the likelihoods and the priors, as shown in Eqs. (6.20) and (6.21). Equation (6.21) becomes the general form of the Bayes rule which considers the mutual exclusive events, $A_1, A_2, \ldots, A_n$.

We will use the Bayes rule which is expressed in Eqs. (6.18) and (6.21) for the classification tasks. Given the input vector, $\mathbf{x}$, we compute the conditional probabilities of categories, $C_1, \ldots, C_{|C|}$, $P(C_1|\mathbf{x}), \ldots, P(C_{|C|}|\mathbf{x})$. The probability, $P(C_i|\mathbf{x})$, is expressed by the Bayes rule which is given in Eq. (6.18), into Eq. (6.22),

$$P(C_i|\mathbf{x}) = \frac{P(\mathbf{x}|C_i)P(C_i))}{P(\mathbf{x})} \tag{6.22}$$

Because the ultimate goal of classifying a data item is to find the maximum of the conditional probabilities, $P(C_1|\mathbf{x}), \ldots, P(C_{|C|}|\mathbf{x})$, rather than their values, we do not need the constant, $P(\mathbf{x})$. So, Eq. (6.22) is transformed into Eq. (6.23), by omitting the constant,

$$P(C_i|\mathbf{x}) \propto P(\mathbf{x}|C_i)P(C_i) \tag{6.23}$$

If the complete balanced distribution over categories is assumed as $P(C_1) = \cdots = P(C_{|C|})$, Eq. (6.23) is further transformed into Eq. (6.24),

$$P(C_i|\mathbf{x}) \propto P(\mathbf{x}|C_i) \tag{6.24}$$

6.3.2 Bayes Classifier

This section is concerned with the Bayes classifier which is the simplest probabilistic learning algorithm. The relation between the posteriori probability, $P(C_i|\mathbf{x})$, and the likelihood, $P(\mathbf{x}|C_i)$, is proportional under the assumption of the balanced distribution over the predefined categories. In the Bayes classifier, it is assumed that the likelihood, $P(\mathbf{x}|C_i)$ follows a Gaussian distribution, given, the category, C_i. The

Gaussian distribution is defined for each category from the training examples, with the parameters: mean vector and covariance matrix. In this section, we describe the Bayes classifier with respect to its learning process.

It is assumed that the probability distribution over examples in each category follows a Gaussian distribution. The parameters which represents the Gaussian distribution, the mean and the variance, are notated by μ and σ. The Gaussian distribution is expressed by Eq. (6.25),

$$p(x) = \frac{1}{\sigma\sqrt{2\pi}} \exp\left(\frac{-(x-\mu)^2}{2\sigma^2}\right) \tag{6.25}$$

if a data item is given as a scalar value. If the data item is given as a dimensional vector, it is expressed by Eq. (6.26),

$$p(\mathbf{x}) = \frac{1}{(2\pi)^{\frac{d}{2}}|\Sigma|^{\frac{1}{2}}} \exp\left(-\frac{1}{2}(\mathbf{x}-\mu)^t \Sigma^{-1}(\mathbf{x}-\mu)\right) \tag{6.26}$$

where μ is the mean vector and Σ^{-1} is the covariance matrix [5]. Because the process of classifying data items focuses on finding the maximum likelihood, Eq. (6.26) is converted into the discriminant function, Eq. (6.27), by omitting the constant, $\frac{1}{(2\pi)^{\frac{d}{2}}}$ and putting the logarithm on it,

$$g(\mathbf{x}) = -\frac{1}{2}(\mathbf{x}-\mu)^t \Sigma^{-1}(\mathbf{x}-\mu) - \frac{1}{2}ln|\Sigma| \tag{6.27}$$

Based on Eq. (6.27), the discriminant function of the category, C_i, is expressed as Eq. (6.28)

$$g_i(\mathbf{x}) = -\frac{1}{2}(\mathbf{x}-\mu_i)^t \Sigma_i^{-1}(\mathbf{x}-\mu_i) - \frac{1}{2}ln|\Sigma_i| \tag{6.28}$$

where the vector μ_i is the mean vector of the category C_i and the matrix Σ_i is the covariance matrix of the category C_i. Equation (6.28) is transformed into Eq. (6.29) with the assumption that the covariance matrix Σ_i is absolutely the identity matrix,

$$g_i(\mathbf{x}) = (\mathbf{x}-\mu_i)^t(\mathbf{x}-\mu_i) \tag{6.29}$$

We compute the mean vectors, $\mu_1, \ldots, \mu_{|C|}$, of the predefined categories, $C_1, \ldots, C_{|C|}$. The discriminant function values, $g_1(\mathbf{x}), \ldots, g_{|C|}(\mathbf{x})$, to the data item, $\mathbf{x}$ by Eq. (6.29) are classified into the category whose discriminant function value is minimum, as expressed in Eq. (6.30),

$$C_{\max} = \operatorname{argmin}_{i=1}^{|C|} g_i(\mathbf{x}) \tag{6.30}$$

We need the three cases, $\Sigma_i = \sigma^2 I$, $\Sigma_i = \Sigma$, and $\Sigma_i =$ arbitrary for defining the discriminant functions [14].

Let us mention the scheme of applying the Bayes classifier to the text categorization task. The labeled sample texts are encoded into numerical vectors by the process which was described in Chaps. 2 and 3. Category by category, mean vectors are computed as the parameters for representing Gaussian distributions. Novice texts are encoded into numerical vectors, have similarities with the mean vectors, and are classified into the category whose similarity is maximum. In this scheme, only the mean vector is considered for representing the Gaussian distribution; because texts are usually encoded into several hundred-dimensional vectors, it very complicated to consider the covariance matrix.

The Bayes classifier is a very simple one which is based on probabilities, but we need to consider its limits. If the covariance matrices are considered together with the mean vectors, it is very complicated to apply the Bayes classifier to the text categorization. The fact that the distribution over data items in each category follows absolutely the Gaussian distribution is only an assumption; it is not true in the reality. If training examples which are labeled identically are scattered as small clusters, the robustness becomes very poor. In computing the similarity between a novice item and a mean vector, discriminations among attributes by their different influences are not considered.

6.3.3   Naive Bayes

This section is concerned with another probabilistic learning which is called Naive Bayes. The assumption which underlies inherently in the learning algorithm is that the attributes of data items are independent of each other. Because the strong semantic relations among words which are selected as features exist, the assumption actually violates against the reality, especially in the text categorization. In spite of that, the approach shown its feasible and good performance in applying it to the text categorization in 1997 [70]. In this section, we describe the Naive Bayes in detail as a typical approach to the text categorization.

The d attributes are notated by $a_1, \ldots, a_d$ under the assumption that a text is encoded into a d dimensional numerical vector. The probability that the attribute a_i is the value, x_i, given the category, C_k, $P(a_i = x_i | C_k)$ is expressed as the likelihood of the attribute value, x_i, $P(x_i | C_k)$. The input vector is expressed as $\mathbf{x} = [x_1 \ x_2 \cdots x_d]$ and the likelihood of the input vector to the category, C_k expressed as the product of likelihoods of individual values by Eq. (6.31),

$$P(\mathbf{x}|C_k) = P(x_1|C_k) \cdot s P(x_d|C_k) = \prod_{i=1}^{d} P(x_i|C_k) \qquad (6.31)$$

The learning process is to compute the likelihoods of individual values and is explained in the next paragraph. If the two attributes, a_i and a_j, are not independent of each other, the product of the two likelihoods does not work as expressed in Eq. (6.32),

$$P(x_i, x_j | C_k) \neq P(x_i | C_k) P(x_j | C_k) \tag{6.32}$$

and this case will be considered in the next section.

The learning process of Naive Bayes is to compute likelihoods of individual attribute values to the predefined categories from training examples. The number of training examples which are labeled with the category, C_k, and the number of training examples which are labeled with the category, C_k, and whose attribute a_i is x_i are denoted by N_k and N_{ki}, respectively. The likelihood of the attribute value, $a_i = x_i$ to the category, C_k is computed by Eq. (6.33),

$$P(a_i = x_i | C_k) = P(x_i | C_k) = \frac{N_{ki}}{N_k} \tag{6.33}$$

For example, if there are two categories, positive class and negative class, there are four attributes, and each attribute has binary value, 0 or 1, the likelihoods are listed as follows:

$$P(a_1 = 0|+), P(a_2 = 0|+), P(a_3 = 0|+), P(a_4 = 0|+)$$
$$P(a_1 = 1|+), P(a_2 = 1|+), P(a_3 = 1|+), P(a_4 = 1|+)$$
$$P(a_1 = 0|-), P(a_2 = 0|-), P(a_3 = 0|-), P(a_4 = 0|-)$$
$$P(a_1 = 1|-), P(a_2 = 1|-), P(a_3 = 1|-), P(a_4 = 1|-)$$

Learning process in the Naive Bayes results in the collection of likelihoods of individual attributes to each category.

Let us classify a novice item which is separated from the training examples, denoted by $\mathbf{x} = [x_1\ x_2 \cdots x_d]$. It is assumed that the posteriori probability depends purely on likelihoods, as expressed in Eq. (6.34),

$$P(C_i | \mathbf{x}) \propto P(\mathbf{x} | C_i) \tag{6.34}$$

The corresponding likelihoods of individual attribute values are gathered from the likelihood collection which is resulted from learning the Naive Bayes as follows:

$$P(x_1 | C_1), \ldots, P(x_d | C_1)$$

$$\vdots$$

$$P(x_1 | C_d), \ldots, P(x_d | C_d)$$

The likelihoods of the input vector $\mathbf{x} = [x_1 \ x_2 \cdots x_d]$ to the categories, $C_1, \ldots, C_{|C|}$, are computed by Eq. (6.35),

$$P(\mathbf{x}|C_1) = P(x_1|C_1) \cdots P(x_d|C_d) = \prod_{i=1}^{d} P(x_i|C_1)$$

$$\vdots \tag{6.35}$$

$$P(\mathbf{x}|C_{|C|}) = P(x_1|C_{|C|}) \cdots P(x_d|C_{|C|}) = \prod_{i=1}^{d} P(x_i|C_{|C|})$$

and the input vector is classified into the category with the maximum likelihood by Eq. (6.30). If at least, one likelihood of individual attribute value is zero, the entire likelihood of the input vector to the category is zero, according to Eq. (6.35).

Because zeros or tiny values of the likelihoods of individual attribute values given a category are very frequent by Eq. (6.32), it is not feasible to use Eq. (6.35) for finding the likelihoods. If at least, one likelihood of an attribute value is zero or a tiny value, the entire vector likelihood becomes a tiny value or zero according to Eq. (6.35). So, in the literature, the smooth version which is expressed as Eq. (6.36) replaces Eq. (6.33),

$$P(x_i|C_k) = \frac{B + N_{ki}}{B + N_k} \tag{6.36}$$

where B is a constant. If replacing Eq. (6.33) by Eq. (6.36) for computing the input vector likelihood by Eq. (6.35), its value avoids the zero value; if there is no example whose attribute is such value, the likelihood is given as $\frac{B}{B+N_k}$ which is the minimum value, instead of zero. In implementing the classification program by adopting the Naive Bayes, Eq. (6.36) is actually used, rather than Eq. (6.33).

6.3.4   Bayesian Learning

This section is concerned with the advanced probabilistic learning which is called Bayesian learning. The learning process of the Naive Bayes is to compute the likelihoods of individual attribute values through training examples. The Bayesian learning proceeds with the two steps: defining causal relations among attributes and computing the likelihoods of individual attribute values. The assumption underlying in the Bayesian learning is that not all attributes are independent of each other. In this section, we describe the Bayesian learning in terms of the learning and the classification, in detail, and compare the three probabilistic learning algorithms which were covered in Sects. 6.3.2, 6.3.3, and this section.

Fig. 6.7 Causal relations among attributes

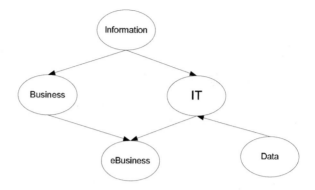

Figure 6.7 illustrates the causal relation among attributes called causal diagram. It is assumed that the causal relations exist among the five attributes: information, business, IT, ebusiness, and data. The two attributes, information and data, are ones which are independent and initial. The attribute, business, depends on the attribute, information; the attribute, IT, depends on the two attributes, information and data; and the attribute, ebusiness, depends on the two attributes, business and IT, as shown in Fig. 6.7. The learning process is to compute the conditional probabilities among the five attributes.

It is assumed that the two attributes, a_i and a_j, have the causal relation from a_i to a_j, which is notated by $a_i \rightarrow a_j$. If the two attributes are not independent of each other, the likelihood of the two attributes to the category, C_k, is not applicable as expressed in Eq. (6.37),

$$P(a_i = x_i, a_j = x_j | C_k) \neq P(a_i = x_i | C_k) P(a_j = x_j | C_k) \tag{6.37}$$

The likelihood is computed by Eq. (6.38),

$$P(a_i = x_i, a_j = x_j | C_k) = P(a_i = x_i | a_j = x_j, C_k) P(a_j = x_j | C_k) \tag{6.38}$$

The term, $P(a_i = x_i | a_j = x_j, C_k)$, is computed by Eq. (6.39),

$$P(a_i = x_i | a_j = x_j, C_k) = \frac{N_{kij}}{N_{ki}} \tag{6.39}$$

where N_{kij} is the number of examples whose attributes a_i and a_j are assigned to the values x_i and x_j, respectively, under the category C_k, and N_{ki} is the number of examples whose attribute a_i is assigned to value x_i under the category C_k. Equation (6.39) may be modified into Eq. (6.40) for preventing the likelihood, $P(x_j | x_i, C_k)$, from being a zero or a tiny value.

$$P(a_i = x_i | a_j = x_j, C_k) = \frac{B + N_{kij}}{B + N_{ki}} \tag{6.40}$$

Table 6.2 KNN vs RNN

	Bayes classifier	Naive Bayes	Bayesian
Attribute	Group	Independent relation	Conditionally independent relation
Classification criteria	Maximum likelihood		
Learning	Distribution parameter estimation	Likelihood estimation	Causal relation likelihood estimation
Complexity	Simple	Medium	Complexity

There are the two steps of learning process in the Bayesian learning; the first step is to define the causal relations among attributes and the second step is to compute their likelihoods. The second step was mentioned above and let us consider the first step. All possible pairs of attributes are generated and the causal relation from a_i to a_j is decided based on the following two factors: $\max_{k=1}^{|C|} P(x_i, x_j | C_k)$ which is called the support in the area of data mining and $\max_{k=1}^{|C|} P(x_i | x_j, C_k)$ which is called the confidence in the area of data mining. As the external parameters, the support threshold and the confidence threshold are needed; the causal relation between the two attributes is decided when both its confidence and support are equal to or greater than the thresholds. Because it takes the quadratic complexity for defining the causal relations to the number of attributes, the Bayesian learning is not feasible to the text categorization, where texts tend to be encoded into very high dimensional numerical vectors.

Table 6.2 illustrates the comparisons of the three probability learning algorithms which are covered in Sect. 10.3. The Bayes classifier starts with the likelihoods of the input vector, the Naive Bayes does with those of individual attribute values, and the Bayesian learning defines the causal relations among attributes before computing their likelihoods. The three machine learning algorithms classify data items based on their likelihoods; they are called commonly maximum likelihood learning. In the learning process, the Bayes classifier defines the probability distributions over the training examples, the Naive Bayes computes the likelihoods of individual attribute values, and the Bayesian learning adds casual relations. To the number of attributes, it takes constant complexity, linear complexity, and quadratic complexity in the Bayes classifier, the Naive Bayes, and the Bayesian learning, respectively.

6.4   Kernel Based Classifier

This section is concerned with the support vector machine which is the most popular approach to the text categorization. In Sect. 6.4.1, we study the Perceptron which is a linear model, in order to provide the background for understanding the support vector machine. In Sect. 6.4.2, we describe and characterize mathematically the kernel functions. In Sect. 6.4.3, we describe the process of classifying an item by the support vector machine. In Sect. 6.4.4, we explain the process of deriving the constraints as the dual problem and optimizing them.

6.4.1 Perceptron

This section is concerned with the Perceptron as the early neural network. The neural networks was invented in 1958 by Rosenblatt as the first learnable neural networks [80]. Because it has no ability to solve even XOR problem, it was criticized by Minsty and Papert in 1969 [69]. In 1988, the Perceptron was expanded into the MLP (Multiple Layer Perceptron) by Rumelhart and McClelland, in order to solve the limit [81]. In this section, we study the Perceptron as the background for understanding the support vector machine.

The classification boundary which is defined by the Perceptron is presented in Fig. 6.8. The training examples which are labeled with the positive class or the negative class are represented into the two-dimensional vectors and are plotted into the dimensional space as shown in Fig. 6.8. The classification boundary is expressed as a line in the two- dimensional space; the classification boundary is given as a hyperplane in more than three-dimensional spaces. The classification boundary is expressed as Eq. (6.41),

$$\mathbf{w} \cdot (x) + b = 0 \qquad (6.41)$$

where $\mathbf{w}$ is the weight vector and b is the bias. If the data point in $\mathbf{w} \cdot (x) + b \geq 0$, it is classified into the positive class, otherwise it is done into the negative class. The

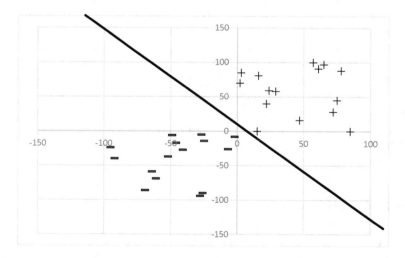

Fig. 6.8 Linear classification boundary

learning in the Perceptron is the process of finding the weight vector, **w**, and the bias, b for classifying the training examples correctly as much as possible.

The training examples are denoted into $(\mathbf{x}_1, t_1), (\mathbf{x}_2, t_2), \ldots, (\mathbf{x}_N, t_N)$, $\mathbf{x}_i \in \mathbf{R}^d, t \in \{-1, 1\}$ and the linear equation is expressed as Eq. (6.42),

$$y = \begin{cases} 1 & \text{if } \mathbf{w} \cdot (x) + b \geq 0 \\ -1 & \text{otherwises} \end{cases} \tag{6.42}$$

The weight vector, **w** and the bias, b are initialized at random around zero. Each training example is classified with the initialized ones; if it is classified correctly, it proceeds to the next example, otherwise the weights are updated. The update rules of the weight vector and the bias are expressed as Eqs. (6.43) and (6.44),

$$\mathbf{w}(t + 1) = \mathbf{w}(t) + \eta(t_i - y_i)x_i \tag{6.43}$$

$$b(t + 1) = b(t) + \eta(t_i - y_i)x_i \tag{6.44}$$

The algorithm iterates classifying each training example and updating the weight vector and the bias until no misclassification, under the assumption of the linearly separable distribution like that in Fig. 6.8.

It is assumed that the task is given as a binary classification which is decomposed from a text classification, and the training examples which are prepared are labeled with the positive class or the negative class. The training examples are encoded into dimensional numerical vectors and they are labeled with -1 and $+1$. The Perceptron learns the training example by the process which was mentioned above. A novice text is encoded into a numerical vector and its output is computed by Eq. (6.42). The output value, $+1$, indicates the positive class and -1 does the negative class.

In 1969, the Perceptron was criticized by Papert and Minsky by its own limits, as mentioned above [69]. It is applicable to linearly separable classification problems; the errors on training examples are allowed somewhat in case of nonlinear separable problems. The reason of criticizing the Perceptron is no ability to solve even the trivial problem, exclusive or problem. However, in 1988, Winter and Widrow proposed the solution to the limit by combining multiple Perceptrons with each other [96]. The Perceptron was expanded into the MLP (Multiple Layer Peceptron) by Rumelhart in 1986 [81], and expanded into the Support Vector Machine by Corte and Vapnik, in 1995 [12].

6.4.2   Kernel Functions

This section is concerned with the kernel functions which become the core operations in performing the support vector machines. Their idea is to map a vector space which is non-linearly separable into one which is linear separable. The inner product of two vectors in any dimension is always given as a scalar value. So, the inner product of two vectors even in the mapped space may be computed by a kernel function, without mapping the vectors in the original space. In this section, we describe kernel functions in detail, before discussing the support vector machine.

As mentioned above, the inner product value of two vectors in any dimension space is absolutely given as a scalar value. The two vectors, $\mathbf{x}$ and $\mathbf{y}$, are mapped into ones in another space, $\phi(\mathbf{x})$ and $\phi(\mathbf{y})$. The inner product of the two vectors, $\phi(\mathbf{x})$ and $\phi(\mathbf{y})$, is given as $\phi(\mathbf{x}) \cdot \phi(\mathbf{y})$ which is a scalar value. The inner product of the two mapped vectors is expressed as the kernel function of the two original vectors, $\mathbf{x}$ and $\mathbf{y}$, as expressed in Eq. (6.45),

$$\phi(\mathbf{x}) \cdot \phi(\mathbf{y}) = K(\mathbf{x}, \mathbf{y}) \tag{6.45}$$

Therefore, we do not need to map the training examples explicitly into another space, in executing the learning process of the support vector machine.

Let us mention some representative types of kernel functions. The inner product of two vectors is mentioned as the basic kernel function as expressed in Eq. (6.46),

$$K(\mathbf{x}, \mathbf{y}) = \mathbf{x} \cdot \mathbf{y} \tag{6.46}$$

Another type of kernel function which is called polynomial kernel function is expressed in Eq. (6.47),

$$K(\mathbf{x}, \mathbf{y}) = (\mathbf{x} \cdot \mathbf{y} + c)^p \tag{6.47}$$

The type of kernel function, Radial Basis Function, is expressed based on the Gaussian distribution as Eq. (6.48),

$$K(\mathbf{x}, \mathbf{y}) = \exp\left[-\frac{||\mathbf{x} - \mathbf{y}||}{\sigma^2}\right] \tag{6.48}$$

The kernel function which is sigmoidal function is expressed in Eq. (6.49),

$$K(\mathbf{x}, \mathbf{y}) = \tanh(2\mathbf{x} \cdot \mathbf{y} + c) \tag{6.49}$$

The kernel function begins with the inner production which is expressed in Eq. (6.46), in defining it. The addition of two kernel functions which is expressed in Eq. (6.50) and the multiplication of a kernel function by a scalar constant which is expressed in Eq. (6.51) are also a kernel function.

$$K(\mathbf{x}, \mathbf{y}) = K_1(\mathbf{x}, \mathbf{y}) + K_2(\mathbf{x}, \mathbf{y}) \tag{6.50}$$

$$K(\mathbf{x}, \mathbf{y}) = \alpha_1 K_1(\mathbf{x}, \mathbf{y}) \tag{6.51}$$

The product of two kernel functions or vector functions which is expressed in Eqs. (6.52) and (6.53), respectively, is also a kernel function.

$$K(\mathbf{x}, \mathbf{y}) = K_1(\mathbf{x}, \mathbf{y}) K_2(\mathbf{x}, \mathbf{y}) \tag{6.52}$$

$$K(\mathbf{x}, \mathbf{y}) = f(\mathbf{x}) f(\mathbf{y}) \tag{6.53}$$

The kernel function of the two vectors which are mapped ones in another space, instead of original vectors, which is expressed in Eq. (6.54) is also a kernel function.

$$K(\mathbf{x}, \mathbf{y}) = K_3(\phi(\mathbf{x}), \phi(\mathbf{y})) \tag{6.54}$$

The product of a matrix between the two vectors which is expressed by Eq. (6.55) is also a kernel function.

$$K(\mathbf{x}, \mathbf{y}) = \mathbf{x}\mathbf{B}\mathbf{y} \tag{6.55}$$

The kernel functions which are covered in this section imply the similarity between two. Refer to Eqs. (6.1)–(6.3) for computing the similarity. From the equations, it is discovered that the inner product is the base kernel function which is proportional to the similarity. The assumption underlying in defining it is that the kernel function means the inner product between the two mapped vectors. By computing the inner product of the mapped vectors, using the kernel function, we can avoid mapping individual vectors, explicitly.

6.4.3 Support Vector Machine

The goal of learning in the support vector machine is to define the dual parallel linear boundaries with their maximal margin in the mapped space. In the Perceptron, there exist infinitely many boundaries for separating the training examples into the two groups, according to their labels. In the support vector machine, it is assumed that vectors are mapped into ones in the new space, and dual boundaries which are parallel to each other are decided in the new space. The constraint for making the decision boundaries in the support vector machine is to maximize the margin which is the distance between the two parallel linear boundaries. In this section, we describe the linear equations which are involved in the support vector machine, and its classification process.

The linear equations which represent the dual parallel boundaries are defined under the assumption of the base kernel function which is expressed in Eq. (6.46). The two linear functions which are involved in modeling the support vector machine are expressed as Eqs. (6.56) and (6.57),

$$1 = \mathbf{w} \cdot \mathbf{x} + b \tag{6.56}$$

$$-1 = \mathbf{w} \cdot \mathbf{x} + b \tag{6.57}$$

The positive class satisfies the inequation, $\mathbf{x} \cdot \mathbf{w} + b \geq 1$ and the negative class satisfies the inequation, $\mathbf{x} \cdot \mathbf{w} + b \leq -1$. The area between the two parallel equations which is the margin between the two hyperplanes is expressed in Eq. (6.58),

$$-1 < \mathbf{w} \cdot \mathbf{x} + b < 1 \qquad (6.58)$$

Equations (6.56) and (6.57) are combined into Eq. (6.59), as the equation for modeling the support vector machine,

$$f(\mathbf{x}) = sign(\mathbf{w} \cdot \mathbf{x} + b) \qquad (6.59)$$

where $sign(\cdot)$ generates -1 as the numerical code of the negative class, or $+1$ as the that of the positive class.

It is assumed that the training examples are given as $\mathbf{x}_1, \mathbf{x}_2, \dots, \mathbf{x}_N$. Because the weight vectors are determined by training examples, the weight vector in Eqs. (6.56) and (6.57) is expressed as a linear combination of training examples which is expressed in Eq. (6.60),

$$\mathbf{w} = \sum_{i=1}^{N} \alpha_i \mathbf{x}_i \qquad (6.60)$$

The coefficient in Eq. (6.60), α_i, is called Lagrange multiplier, and Eq. (6.61) which is the equation of the support vector machine classifier is derived by substituting Eqs. (6.60)–(6.59),

$$f(\mathbf{x}) = sign\left(\sum_{i=1}^{N} \alpha_i (\mathbf{x}_i \cdot \mathbf{x}) + b\right) \qquad (6.61)$$

The Lagrange multipliers are optimized for making the dual linear boundaries with the maximal margin, as the learning process. After optimizing so, the training examples which correspond to nonzero optimized Lagrange multipliers are called support vectors and influence on classifying novice examples.

Equation (6.61) expresses the support vector machine where any input vector is not mapped into one in another space. The training examples, $\{\mathbf{x}_1, \mathbf{x}_2, \dots, \mathbf{x}_N\}$, are mapped into $\{\phi(\mathbf{x}_1), \phi(\mathbf{x}_2), \dots, \phi(\mathbf{x}_N)\}$. Equation (6.61) is modified into Eq. (6.62), expressing the support vector machine which deals with the mapped input vectors,

$$f(\mathbf{x}) = sign\left(\sum_{i=1}^{N} \alpha_i (\phi(\mathbf{x}_i) \cdot \phi(\mathbf{x})) + b\right) \qquad (6.62)$$

If the inner product between the mapped vectors is replaced by the kernel function, Eq. (6.62) is changed into Eq. (6.63), as the general form of the support vector machine,

$$f(\mathbf{x}) = sign\left(\sum_{i=1}^{N} \alpha_i K(\mathbf{x}_i, \mathbf{x}) + b\right) \tag{6.63}$$

where $\phi(\mathbf{x}_i) \cdot \phi(\mathbf{x}) = K(\mathbf{x}_i, \mathbf{x})$. The strategy of optimizing the Lagrange multipliers based on Eq. (6.63) will be mentioned in Sect. 6.4.4.

Let us introduce the previous cases of applying the support vector machine to the text categorization. In 1998, the support vector machine was initially applied to the text categorization [47]. In 2001, Tong and Koller applied the support vector machine to the text categorization, following the learning paradigm, active learning [92]. In 2002, Joachims published his book on applying the support vector machine to the text categorization [48]. Sebastiani mentioned the support vector machine as the best approach to the text categorization, in his survey paper [85].

6.4.4   Optimization Constraints

This section is concerned with the constraints for optimizing the Lagrange multipliers. The support vectors are expressed from the dual equations of the weights, Eqs. (6.56) and (6.57), into the linear equations of the Lagrange multipliers, Eq. (6.63). From Eqs. (6.56) and (6.57), we derive constraints before optimizing the Lagrange multipliers. The constraints are transformed into another ones as the optimization direction. In this section, we describe the process of deriving the constraints and optimizing the Lagrange multipliers.

The constraints for optimizing weights are derived from Eqs. (6.56) and (6.57). The constraint, Eq. (6.64), is derived for minimizing the weight scale in order to maximize the margin between the dual hyperplanes,

$$\Phi(\mathbf{x}) = \frac{1}{2}\mathbf{w}^{\mathbf{T}} \cdot \mathbf{w} \tag{6.64}$$

From Eq. (6.58), we derive Eq. (6.65), as one more constraint,

$$d_i(\mathbf{w}^{\mathbf{T}}x_i + b) \geq 1 \tag{6.65}$$

The two constraints, minimizing Eqs. (6.64) and (6.65), should be satisfied for optimizing the weight vectors. However, in the learning process of the support vector machine, not weight vectors but the Lagrange multipliers are optimized.

The two constraints which are mentioned above are mapped into Eq. (6.66) which we minimize,

$$J(\mathbf{w}, b, \alpha) = \frac{1}{2}\mathbf{w}^{\mathbf{T}} \cdot \mathbf{w} - \sum_{i=1}^{N} \alpha_i [d_i(\mathbf{w}^{\mathbf{T}} \cdot \mathbf{x}_i + b)] \tag{6.66}$$

By the partial differentiation of Eq. (6.66) by the bias, b, we derive Eq. (6.67),

$$\frac{\partial J(\mathbf{w}, b, \alpha)}{\partial b} = \sum_{i=1}^{N} \alpha_i d_i \tag{6.67}$$

By the partial differentiation of Eq. (6.66) by the weight vector, $\mathbf{w}$, Eq. (6.68) is induced,

$$\frac{\partial J(\mathbf{w}, b, \alpha)}{\partial \mathbf{w}} = \mathbf{w} - \sum_{i=1}^{N} \alpha_i d_i \mathbf{x}_i \tag{6.68}$$

Equations (6.67) and (6.68) should become zero for minimizing Eq. (6.66), as expressed in Eqs. (6.69) and (6.70),

$$\sum_{i=1}^{N} \alpha_i d_i = 0 \tag{6.69}$$

$$\mathbf{w} - \sum_{i=1}^{N} \alpha_i d_i \mathbf{x}_i = 0 \tag{6.70}$$

Therefore, Eqs. (6.69) and (6.70) are constraints for optimizing the weight vectors and the bias; it is called primal problem.

Equation (6.64) is expanded into Eq. (6.71)

$$J(\mathbf{w}, b, \alpha) = \frac{1}{2}\mathbf{w}^{\mathbf{T}} \cdot \mathbf{w} - \sum_{i=1}^{N} \alpha_i d_i \mathbf{w}^{\mathbf{T}}\mathbf{x}_i - b\sum_{i=1}^{N} \alpha_i d_i + \sum_{i=1}^{N} \alpha_i \tag{6.71}$$

Equation (6.71) is expressed into Eq. (6.72) by Eq. (6.69),

$$J(\mathbf{w}, b, \alpha) = \frac{1}{2}\mathbf{w}^{\mathbf{T}} \cdot \mathbf{w} - \sum_{i=1}^{N} \alpha_i d_i \mathbf{w}^{\mathbf{T}}\mathbf{x}_i + \sum_{i=1}^{N} \alpha_i \tag{6.72}$$

We derive Eq. (6.73) by Eq. (6.70),

$$\mathbf{w}^{\mathbf{T}} \cdot \mathbf{w} = \sum_{i=1}^{N} \alpha_i d_i \mathbf{w}^{\mathbf{T}}\mathbf{x}_i = \sum_{i=1}^{N}\sum_{j=1}^{N} \alpha_i \alpha_j d_i d_j \mathbf{x}_i \mathbf{x}_j \tag{6.73}$$

Equation (6.72) is mapped into the function of the Lagrange multipliers which is expressed in Eq. (6.74),

$$Q(\alpha) = \sum_{i=1}^{N} \alpha_i - \frac{1}{2} \sum_{i=1}^{N} \sum_{j=1}^{N} \alpha_i \alpha_j d_i d_j \mathbf{x}_i \mathbf{x}_j \qquad (6.74)$$

The Lagrange multipliers are optimized by maximizing Eq. (6.74), keeping the constraints, $\sum_{i=1}^{N} \alpha_i d_i = 0$ and $\alpha_i \geq 0$; this is called dual problem.

We may consider the SMO (Sequential Minimization Optimization) for optimizing the Lagrange multipliers as the learning process of the support vector machine. It was invented by Platt in 1998 [76]. The Lagrange multipliers range between zero and positive constants which are given as external parameters. In the SMO algorithm, until no violation of the KKT (Karush–Kuhn–Tucker) condition, it iterates finding the first Lagrange multiplier which violates the KKT condition and picking the second one to optimize the pair of the first and the second. Refer to the literatures for detailed explanation about the SMO algorithm.

6.5 Summary and Further Discussions

This chapter is summarized as some typical machine learning algorithms which are used as the approach to text categorization. We described the KNN algorithm and its variants as the simple and practical approaches to any classification task, as well as the text categorization. Based on the Bayes rule which was covered in Sect. 6.3.1, we studied the three probabilistic learning algorithms, Bayes Classifier, Naive Bayes, and Bayesian Learning. Together with the Perceptron, we explained the support vector machine which is the most popular classification algorithm. In this section, we make further discussion about what we studied in this chapter.

When applying the KNN algorithm to the real task, it should be customized into its more suitable version. In Sect. 6.1, we mentioned some variants as well as its initial version. The Radius Nearest Neighbor, the distance criminated version, and the attribute discriminated version were described as the variants from Sects. 6.2.2–6.2.4. Customizing the KNN algorithm to real applications means to modify it into the discriminated version or to combine it with other machine learning algorithms. Note that it costs no ignorable time for tuning the KNN algorithm to the real task by customizing it.

In the Naive Bayes and the Bayesian learning, it is assumed that the attribute values are given as discrete ones. When any attribute value is given as a continuous one, the number of all possible values in the attribute becomes infinite. So, it is required to discretize each continuous attribute into a finite number of values for using one of them. Their performances depend strongly on how many intervals the value is discretized with. We may try to apply fuzzy concepts to attribute values for computing likelihoods in using one of them.

In 2002, Lodhi et al. proposed the string kernel in applying the support vector machine to the text categorization [61]. We may consider the issues in encoding texts into numerical vectors: huge dimensionality where the dimension of numerical vectors which represent texts is at least several hundreds and the sparse distribution where more than 95% of elements are zero values. In using the string kernel, the similarity between two raw texts is computed based on character sequences without encoding texts into numerical vectors. However, using the string kernel takes much more time for learning and improves very little performance. It was successful in protein classification in bioinformatics.

The neural networks, MLP (multiple-layer Perceptron), was initially applied to the text categorization by Winer in 1995 [95]. The neural networks showed better results on the text collection, Reuter21578, compared with the KNN. Difficulties in applying the neural networks to the task are too much time being taken for the learning process and requirement of many features and many training examples. In spite of its better performance, the decision tree, the Naive Bayes, and the KNN are preferred to the neural networks. In 2002, Sebastiani recommended the support vector machine as the approach, in his survey paper on the text categorization [85].

Chapter 7
Text Categorization: Implementation

This chapter is concerned with the implementation of the prototype version of the text categorization system in Java. We present the system architecture in Sect. 7.1, and define the classes which are involved in implementing the system, in Sect. 7.2. We present the implementations of methods in the classes in Sect. 7.3. We demonstrate the implemented text categorization system in terms of its execution in Sect. 7.4, and make the summarization and further discussions in Sect. 7.5. In this chapter, we implement the text categorization system in Java, and demonstrate it, in order to provide a guide for implementing its real version.

7.1 System Architecture

The architecture of the text categorization system which we implement in this chapter is illustrated in Fig. 7.1. The system gathers sample texts and generates the features as the attributes of numerical vectors. Sample texts are encoded into numerical vectors which are given as training examples. Test texts are also encoded into numerical vectors and classified into one of the predefined categories. In this section, we describe the system architecture which is illustrated in Fig. 7.1 with respect to its module functions.

Let us explain the modules in Fig. 7.1, the feature generation and the text encoder. The categories are predefined as a flat list and sample texts are allocated to each category. In the module, feature generation, the sample texts are indexed into a list of words called feature candidates, and the words with highest frequencies are selected among them as the features. The sample texts are encoded into numerical vectors each of which consists of frequencies of the selected words, in the module, text encoding. The role of modules, feature generation and text encoding, is to represent the sample texts which are allocated to the predefined categories, into the numerical vectors.

© Springer International Publishing AG, part of Springer Nature 2019
T. Jo, *Text Mining*, Studies in Big Data 45,
https://doi.org/10.1007/978-3-319-91815-0_7

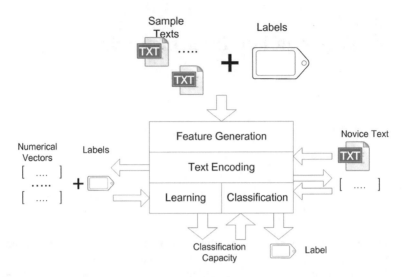

Fig. 7.1 Architecture of text categorization system

Let us explain the learning module below the text encoding module, in Fig. 7.1. We adopt the 1 Nearest Neighbor as the approach for implementing the version of the text categorization system. Once novice texts are given as inputs, they are also encoded into numerical vectors based on the features which are defined by the above process. For each test example, it is encoded into numerical vector, and its similarities with the training examples are computed. Because the adopted approach belongs to lazy learning, no learning happens in advance in this system.

The role of the classification module is to decide which category or categories of the predefined ones are assigned to each text. The training examples are sorted by their similarities with a given novice text. Most similarity training examples are selected as its nearest neighbor by the 1 Nearest Neighbor algorithm. The given text is classified by the label of its nearest neighbor. Although we adopt the 1 Nearest Neighbor as the classification tool for implementing the system easily and fast, it will be expanded into k Nearest neighbor in its next version.

We adopt Java as the tool of implementing the text classification system. Because the program which is developed by Java is executed by the virtual machine; its execution may be slow. In spite of that, the reason of adopting Java is that the performance of CPU is increasing continually and debugging is very easy. In Java programming which keeps the object oriented programming principle, it is very easy to be reused the source code in other programs for implementing a program. In this chapter, we present the implementations of the text categorization system in Java source codes which are given in the Eclipse.

7.2 Class Definitions

This section is concerned with the class definitions which are involved in implementing the text categorization system in Java with their properties and methods. In Sect. 7.2.1, we define the three classes and explain their properties and methods. In Sect. 7.2.2, we define 'Classifier' as an interface and 'KNearestNeighbor' as a class which is specific to the interface. In Sect. 7.2.3, we define the API (Application Programming Interface) class, 'TextClassificationAPI,' whose methods are involved in the main program. Therefore, in this section, we explain the class definitions which are involved in developing the text categorization system, with respect to their properties, methods, and relations with other classes.

7.2.1 Classes: Word, Text, and PlainText

This section is concerned with defining the three classes: Word, Text, and PlainText. The class, 'Word,' is defined, assuming that a text consists of words. The class, 'Text,' which indicates the frame of individual data items, is defined as an abstract class for supporting other types of texts such as XML documents and HTML ones, as well as plain texts, in subsequent versions. We define the class, 'PlainText,' as the class which is specific to the class, 'Text'; the class, 'Text,' is a super class, and the class, 'PlainText,' is a subclass. Therefore, in this section, we explain the three class definitions with respect to their properties, methods, and relations with other classes.

The class, 'Word,' is defined in Fig. 7.2. The properties of this class are 'wordName' which is the symbolic identifier and 'wordFrequency' which is its occurrences in a given text. An object is created with its word name by the following declaration:

$$\text{Word wordItem} = \text{new Word("computer");}$$

Its frequency is initialized to zero. The method 'isPresent' returns a Boolean value depending on whether the word appears in the text which is given as an argument. The methods, 'computeWordFrequency' and 'computeWordWeight,' return a numerical value which indicates a frequency or a weight of the word in the given text, respectively. The method, 'computeWordSimilarity,' returns its semantic similarity with another word. Among the methods in this class, the method, 'isPresent,' is called a predicate method which returns true or false depending on the program status.

The class, 'Text,' is defined as an abstract class, as shown in Fig. 7.3. In this class, the properties are 'textFileName' which is the text identifier given as a file name together with its directory path, 'wordList' which is the list of words as its indices, "fullText" which is the entire text content, 'textLabel' which is its

```
package TextClassification;
import java.util.*;
public class Word {
    String wordName;
    int wordFrequency;
    public Word(String wordName){
        this.wordName = wordName;
        this.wordFrequency = 0;
    }

    void setWordFrequency(int wordFrequency){
        this.wordFrequency = wordFrequency;
    }

    String getWordName(){
        return this.wordName;
    }

    int getWordFrequency(){
        return this.wordFrequency;
    }

    boolean isPresent(String fullText){
        return true;
    }

    int computeWordFrequency(String fullText){
        return 0;
    }

    double computeWordWeight(String fullText, Vector textList){
        return 0.0;
    }

    double computeWordSimilarity(Word another){
        return 0.0;
    }
}
```

Fig. 7.2   The class: word

Fig. 7.3   The class: text

```
package TextClassification;
import java.util.*;
public abstract class Text {
    String textFileName;
    Vector<Word> wordList;
    String fullText;
    String textLabel;
    double featureVector[];

    abstract void setTextLabel(String textLabel);
    abstract void setFullText(String fullText);
    abstract String getTextFileName();
    abstract String getTextLabel();
    abstract String getFullText();
    abstract void loadFullText();
    abstract void indexFullText();
    abstract void encodeFullText(int dimension);
    abstract double computeSimilarity(Text another);
}
```

Fig. 7.4 The class: PlainText

```
package TextClassification;
import java.util.*;
public class PlainText extends Text {
    public PlainText(String textFileName){
        this.textFileName = textFileName;
    }
    void setTextLabel(String textLabel) {
        this.textLabel = textLabel;
    }
    void setFullText(String fullText){
        this.fullText = fullText;
    }
    String getTextFileName() {
        return this.textFileName;
    }
    String getTextLabel() {
        return this.textLabel;
    }
    String getFullText() {
        return this.fullText;
    }
    void loadFullText() {}
    void indexFullText() {}
    void encodeFullText(int dimension) {}
    double computeSimilarity(Text another) {
        return 0.0;
    }
}
```

label, and 'featureVector' which is the numerical vector representing the text. The methods are declared as abstract ones without any implementation as shown in Fig. 7.3, and their implementations will be provided by its subclass. In this class, methods are 'loadFullText' which loads a full text from a file and assigns it to the property, 'fullText,' 'indexFullText' which indexes the full text into a list of words, 'encodeFullText' which encodes the full text into a numerical vector, and 'computeSimilarity' which computes its similarity with another text. We may add more text formats such as XML document, HTML one, and any other customized texts as subclasses derived from this class, in developing subsequent versions of the text categorization system.

The class, PlainText, is defined as a subclass which is inherited from the class, Text, in Fig. 7.4. An object is created with its text file name which is its identifier. Its properties are inherited from its superclass, Text. The method implementations in the class, Text, are provided in this class. Even if the abstract class looks similar as the interface, properties are defined as well as methods in the abstract class.

The category which indicates a label which is assigned to a text is defined as a separated class in Fig. 7.5. An object of this class is created with its name and its empty list of texts. The properties of this class are 'categoryName' which is the symbolic category name and 'sampleTextList' which is the list of sample texts

Fig. 7.5 The class: category

```
package TextClassification;
import java.util.*;
public class Category {
    String categoryName;
    Vector<Text> sampleTextList;

    public Category(String categoryName){
        this.categoryName = categoryName;
        this.sampleTextList = new Vector<Text>();
    }

    void addSampleTextItem(Text textItem){
        this.sampleTextList.addElement(textItem);
    }

    String getCategoryName(){
        return this.categoryName;
    }

    Text getSampleTextItem(int index){
        return this.sampleTextList.elementAt(index);
    }

    int getSize(){
        return this.sampleTextList.size();
    }
}
```

which belong to the current category. The category manages a list of texts through the methods which are defined in Fig. 7.5. The reason of defining a category as a separated class is that the categories are predefined and sample texts should be allocated to each of them, as preliminary tasks.

7.2.2   Interface and Class: Classifier and KNearestNeighbor

This section is concerned with the interface, Classifier, and its specific class, 'KNearestNeighbor.' The interface is similar as the abstract class which is mentioned in Sect. 7.2.1. Unlike the abstract class, in the interface, the methods are only listed without their implementations. The relation between an abstract class and its specific classes is based on their inheritance, whereas that between an interface and its specific classes is based on the polymorphism which will described below. In this section, we explain the polymorphism and describe the interface, Classifier, and the class, 'KNearestNeighbor.'

Defining the polymorphism as a concept of the object-oriented design and programming is intended to allow the variety of objects under their same class with respect their method implementations. The identical type of objects is required for implementing an array of objects. The polymorphism is used for the case of identical methods on objects but their different implementations. In the Java programming, the methods which are applied identically to objects are declared in the interface,

```
package TextClassification;
import java.util.*;
public interface Classifier {
    void setSampleTextList(Vector<Text> sampleTextList);
    void setNoviceTextList(Vector<Text> noviceTextList);
    Vector<Text> getNoviceTextList();
    void learnSampleTextList();
    void classifyNoviceTextList();
}
```

Fig. 7.6 The interface: classifier

and various classes which implement the methods differently are defined as its specific ones. The KNN algorithm is implemented as an object of the specific class, 'KNearestNeithbor,' and the object is declared in Java as follows:

Classifier knn = new KNearestNeighbor(3);

In Fig. 7.6, the interface, 'Classifier,' is defined. In the interface, only methods which are involved in executing the classifier are declared without their implementations. The interface definition is intended to expand the program easily by adding more classification algorithms. Various kinds of classification algorithms are treated as objects under the class, Classifier. Implementations of the declared methods are provided in the specific class, 'KNearestNeighbor.'

The class, 'KNearestNeighbor,' is defined as the class which is specific to the interface, Classifier, by the reserved statement, "implements Classifier." The object is created by the above statement. The object, 'knn,' in the above statement is treated as an instance of the class, Classifier. Its methods are actually implemented in the class, 'KNearestNeighbor.' If adding more specific classes, 'RadiusNearestNeithbor,' one more object is created as another classification algorithm by the following statement:

Classifier rnn = new RadiusNearestNeighbor(0.6);

Declaring the classifier 'KNearestNeighbor' as the interface and its specific class is intended for easy addition of more machine learning algorithms and each combination of different ones. The behaviors which are shared by any kind of machine learning algorithm are to assign training and test examples, to learn training examples, and to classify test ones. More machine learning algorithms such as Radius Nearest Neighbor, Naive Bayes, and Support Vector Machine may be added by defining more classes which are specific to the interface, 'Classifier,' subsequently. After adding the classes, the three machine learning algorithms are declared as the three objects:

$$\text{Classifier } c1 \; = \; \text{new KNearestNeighbor(3)};$$

$$\text{Classifier } c2 \; = \; \text{new RadiusNearestNeighbor(0.6)};$$

$$\text{Classifier } c3 \; = \; \text{new NaiveBayes()};$$

and the three objects, c1, c2, and c3, are treated as identical objects of the class Classifier. We may implement the multiple classifier combinations such as boosting, voting, and expert gates by doing so, easily.

7.2.3   Class: TextClassificationAPI

This section is concerned with defining the class, 'TextClassificationAPI,' as illustrated in Fig. 7.7. The class, 'TextClassificationAPI,' is called API (Application Program Interface) class which is created in the main program, and whose methods are called there. The text categorization system which is implemented in this chapter belongs to the GUI (Graphic User Interface) program. We mention creation of objects and invoking of methods in the GUI class in Sect. 7.4.4. In this section, we explain the class definition which is presented in Fig. 7.7, in terms of its properties, constructors, and methods.

The API class is explained as the scheme of implementing the program in Java. The API class means the public class which is touched by the main program in the console environment. In the API class, the API is attached to a class name as its postfix, and its role is to interface with other classes directly and indirectly. The API class is actually touched by the GUI class in this program with its graphical user interface. Only one API class is usually defined in implementing an application program.

Let us mention the properties which are defined in the class, 'TextClassifica-tionAPI.' The property, 'sampleTextList,' means the list of sample texts which are labeled and used for building the classification capacity. The property, 'novice-TextList,' is the list of texts which are prepared as the classification targets. The property, 'featureList,' means the list of attributes which are given as words for encoding texts into numerical vectors, and the property, dimension, means the number of features in encoding texts into numerical vectors. The property, 'categoryList,' is the list of categories which are predefined as the classification frame.

We explain the methods which are defined in Fig. 7.7 in their functional view. The contents of both sample and novice texts are loaded as a list of text objects by the methods, 'loadSampleTextList' and 'loadNoviceTextList.' Text objects are encoded into numerical vectors by the methods, 'encodeSampleTextList' and 'encodeNoviceTextList.' The novice texts in the list are classified by the KNN through the method, 'classifyNoviceTextList.' The method, 'generateFeatureList,' extracts features as the attributes for implementing both the methods, 'encodeSam-

```
public class TextClassificationAPI {
    Vector<Text> sampleTextList;
    Vector<Text> noviceTextList;
    Vector<String> featureList;
    Vector<String> categoryList;
    int dimension;
        public TextClassificationAPI(int dimension){
        this.dimension = dimension;
    }
    public void setSampleTextList(Vector<Text> sampleTextList){
        this.sampleTextList = sampleTextList;
    }
    public void setNoviceTextList(Vector<Text> noviceTextList){
        this.noviceTextList = noviceTextList;
    }
    public void setCategroyList(Vector<String> categoryList){
        this.categoryList = categoryList;
    }
    public Vector<Text> getNoviceTextList(){
        return this.noviceTextList;
    }
    public void loadSampleTextList(){}
    public void loadNoviceTextList(){}
    public void encodeSampleTextList(){}
    public void encodeNoviceTextList(){}
    public void classifyNoviceTextList(){}
    public void generateFeatureList(){}
    public void flatSampleTextList(){}
}
```

Fig. 7.7 The class: TextClassificationAPI

pleTextList' and 'encodeNoviceTextList,' and the method, 'flatSampleTextList,' converts the group of categories into the list of sample texts which are labeled with their own categories.

The GUI class is necessary for implementing the graphic user interface of this program, separately from the API class. The object of the API class is created, and its methods are invoked in the main program in the console running environment. The GUI class is required for expanding the console program into the GUI program. In the GUI class, the object of the class, TextClassificationAPI, is created and its methods are called. The properties and methods in the GUI class will be explained in detail, in Sect. 7.4.1.

7.3 Method Implementations

In this section, we present the method implementations which are involved in developing the text categorization system. In Sect. 7.3.1, we explain the implementations of the methods which are defined in class, 'Word.' In Sect. 7.3.2, we cover those methods which are included in class 'PlainText.' In Sect. 7.3.3, we describe the methods in the class, 'KNearestNeighbor,' with respect to their implementations. In Sect. 7.3.4, we mention the method implementations, concerned with the class, 'TextClassificationAPI.'

```
boolean isPresent(String fullText){
    int position = fullText.indexOf(this.wordName);
    if(position != -1)
        return true;
    return false;
}
```

Fig. 7.8 The method: isPresent

```
int computeWordFrequency(String fullText){
    if(fullText.length() == 0){
        return 0;
    }
    int offset = fullText.indexOf(this.wordName);
    if(offset == -1){
        return 0;
    }
    return 1 + computeWordFrequency(fullText.substring(offset + 1));
}
```

Fig. 7.9 The method: computeFrequency

7.3.1 Class: Word

This section is concerned with the method implementations in the class, 'Word.' The class, Word, was defined in Fig. 7.2 with its properties and methods. An object of the class, Word, is created by its symbolic name. The property, 'wordFrequency,' is initialized to zero in creating the object. In this section, we explain the implementations of individual methods which are members of the class, 'Word.'

Figure 7.8 illustrates the implementation of the method, 'isPresent.' The method belongs to the predicate method which checks the program status and returns a Boolean value, and 'is' used as the prefix of its name. This method checks whether the current object appears in the full text which is given as its argument, or not; if so, it returns true. The method, 'indexOf,' which is called in Fig. 7.9, is the method of a string object which returns the position where the argument, substring, occurs first time. If the substring does not appear in the string, it returns -1.

The implementation of the method, 'computeFrequency,' is illustrated in Fig. 7.9. The full text is given as the argument, like the method, 'isPresent,' and the method, 'computeFrequency,' is implemented recursively; the method is invoked in itself. It checks whether the property, 'wordName,' occurs in the current full text; if it is so, it returns zero. Otherwise, it updates the argument, 'fullText,' by deleting its prefix which ends with the word occurrence, and invokes the method with the updated full text, recursively by incrementing one. This method counts the occurrences of the word, which is given the property, 'wordName,' in the full text which is given the argument, by the recursive call.

Figure 7.10 illustrates the implementation of the method, 'computeWordWeight.' The two arguments, 'fullText' and corpus, are given in the method, and the method, 'computeWordFrequency,' whose implementation is presented in Fig. 7.9

```
double computeWordWeight(String fullText, Vector<String> corpus){
    int corpusSize = corpus.size();
    int termFrequency = this.computeWordFrequency(fullText) ;
    int documentFrequency = 0;
    for(int i = 0;i < corpusSize ;i++){
        String corpusFullTextItem = corpus.elementAt(i);
        if(this.isPresent(corpusFullTextItem))
            documentFrequency++;
    }
    double wordWeight = Math.log((termFrequency + 1)*(corpusSize/documentFrequency));
    return wordWeight;
}
```

Fig. 7.10 The method: computeWordWeight

```
double computeWordSimilarity(Word another,Vector<String> corpus){
    int corpusSize = corpus.size();
    int documentFrequency1 = 0;
    int documentFrequency2 = 0;
    int collocationFrequency = 0;
    for(int i = 0;i < corpusSize ;i++){
        String corpusFullTextItem = corpus.elementAt(i);
        if(this.isPresent(corpusFullTextItem))
            documentFrequency1++;
        if(another.isPresent(corpusFullTextItem))
            documentFrequency2++;
        if((this.isPresent(corpusFullTextItem)) && another.isPresent(corpusFullTextItem))
            collocationFrequency++;
    }
    return (2 * collocationFrequency)/(documentFrequency1 + documentFrequency2);
}
```

Fig. 7.11 The method: computeWordSimilarity

is involved. The frequency of the word in the full text is computed by invoking the method, 'computeWordFrequency,' and the number of total texts and the number of texts including the word are counted in the corpus. The word is weighted by the TF-IDF (Term Frequency-Inverse Document Frequency).

Figure 7.11 illustrates the implementation of the method, 'computeWordSimilarity.' The method computes the semantic similarity with another word, and the opposite word and corpus which is the reference for computing the similarity are given as arguments. In the corpus, the number of documents which include either of the two words and the number of documents which include both of them are counted. The semantic similarity between the two words is computed as the ratio of the documents including both words to the summation of ones including either of them, as shown in the return statement in Fig. 7.11. The semantic similarity which is returned from this method is given as a normalized value between zero and one.

7.3.2 Class: PlainText

This section is concerned with the method implementations which are involved in the class, 'PlainText.' The class is the subclass which receives the inheritances from the super class, 'Text.' The implementations of the methods which are defined in the

```
void loadFullText() {
    FileString fs = new FileString(this.textFileName);
    fs.loadFileString();
    this.fullText = fs.getFileString();
}
```

Fig. 7.12 The method: loadText

```
void indexFullText() {
    this.wordList = new Vector<String>();
    System.out.println("PlainText::indexFullText();");
    Vector<String> tokenList = this.tokenizeTextContent();
    Vector<String> stopWordList = this.loadStopWordList("stopWordList.ini");
    Vector<String> wordNameList = this.filterStopWordList(tokenList,stopWordList);
    this.wordList = this.filterSimilarWordList(wordNameList);
    int size = wordNameList.size();
    for(int i = 0;i < size; i++){
        String wordName = wordNameList.elementAt(i);
        Word wordItem = new Word(wordName);
        wordItem.computeWordFrequency(this.fullText);
    }
}
```

Fig. 7.13 The method: indexText

class, Text, are provided in this class. The methods are implemented differently in other subclasses which will be added in the future. In this section, we explain the method implementations in the class, 'PlainText.'

Figure 7.12 illustrates the implementation of the method, 'loadText.' This method loads a text as a string from a file. An object of the class, 'FileString,' is created and the method, 'loadFileString,' is invoked. It takes the full text by invoking the method, 'getFileString.' The exceptional handling for access to the file should be embedded in the methods in the class, 'FileString.'

The implementation of the method, 'indexText,' is illustrated in Fig. 7.13. This method indexes a string which is the value of the property, 'fullText,' into a list of words. In Chap. 2, we explained the basic three steps of text indexing, tokenization, stemming, and stop word removal. A list of words is generated and for each word, its frequency is counted. A list of objects of the class, 'Word,' is generated as the output from this method.

Figure 7.14 illustrates of the method, 'encodeText.' It encodes a text into a numerical vector with the two arguments, dimension and 'featrueList.' It checks whether dimension and the actual number of features match with each other, and if they do not, it terminates. If they match with each other, for each feature, its frequency is counted in the given text. The numerical vector which represents a text consists of integers which indicate the frequencies of features.

Figure 7.15 illustrates the implementation of the method, 'computeSimilarity.' It computes the similarity between two texts under the assumption of encoding them into numerical vectors. We adopt Eq. (6.3) which is a variant of the cosine similarity, for doing so. The inner product and the norm of the two vectors are computed and the similarity is generated by applying Eq. (6.3). The method returns a normalized value between zero and one.

```
void encodeFullText(int dimension, Vector<String> featureList){
    int featureSize = featureList.size();
    if(dimension != featureSize){
        System.out.println("Mismatch between dimension and feature size!!");
        return;
    }
    this.featureVector = new int[dimension];
    for(int i = 0;i < dimension;i++){
        String featureName = featureList.elementAt(i);
        this.featureVector[i] = this.computeWordFrequency(featureName, this.fullText);
    }
}
```

Fig. 7.14   The method: encodeText

```
double computeSimilarity(Text another) {
    int[] anotherFeatureVector = another.getFeatureVector();
    int dimension = this.featureVector.length;
    int thisNorm = 0;
    int anotherNorm = 0;
    int innerProduct = 0;
    for(int i = 0;i < dimension;i++){
        thisNorm = thisNorm + (this.featureVector[i] * this.featureVector[i]);
        anotherNorm = anotherNorm + (anotherFeatureVector[i] * anotherFeatureVector[i]);
        innerProduct = innerProduct + (anotherFeatureVector[i] * this.featureVector[i]);
    }
    double similarity = (2 * (double)innerProduct)/((double)thisNorm + (double)anotherNorm);
    return similarity;
}
```

Fig. 7.15   The method: computeSimilarity

7.3.3   Class: KNearestNeighbor

This section is concerned with the method implementations in the class, 'KNearestNeighbor.' The methods which are defined in the interface, Classifier, are implemented in this class which is a specific class. We adopt the KNN which was described in Sect. 6.1, in implementing the text categorization system. More classifiers such as Naive Bayes, Bayes Classifier, and its variants are added in upgrading the system in the future. In this section, we explain the method implementations and present the direction of upgrading the system.

Lazy learning is the learning paradigm where training examples had never been learned in advance until novice examples were given. So, there is no implementation in the method, 'learnSampleTextList,' in the class; that means that the adopted learning algorithm learns no training example in advance. It learns training examples interactively for each test example, so the interactive learning process is implemented in the method, 'classifyNoviceTextList.' Eager learning is one where the classification capacity is built by learning training examples in advance, as the opposite one to the lazy learning. However, we need to implement the method, 'learnSampleTextList,' in the class of eager learning algorithms.

Figure 7.16 illustrates the method implementation, 'classifyNoviceTextList.' Because the KNN algorithm belongs to the lazy learning, the process of learning the training examples and classifying a novice example is given in this method. For each novice text, its similarities with the training examples are computed and

```
public void classifyNoviceTextList() {
    int noviceTextSize = this.noviceTextList.size();
    int sampleTextSize = this.sampleTextList.size();
    for(int i = 0;i < noviceTextSize; i++){
        Text noviceTextItem = this.noviceTextList.elementAt(i);
        double maxSimilarity = 0.0;
        int maxIndex = 0;
        for(int j = 0;j < sampleTextSize; j++){
            Text sampleTextItem = this.sampleTextList.elementAt(j);
            double similarity = noviceTextItem.computeSimilarity(sampleTextItem);
            if(maxSimilarity < similarity){
                maxSimilarity = similarity;
                maxIndex = j;
            }
        }
        Text nearestTextItem = this.sampleTextList.elementAt(maxIndex);
        String classifiedLabel = nearestTextItem.getTextLabel();
        noviceTextItem.setTextLabel(classifiedLabel);
        this.noviceTextList.setElementAt(noviceTextItem, i);
    }
}
```

Fig. 7.16   The method: classifyNoviceTextList in class, K Nearest Neighbor

the most similar training example is selected as its nearest neighbor. The novice example is classified into the label of the nearest neighbor. In implementing the text categorization system which is the initial demonstration version, we adopt the 1 Nearest Neighbor, as the simplest classification algorithm.

The requirement for expanding the 1 Nearest Neighbor into the KNN is to modify the implementation of the method, 'classifyNoviceTextList.' In order to rank training examples by their similarities, we need a sorting algorithm. The linear logarithmic complexity is the least one for sorting data items; the quick sort or the heap sort is recommendable. If N and M are the number of training examples and test examples, respectively, the complexity of the current version, $O(MN)$, is increased to $O(MNlogN)$, for expanding so. However, the complexity, $O(MN)$, is fixed in expanding the 1 Nearest Neighbor to the Radius Nearest Neighbor.

It is possible to combine multiple KNN algorithms with each other as an approach to text categorization. We need to discriminate the KNN algorithm by some factors. They may be discriminated by different numbers of nearest neighbors. We may consider some variants of KNN which were described in Sect. 6.3. We may consider schemes of combining discriminated KNN algorithms with each other in upgrading the text categorization system in the future.

7.3.4   Class: TextClassificationAPI

This section is concerned with the methods which are involved in the class, 'TextClassificationAPI.' The API class is mentioned as the class whose objects and methods are created and invoked in the main program in the console environment.

```
public void flatSampleTextList(){
    int size = this.categoryList.size();
    System.out.println("size = " + size);
    for(int i = 0;i < size; i++){
        Category categoryItem = this.categoryList.elementAt(i);
        int categorySize = categoryItem.getSize();
        String categoryName = categoryItem.getCategoryName();
        for(int j = 0;j < categorySize; j++){
            Text textItem = categoryItem.getSampleTextItem(j);
            textItem.setTextLabel(categoryName);
            this.sampleTextList.addElement(textItem);
        }
    }
}
```

Fig. 7.17 The method: flatSampleTextList

```
public void generateFeatureList(){
    if(this.sampleTextList == null){
        System.out.println("Null Pointer!!");
        return;
    }
    int size = this.sampleTextList.size();
    String integratedFullText = "";
    for(int i = 0;i < size; i++){
        Text textItem = this.sampleTextList.elementAt(i);
        integratedFullText = integratedFullText + textItem.getFullText() + "\n";
    }
    Text integratedText = new PlainText("dummyFileName");
    integratedText.setFullText(integratedFullText);
    this.featureList = integratedText.generateFeatureList(this.dimension);
}
```

Fig. 7.18 The method: generateFeatureList

The object of this class created and its methods are invoked in the GUI class, in this program. The definition of the API class was covered in Sect. 7.2.3. In this section, we will explain the method implementations in detail.

Figure 7.17 illustrates the implementation of the method, 'flatSampleTextList.' A list of categories each of which contains texts is obtained from the GUI interface. The list of categories is mapped into a list of texts which are labeled with their own categories through this method. For each category, each text is labeled by its own category name and inserted into the property, 'sampleTextList.' The property, 'categoryList,' is the list of categories which is obtained from the GUI interface, and the property, 'sampleTextList,' is the list of labeled sample texts.

Figure 7.18 illustrates the implementation of the method, 'generateFeatureList.' The sample texts are concatenated into a single text. Because no file name is given to the integrated text, its object is created by its dummy file name. The integrated text is set as an object and the method, 'generateFeatureList,' is invoked by this object. The number of features is fixed to the dimension as the size of property, 'featureList.'

```
public void encodeSampleTextList(){
    if(this.sampleTextList == null){
        System.out.println("Null Pointer!!");
        return;
    }
    int size = this.sampleTextList.size();
    for(int i = 0;i < size; i++){
        Text textItem = this.sampleTextList.elementAt(i);
        textItem.encodeFullText(this.dimension, this.featureList);
    }
}

public void encodeNoviceTextList(){
    if(this.noviceTextList == null){
        System.out.println("Null Pointer!!");
        return;
    }
    int size = this.noviceTextList.size();
    for(int i = 0;i < size; i++){
        Text textItem = this.noviceTextList.elementAt(i);
        textItem.encodeFullText(this.dimension, this.featureList);
    }
}
```

Fig. 7.19   The method: encodeSampleTextList and encodeNoviceTextList

```
public void classifyNoviceTextList(){
    Classifier textClassifier = new KNearestNeighbor();
    textClassifier.setSampleTextList(this.sampleTextList);
    textClassifier.setNoviceTextList(this.noviceTextList);
    textClassifier.classifyNoviceTextList();
    this.noviceTextList = textClassifier.getNoviceTextList();
    int size = this.noviceTextList.size();
    for(int i = 0; i < size; i++){
        Text textItem = this.noviceTextList.elementAt(i);
        String textFileName = textItem.getTextFileName();
        String textLabel = textItem.getTextLabel();
        System.out.println(textFileName + ": " + textLabel);
    }
}
```

Fig. 7.20   The method: classifyNoviceTextList in the Class, TextClassificationAPI

Figure 7.19 illustrates the implementations of the methods, 'encodeSample-TextList' and 'encodeNoviceTextList.' If a list of texts is given as a null, the methods do nothing as their exception handlings. The method, 'encodeSampleTextList,' encodes a list of sample texts into numerical vectors and the method, 'encodeN-oviceTextList,' does a list of novice texts so. The method, 'generateFeatureList,' should be executed before executing the methods. The two methods need to be merged into the method, 'encodeTextList,' by adding one more argument, a list of texts.

Figure 7.20 illustrates the implementation of the final method, 'classifyNovice-TextList.' It is required for executing the method to execute the above methods.

A classifier is created as an object of the class, 'KNearestNeighbor,' sample texts and novice texts are assigned to the object, and the method, 'classifyNoviceText,' is invoked. The reason of not invoking the method, 'learnSampleText,' is that the KNN which is adopted in implementing the system does not learn training examples in advance, as a lazy learning algorithm. The classifier, KNN, is implemented as a 1 Nearest Neighbor in this program.

7.4 Graphic User Interface and Demonstration

This section is concerned with the graphical user interface and demonstration of the developed text categorization system. In Sect. 7.4.1, we explain the final class, 'TextClassificationGUI,' together with the graphic user interface. In Sect. 7.4.2, we demonstrated the program in the process of doing preliminary tasks and learning sample texts. In Sect. 7.4.3, we present the process of classifying texts by the program. In Sect. 7.4.4, we mention the guides for upgrading the developed program.

7.4.1 Class: TextClassificationGUI

Figure 7.21 illustrates the initial screen of the implemented text categorization system. On top of the initial screen, the drop-down box contains predefined categories, and the two buttons are for adding more categories and for deleting some categories, respectively. In the middle of the initial screen, there are the three buttons: one for adding sample texts under the category which is selected in the above drop-down box, another for showing a list of sample texts in the area which is labeled 'sample texts,' the other for encoding the sample texts into numerical vectors. In the bottom of the initial screen, there are the two buttons: one for providing texts to classify and the other for classifying them. In this section, we study the properties and methods of class, 'TextClassificationGUI,' for activating the initial screen in Fig. 7.22.

The properties in the class, 'TextClassificationGUI,' are presented in Fig. 7.22. The objects of the class, 'JButton,' among the properties, correspond to buttons in the interface which is shown in Fig. 7.21. The objects of the class, 'JList,' and those of the class, 'JCombobox,' correspond to the text areas and the drop-down box, respectively. The objects of the class, 'JLabel,' are the labels which are nearby the two text areas in the interfaces. The properties 'categoryList,' 'noviceTextListVector,' and 'textClassifier' are the category list, the novice text list, and the object of text classifier algorithm, respectively.

Figure 7.23 illustrates the implementation of the constructor of this class. The components which are involved in the interface are created as objects. If necessary, their options should be configured. The method, 'addActionLister,' is invoked for

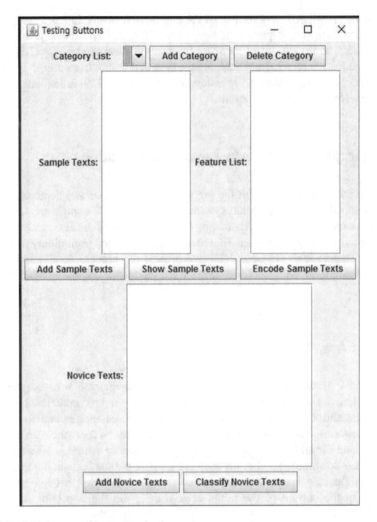

Fig. 7.21 Initial screen of text categorization system

each object of the class, 'JButton,' for implementing the event-driven programming. Refer to the literature, for the detailed contents about the event-driven programming.

Figure 7.24 shows the implementation of the method, 'ActionPerformed.' The method is defined in the class, "ButtonHandler" which is specific to the interface, 'ActionListener,' and nested in the class, 'TextClassificationGUI.' The implementation consists of actions which are corresponding to buttons which are pressed in the interface in Fig. 7.21. The statements within a block of if-statement indicates that when pressing the corresponding button, the program runs, accordingly. The method, 'ActionPerformed,' is used for implementing the event-driven programming, in Java.

```
public class TextClassificationGUI extends JFrame{
    Vector<Category> categoryList;
    Vector<String> noviceTextListVector;
    JLabel categoryListLabel;
    JComboBox categoryListCombo;
    JButton addCategoryButton;
    JButton deleteCategoryButton;

    JLabel sampleTextLabel;
    JList sampleTextList;
    JLabel featureListLabel;
    JList featureList;
    JButton addSampleTextListButton;
    JButton showSampleTextListButton;
    JButton encodeSampleTextListButton;

    JLabel noviceTextLabel;
    JList noviceTextList;
    JButton addNoviceTextListButton;
    JButton classifyNoviceTextListButton;

    TextClassificationAPI textClassifier;
```

Fig. 7.22 Class: TextClassificationGUI

The main program is illustrated in Fig. 7.25. The object of the class, 'TextClassificationGUI,' is created and its methods are invoked. The configurations of the GUI (Graphic User Interface) are made by invoking the methods, 'setDefaulCloseOperation' and 'setSize,' and a window is shown by invoking the method, 'setVisible.' The object of the class, 'TextClassificationAPI,' is created in the class, 'TextClassificationGUI,' and the involved methods are invoked in the method, 'ActionPerformed.' The class, 'TextClassificationGUI,' becomes the final one in this program.

7.4.2 Preliminary Tasks and Encoding

This section is concerned with the process of demonstrating preliminary tasks and the encoding process of the text categorization system. By adding categories to the drop-down box, they are defined as a list. For each category in the drop-down box, texts which are given as plain text files are added as sample ones. The added texts are encoded into numerical vectors. In this section, we present the demonstration of the system in doing the preliminary tasks and encoding into numerical vectors.

```
public TextClassificationGUI(){
    super("Testing Buttons");
    setLayout(new FlowLayout());
    this.categoryList = new Vector();
    this.noviceTextListVector = new Vector();
    textClassifier = new TextClassificationAPI(50);

    this.categoryListLabel = new JLabel("Category List:    ");
    this.add(this.categoryListLabel);

    String categoryNameList[] = {};
    this.categoryListCombo = new JComboBox(categoryNameList);
    this.categoryListCombo.setMaximumRowCount(5);
    this.add(this.categoryListCombo);

    this.addCategoryButton = new JButton("Add Category");
    this.add(this.addCategoryButton);

    this.deleteCategoryButton = new JButton("Delete Category");
    this.add(this.deleteCategoryButton);

    this.sampleTextLabel = new JLabel("Sample Texts:");
    this.add(sampleTextLabel);
    this.sampleTextList = new JList();
    sampleTextList.setFixedCellHeight(30);
    sampleTextList.setFixedCellWidth(120);
    this.add(new JScrollPane(sampleTextList));

    this.featureListLabel = new JLabel("Feature List:");
    this.add(featureListLabel);
    this.featureList = new JList();
    featureList.setFixedCellHeight(30);
    featureList.setFixedCellWidth(120);
    this.add(new JScrollPane(featureList));

    this.addSampleTextListButton = new JButton("Add Sample Texts");
    this.add(addSampleTextListButton);
    this.showSampleTextListButton = new JButton("Show Sample Texts");
    this.add(showSampleTextListButton);
    this.encodeSampleTextListButton = new JButton("Encode Sample Texts");

    this.noviceTextLabel = new JLabel("Novice Texts:");
    this.add(noviceTextLabel);
    this.noviceTextList = new JList();
    noviceTextList.setFixedCellHeight(30);
    noviceTextList.setFixedCellWidth(250);
    this.add(new JScrollPane(noviceTextList));

    this.addNoviceTextListButton = new JButton("Add Novice Texts");
    this.add(addNoviceTextListButton);
    this.classifyNoviceTextListButton = new JButton("Classify Novice Texts");
    this.add(classifyNoviceTextListButton);

    ButtonHandler handler = new ButtonHandler();
    this.addCategoryButton.addActionListener(handler);
    this.deleteCategoryButton.addActionListener(handler);
    this.addSampleTextListButton.addActionListener(handler);
    this.showSampleTextListButton.addActionListener(handler);
    this.encodeSampleTextListButton.addActionListener(handler);
    this.addNoviceTextListButton.addActionListener(handler);
    this.classifyNoviceTextListButton.addActionListener(handler);
}
```

Fig. 7.23 Constructor implementation in TextClassificationGUI

```
private class ButtonHandler implements ActionListener{

    public void actionPerformed(ActionEvent event){
        String eventName = event.getActionCommand();
        if(eventName == "Add Category"){
            String categoryName = JOptionPane.showInputDialog("Category Name:");
            TextClassificationGUI.this.categoryListCombo.addItem(categoryName);
            Category categoryItem = new Category(categoryName);
            TextClassificationGUI.this.categoryList.addElement(categoryItem);
        }
        if(eventName == "Delete Category"){
            Object item = TextClassificationGUI.this.categoryListCombo.getSelectedItem();
            int selectedIndex = TextClassificationGUI.this.categoryListCombo.getSelectedIndex();
            TextClassificationGUI.this.categoryListCombo.removeItem(item);
            TextClassificationGUI.this.categoryList.removeElementAt(selectedIndex);
        }
        if(eventName == "Add Sample Texts"){
            JFileChooser fileChooser = new JFileChooser();
            fileChooser.setMultiSelectionEnabled(true);
            int result = fileChooser.showOpenDialog(TextClassificationGUI.this);
            File[] fileList = fileChooser.getSelectedFiles();
            int selectedIndex = TextClassificationGUI.this.categoryListCombo.getSelectedIndex();
            Category selectedCategory = (Category)TextClassificationGUI.this.categoryList.elementAt(selectedIndex);
            String currentDirectory = fileChooser.getCurrentDirectory().getAbsolutePath();
            for(int i = 0; i < fileList.length; i++){
                String fileName = currentDirectory + fileList[i].getName();
                System.out.println(fileName);
                selectedCategory.addSampleTextItem(fileName);
            }
            Vector<String> sampleTextFileNameList = selectedCategory.getSampleTextFileNameList();
            TextClassificationGUI.this.sampleTextList.setListData(sampleTextFileNameList);
        }
        if(eventName == "Show Sample Texts"){
            int selectedIndex = TextClassificationGUI.this.categoryListCombo.getSelectedIndex();
            Category selectedCategory = (Category)TextClassificationGUI.this.categoryList.elementAt(selectedIndex);
            Vector<String> sampleTextFileNameList = selectedCategory.getSampleTextFileNameList();
            TextClassificationGUI.this.sampleTextList.setListData(sampleTextFileNameList);
        }
        if(eventName == "Encode Sample Texts"){
            TextClassificationGUI.this.textClassifier.setCategoryList(TextClassificationGUI.this.categoryList);
            TextClassificationGUI.this.textClassifier.flatSampleTextList();
            TextClassificationGUI.this.textClassifier.loadSampleTextList();
            TextClassificationGUI.this.textClassifier.generateFeatureList();
            TextClassificationGUI.this.textClassifier.encodeSampleTextList();
            Vector<String> featureList = TextClassificationGUI.this.textClassifier.getFeatureList();
            TextClassificationGUI.this.featureList.setListData(featureList);
        }

        if(eventName == "Add Novice Texts"){
            JFileChooser fileChooser = new JFileChooser();
            fileChooser.setMultiSelectionEnabled(true);
            int result = fileChooser.showOpenDialog(TextClassificationGUI.this);
            File[] fileList = fileChooser.getSelectedFiles();
            String currentDirectory = fileChooser.getCurrentDirectory().getAbsolutePath();
            for(int i = 0; i < fileList.length; i++){
                String fileName = currentDirectory + fileList[i].getName();
                noviceTextListVector.addElement(fileName);
                TextClassificationGUI.this.textClassifier.addNoviceTextFileName(fileName);
            }
            TextClassificationGUI.this.noviceTextList.setListData(noviceTextListVector);
        }

        if(eventName == "Classify Novice Texts"){
            TextClassificationGUI.this.textClassifier.loadNoviceTextList();
            TextClassificationGUI.this.textClassifier.encodeNoviceTextList();
            TextClassificationGUI.this.textClassifier.classifyNoviceTextList();
            String resultStream = "";
            Vector<Text> classifiedNoviceTextList = TextClassificationGUI.this.textClassifier.getNoviceTextList();
            int size = classifiedNoviceTextList.size();
            for(int i = 0;i < size;i++){
                Text textItem = classifiedNoviceTextList.elementAt(i);
                String textFileName = textItem.getTextFileName();
                String textLabel = textItem.getTextLabel();
                resultStream = resultStream + textFileName + ": " + textLabel +"\n";
            }
            JOptionPane.showMessageDialog(TextClassificationGUI.this, resultStream);
        }
    }
}
```

Fig. 7.24 ActionPerformed implementation in TextClassificationGUI

```
import TextClassification.*;
public class Main {
    public static void main(String[] args) {
        TextClassificationGUI gui = new TextClassificationGUI();
        gui.setDefaultCloseOperation(JFrame.EXIT_ON_CLOSE);
        gui.setSize(475,650);
        gui.setVisible(true);
    }
}
```

Fig. 7.25 Main program

Fig. 7.26 Adding categories in the developed text categorization system

Figure 7.26 illustrates the process of adding categories in the developed text categorization system. The button, 'Add Category,' is pressed in the interface. The category name is entered by the keyboard, and the left button, OK, is pressed.[1] The category name is added to the drop-down box. By doing so, the category predefinition which is a preliminary task is achieved.

---

[1]Because the program runs on the Korean version window, the buttons are labeled in Korean.

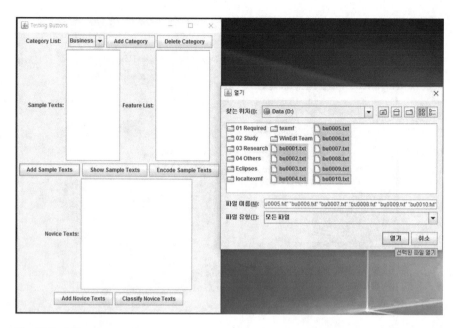

Fig. 7.27 Allocating sample texts to category

Figure 7.27 illustrates the process of allocating sample texts to a particular category. The category, business, is selected in the drop-down box. If the button, 'add sample texts,' is pressed, the pop-up dialog box is given in the left part of Fig. 7.27. Through the browsing, the directory is located in the left dialog box, text files are selected as sample texts. If the left button in the dialog box is pressed, the selected files are added to the sample texts in the category, business.

Figure 7.28 illustrates the results from adding the sample texts to the category, business. The category, business, is selected in the drop-down box, and the sample texts are presented by pressing the button, 'show sample texts.' The list of words in the area are attributes of numerical vectors which represent texts. The text indexing and text encoding which were covered in Chaps. 2 and 3 are executed by pressing the button, 'Encode Sample Texts.'

In the current version of the text categorization system, there is no learning of sample texts in advance. We adopted the 1 nearest neighbor which is the simplest lazy learning algorithm as the approach to the text categorization in this system. More learning algorithms are added in upgrading the system in Fig. 7.27. Accordingly, one more button which is labeled with 'learning' should be added for learning sample texts before classifying novice texts. In the interface, we need to add the option for selecting a machine learning algorithm as a classification tool.

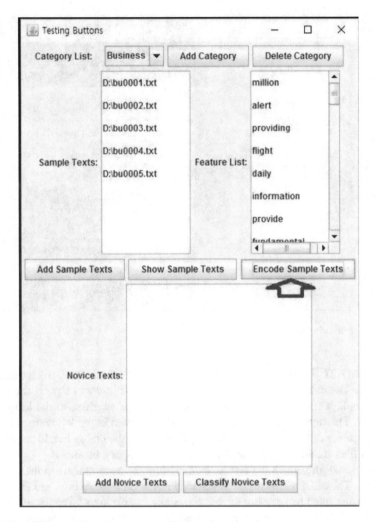

Fig. 7.28 Adding sample texts to category, 'Business'

7.4.3 Classification Process

This section is concerned with the demonstration of classifying texts automatically in the developed system. In the previous section, the categories are predefined, sample texts are allocated to each category, and they are encoded into numerical vectors. The tasks of demonstrating the developed system is to nominate novice texts and to classify them into one of the predefined categories. The classified texts are displayed in the message box as shown in Fig. 7.31. In this section, we describe the process of demonstrating the classification process.

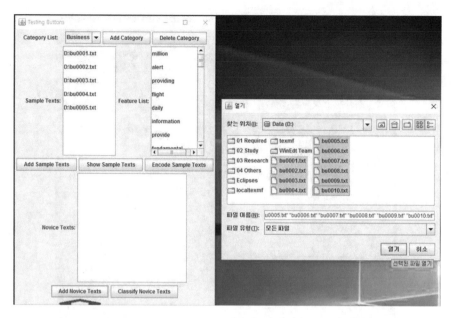

Fig. 7.29 Preparing novice texts

Figure 7.29 illustrates the process of preparing novice texts as classification targets. First, we press the button, 'Add Novice Texts,' in the interface. The dialog box for opening files is popped up, and we select text files as novice texts. In the dialog box, we press the left button and the selected text files are displayed in the area which is labeled, 'Novice Texts.' The novice texts will be classified in the text step.

Figure 7.30 illustrates the process of completing the preparation for classifying texts. The list of categories is stored in the drop-down box, and the sample texts are presented in the area which is labeled with 'Sample Texts.' The list of features for encoding texts into numerical vectors is presented in the area which is labeled with 'Feature List,' and novice texts which are selected by the above process in the area which is labeled with 'Novice Texts.' By pressing the button, 'Classify Novice Texts,' the novice texts are encoded into numerical vectors and are classified by the 1 Nearest Neighbor. The results from classifying the novice texts are displayed in Fig. 7.31.

Figure 7.31 illustrates the results from classifying the texts which are given as a message box. The ten files are selected as the classification targets, as shown in Fig. 7.30. Each line consists of the file name with its path which identifies the text and its classified category, in the message box. For example, the first line, D:\bu0001.txt:Business, consists of 'D:\bu0001.txt' as the file name with its path and 'Business' as its category. We may consider stamping its label on the text by concatenating it to its contents.

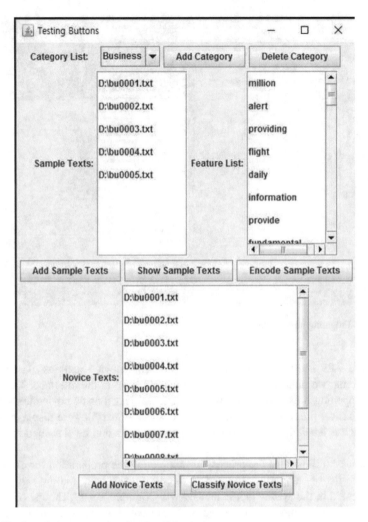

Fig. 7.30 Completing preparation for classifying texts

The text categorization system was implemented as only a prototype program, in this chapter. The first step of improving the system is to expand the 1 Nearest Neighbor into the KNN. We may add the KNN variants such as the RNN, the distance discriminated version, and the attribute discriminated one which are mentioned from Sects. 6.2.2–6.2.4. Other kinds of machine learning algorithms, Bayes Classifier, Naive Bayes, and Bayesian Classifier, which were mentioned in Sect. 6.3, are also added. So, we need to add the option for selecting a text categorization approach in the interface.

Fig. 7.31 Classified texts in
message box

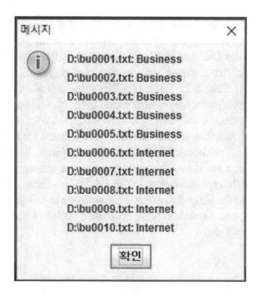

7.4.4   System Upgrading

This section is concerned with the direction of upgrading the text categorization
system which is implemented in this chapter. The text format which is covered by
this system is only the plain text file. In this system, only 1 Nearest Neighbor is
adopted as the approach to the text categorization. Only exclusive classification is
executed, and there is no module for decomposing the multiple classification into
binary classifications, in this version. So, in this section, we present the plans for
upgrading the text categorization system.

Let us consider some types of textual data as the direction of upgrading the
system. What is covered in the current version is the plain text which is given
as the file whose extension is 'txt.' The system will be upgraded into the version
which is able to process XML documents as the standard text format. The PDF
(Portable Document Format) file which is used most popularly is also considered for
upgrading the system. The web documents which are written in HTML are covered
together with the XML formation, in the next version.

Various kinds of machine learning algorithms are available as approaches to
the text categorization. The first step for upgrading the program is to expand
the 1 Nearest Neighbor into the KNN algorithm. After that, we add the KNN
variants such as Radius Nearest Neighbor, distance discriminant version, and
attribute discriminant version which are covered in Sect. 6.2. We may also add
the probabilistic learning algorithms, such as Bayes classifier, Naive Bayes, and
Bayesian learning, which were covered in Sect. 6.3. As a state-of-the-art approach,
we may consider the support vector machine which is the popular classification
algorithm in any application domain.

The overlapping or fuzzy classification may be considered as a direction of upgrading the system. We need to implement the module of decomposing a classification task into binary classification tasks as many as categories. The task is decomposed into binary classifications by the process that were mentioned in Sect. 5.2.3, and classifiers are created as objects as many as categories. The categories which correspond to the classifiers which classify the text into the positive class are assigned to it. However, the reallocation of training examples to the mapped binary classifiers becomes the overhead for doing the decomposition.

As shown in Fig. 7.21, we designed the GUI interface of this program, very simply. We will add the function of creating a text and editing an existing one. We will represent graphically a list of predefined categories, rather than including them in the drop-down box. Sample texts are displayed by clicking one of the predefined ones. Novice texts are organized by their classified labels, as well as displaying them as a list.

7.5   Summary and Further Discussions

This chapter is summarized as developing the text categorization system as an initial prototype version in Java. We presented the classes which are defined by their properties and their methods. We explained the implementations of methods which are included as members in the classes. We explained the GUI class, demonstrated the system, and presented the direction of upgrading the system in future. In this section, we will make further discussion about what we studied in this chapter.

In this chapter, we adopted the 1 Nearest Neighbor as the approach to the text categorization in implementing the system. We mentioned the addition of more machine learning algorithms in upgrading the system as the classes which are specific to the interface, 'TextClassifier.' After adding the complete KNN algorithm, we add further the probabilistic learning, the Perceptron, and the support vector machine. The objects of different machine learning algorithms are treated identically as objects of the class, 'TextClassifier.'

In upgrading the system, we considered XML documents as sample and novice texts as well as plain texts. In the recent version of JDK (Java Development Kit), the module of processing XML documents may be implemented by java library classes. By utilizing the library classes, the system is upgraded into the version which supports the XML documents as the standard format. The class, Text, is set as the abstract class. Therefore, the system is expanded into the version which categorizes the XML documents as well as plain texts.

Let us consider implementing the text categorization system as a web- based application program. In this chapter, we implement the system as an independent application program, in Java. The programming language, JavaScript, is used for developing a web-based program. The graphic user interface which is shown in Fig. 7.31 is activated in the web browser, and the results from classifying texts are displayed in the HTML format. The development of the text categorization as the web-based version is left in the next study.

Chapter 8
Text Categorization: Evaluation

This chapter is concerned with the schemes of evaluating text categorization systems. We make general discussions on the process of evaluating the text categorization systems, in Sect. 8.1, and mention the text collections which are used for evaluating text categorization systems and approaches, in Sect. 8.2. In Sect. 8.3, we describe the F1 measure which is introduced from the area of information retrieval and used as the most popular evaluation metric, in detail. We explain the statistical t test which is introduced from the statistics as the basis for comparing two text classification systems and algorithms, in Sect. 8.4, and make the summarization and further discussions on this chapter, in Sect. 8.5. In this chapter, we describe the text collection for evaluating text categorization systems, evaluation measures, and the schemes of comparing two approaches.

8.1 Evaluation Overview

It is necessary to evaluate text classification performances for adopting an approach for implementing text categorization systems. We prepare the test collection and partition it into the two sets: the training set for learning and the test set for classification and evaluation. The participated approaches learn by examples in the training set and classify those in the test set. We compute the F1 measures of the approaches, and compare them with each other based on the t-test. In this section, we describe briefly the process of evaluating text categorization systems.

Let us mention some test collections which are used for evaluating performances of text categorization systems and approaches. The collection, NewsPage.com, is the small collection of texts which was built by Jo in 2005. The Reuter21578 is mentioned by Sebastiani as the most popular one in 2002 [85]. The 20NewsGroups is one where 20 categories and totally 20,000 texts are composed in a two-level

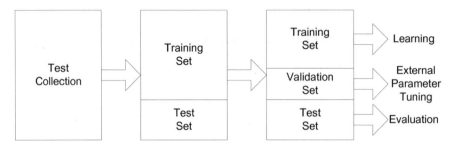

Fig. 8.1   Division of test collection into three sets

Fig. 8.2   *n* cross-validation

hierarchical form and is another popular test collection. The OSHUMED which is the collection of medical documents, each of which is classified by medical areas used for evaluating approaches in the domain, medicine.

Once a test collection is adopted for evaluating text categorization systems and approaches, as shown in Fig. 8.1, it should be divided into the three sets: the training set, the validation set, and the test set. The training set is used for building the classification capacity of the adopted approach. The validation set is separated from the training set, and used for validating the classification capacity which is built by the training set. The test set is used for evaluating finally the text categorization systems and approaches. After installing the text categorization system, a set of real inputs which is called the real set is given for executing it.

In Fig. 8.2 where the white portion indicates the training set and the black one does the validation set, we illustrate the cross-validation which is different from the above simple one. The cross-validation is intended to avoid the dependency and the bias on the fixed validation set. The training set is divided into n portions and we use $n - 1$ portions as the training set and one portion as the validation set. We may use the different portions, rotating over the n portions as the validation set, in each training phase, as shown in Fig. 8.2. By the cross-validation, we expect a more reliable parameter turning in the learning process.

We need the hypothesis inference through the statistical test in comparing two approaches to text categorization with each other. The better performance of the proposed approach than the traditional one which is the results from the experiments is not always such case in the entire population. We need to derive a better case of the proposed approach in the entire population, in order to persuade the results as the confirmed ones. Using the hypothesis inference for doing it is based on the difference between the proposed and traditional ones. Because the number of trials in the experiments is usually limited to less than 30, we had better using the Student's t- distribution, instead of the normal distribution.

8.2   Text Collections

This section is concerned with the text collections which are used for evaluating text categorization systems and approaches. In Sect. 8.2.1, we describe the text collection which is called NewsPage.com as a small collection. In Sect. 8.2.2, we mention 20NewsGroups which has its hierarchical classification frame with the two levels and is the standard collection for evaluating systems and approaches. In Sect. 8.2.3, we explain Reuter21578 which is used for evaluating the overlapping text categorization performances. In Sect. 8.2.4, we cover the text collection, OSHUMED, which is specialized for the medical domain.

8.2.1   NewsPage.com

This section is concerned with the small text collection, NewsPage.com. This collection has the five categories and totally 2000 texts which are news articles copied from the website, www.newspage.com. This collection has been used for comparing text categorization systems with each other in entering evaluation. This collection is manually constructed by copying and pasting individual news articles from newspage.com. Therefore, in this section, we describe the simple text collection which is called, NewsPage.com.

The text collection, NewsPage.com, is specified in Table 8.1. The classification frame is defined as a list of five categories: business, crime, health, Internet, and sports. The set of totally 2500 texts is partitioned into 840 training texts and 360 test texts with the ratio, 7:3. In each category, as shown in Table 8.3, 1500 texts are given; 350 texts are used as the sample texts, and the other are used as the test ones. The texts in this collection are labeled exclusively; each text belongs to only one of the five categories.

Table 8.1 The number of texts in NewsPage.com

Category	#Texts	#Training texts	#Test texts
Business	500	300	75
Health	500	300	75
Internet	500	300	75
Sports	500	300	75
Total	2000	1200	300

The text collection which consists of news articles between 2003 and 2005 is constructed by Jo in 2005. In the website, www.newspage.com, the five sections are selected as the categories, and news articles are gathered section by section. Using a text editor, entire full texts of news articles are copied and pasted into plain text files, individually. A single news article corresponds to a single plain text file; a text file name is given as a text identifier. This collection is useful for the preliminary evaluation of text categorization systems or approaches just after implementing them.

Let us explore previous cases of using the text collection for evaluating the text classification approaches. This collection was used as the source for evaluating the keyword extraction algorithms by Jo in 2003 [23]. In 2008, this collection was used for validating the better performance of the modified version of Support Vector Machine in the text categorization tasks [26]. In 2010, Jo proposed the NTC (Neural Text Categorizer) which is the neural network specialized for the text categorization and used the collection for validating its performance [30]. In 2015, Jo used the text collection for evaluating the table-based matching algorithm in the text categorization [35].

This collection is used for evaluating the hard text categorization where all texts are labeled with only one category. The spam mail filtering which is an instance of text categorization belongs to both binary classification and hard classification. Because each text tends to cover more than one topic, soft text categorization which allows each text to be labeled with more than one topic is more desirable. Nevertheless, the collection is needed for evaluating text categorization systems as the entrance. Soft text categorization tends to be decomposed into hard binary classification tasks.

8.2.2   20NewsGroups

Let us mention the text collection, 20NewsGroups, which is another collection for evaluating text categorization performances. The text collection is published in the website, http://qwone.com/~jason/20Newsgroups/. This collection consists of 20 categories; approximately 1000 texts in each class; there are totally 20,000 texts in the collection. The 20 categories are grouped into the four groups; the collection is organized as the two- level hierarchical structure. In this section, we describe the text collection, 20NewsGroups, with respect to its categorical structure and its evaluation process.

The partition of this collection into the training set and the test set is illustrated in Table 8.2. The collection consists of the 20 categories, and each category consists of 1000 texts. In each category, 700 texts belong to the training set and 300 ones belong to the test one; totally, the training set consists of 14,000 texts and the test set consists of 6000 ones. The text classification on the collection may be decomposed into the 20 binary classification tasks, rather than a single multiple classification task where each text is classified into one of the 20 categories. Based on the prefixes

Table 8.2 The number of texts in 20NewsGroups

Category	#Texts	#Training Texts	#Test Texts
Comp	5000	3500	1500
Rec	4000	2800	1200
Sci	4000	2800	1200
Talk	4000	2800	1200
Others	3000	2100	900
Total	20,000	14,000	6000

of the category names which presented in the first column in Table 8.2, the 20 categories are organized into the five groups: comp, sci, rec, talk, and others.

Let us mention the process of evaluating text categorization systems and algorithms on this text collection. The multiple classification task is decomposed into 20 binary classification tasks, and some of them are selected by the planned experiment scale. The training examples which correspond to the given category and some of others are allocated as positive examples and negative examples, respectively. The F1 measures are computed to the selected binary classification tasks and are averaged into a single value. The process of doing so will be explained in detail in Sect. 8.3.

Let us mention the previous cases of using the collection, 20NewsGroups, for evaluating text categorization system. In 2005, the collection was used by Dong and Han for evaluating the SVM (Support Vector Machine) Boost [13]. It was used in 2007 by Li et al. for evaluating the performances of hierarchical text categorizations [59]. In 2011, it was used by Cai and He for designing text categorization systems experimentally [9]. In 2015, it was used by Jo for comparing his proposed approach to text categorization with the traditional ones [35].

The text collection may be used for evaluating the performances of hierarchical text classifications. In the first level, we set the general multiple classification where each item is classified into one of the four general categories. To each general category, we allocate the specific multiple classification task where items which correspond to their own general category are classified into one of its specific categories. Both the general classification task and the specific ones may be decomposed into binary classifications, and each of them is evaluated by the F1 measure. In integrating the classification results, we may assign higher weights to the specific classification tasks, and a lower weight to the general classification task.

8.2.3 Reuter21578

This section is concerned with another standard text collection, Reuter21578. The two collections, NewsPage.com and 20NewsGroups, are used for evaluating the exclusive classifications, whereas the collection, Reuter21578, is used for evaluating the overlapping classification. In the collection, more than 100 categories are given

Table 8.3 The number of
texts in Reuter21578

Category	#Texts	#Training texts	#Test texts
Acq	2292	1596	696
Crude	374	253	121
Earn	3923	2840	1083
Grain	51	41	10
Interest	271	190	81
Money-FX	293	206	87
Ship	144	108	36
Trade	326	251	75
Total	7674	5485	2189

and a very variable number of texts is allocated to each category. The text may belong to more than one category; the sparse categories with less than ten texts and the dense categories with more than 1000 texts coexist. In this section, we describe the collection, Reuter21578, as the typical standard text collection for evaluating the classification results.

Table 8.3 illustrates the categories and the number of training and test examples in Reuter21578. Even if more than 100 categories are available in the collection, we list only some of them as the representative ones in Table 8.3. Only the two categories have more than 1000 texts as the major ones, but the others have much less than 1000. The number of texts has very unbalanced distribution over categories; for example, the number of texts in the category, corn, has less than a tenth of the number of texts in the category, earn. Texts which belong to more than two categories exist in the collection.

This collection is used for evaluating the overlapping text classification performance, as mentioned above. We select about ten categories as representative categories as displayed in Table 8.3; note that empty categories without any text exist in the collection. The overlapping multiple classification task is decomposed into the ten binary classification tasks by the process which was described in Sect. 5.2.3. The F1 measures are computed in each binary classification, and the micro- and macro-averaged F1 is found, by the process which is described in Sect. 8.3. The decomposition into binary classification tasks may be the option to the exclusive multiple classifications, depending on a number of predefined categories, but is the requirement to the overlapping ones.

Let us mention the typical previous cases of using the collection, Reuter21578, for evaluating text mining systems. In 1999, the collection was used for evaluating various approaches to text categorization by Yang and Liu [99]. In 2002, the collection, Retuer21578, is mentioned as the most standard test collection for evaluating text categorization systems by Sebastiani [85]. In 2006, the collection, Retuer21578, was used for evaluating the text classification systems based on the KNN by Tan [90]. This collection was used for evaluating text categorization systems by Jo in 2015 [35].

The text collection, Reuter21578, is divided into R8, R16, and R52, depending on the number of selected categories. There are actually 126 predefined categories in the collection; 90 categories among them have at least one text. R8 is the version where the top frequent 8 categories are used as the simple collection, and R52 is the version where the top 52 categories are used as the advanced one. The categories are unbalanced in this collection; some categories have no text. This collection is used for evaluating the tolerances of text categorization systems to sparse categories as well as the classification performances.

8.2.4   OSHUMED

Let us mention the text collection which is specialized to the medical domain; it is called OSHUMED. The trials to converge the computer science with the biology and the medicine motivate for processing information in this domain. The collection is used for evaluating text categorization systems, specializing them in the medical domain. The 22 categories about the medicine are given, and more than 20,000 text are contained totally in the collection. In this section, we describe the collection of medical documents with respect to its components and cases of using the collection for evaluating text categorization systems.

The categories, the training set, and the test set of the collection are illustrated in Table 8.4. The 22 medical categories and a various number of medical documents in each category from 135 to 2560 are given in the collection. As shown in Table 8.4, the test set is larger than the training set unlike the previous collection. The collection is used for evaluating tolerances to sparse categories as well as classification performance within the medical domain. The expert knowledge about the medicine is required for building the collection by predefining the categories and allocating sample texts.

The text collection, OSHUMED, is characterized as the one which is specialized to the medical domain. The demand for process information in the domain is increased by the necessity of implementing Ubiquitous Health Care Systems. The text collection is used for evaluating text categorization systems, specializing for the domain. Even if the classification in the collection belongs to the exclusive classification, it is necessary to decompose the task into binary classifications, because of the large number of categories. The specialization to the medical domain makes the collection distinguished from the previous text collections: 20NewsGroups and Reuter21578.

Let us mention some previous cases of using the text collection for performance evaluations. Use of this collection is traced even to 1996 [98]. In 1998, Joachims initially proposed the SVM as an approach to the text categorization and used the collection for evaluating it [47]. In 2011, the collection is also used for evaluating the retrieval strategies by Kreuzthaler et al. [55]. In 2013, Jo used the collection for evaluating his proposed approach to the text categorization and simulating the semantic operations [33, 34].

Table 8.4 Training and test set in the test bed: OSHUMED

Category name	Training set	Test set	Total
Bacterial infections and mycoses	423	506	929
Virus diseases	158	233	391
Parasitic diseases	65	70	135
Musculoskeletal diseases	1163	1467	2693
Digestive system diseases	283	429	712
Stomatognathic diseases	588	632	1220
Respiratory tract diseases	100	146	246
Otorhinolaryngologic diseases	473	600	1073
Nervous system diseases	125	129	254
Eye diseases	621	941	1562
Urologic and male genital diseases	162	202	264
Female genital diseases and and pregnancy complications	491	548	1039
Cardiovascular diseases	281	386	667
Hemic and lymphatic diseases	1259	1301	2560
Neonatal diseases and abnormalities	215	320	535
Skin and connective tissue diseases	200	228	428
Nutritional and metabolic diseases	295	348	643
Endocrine diseases	388	400	788
Immunologic diseases	191	191	382
Disorders of environmental origin	525	695	1220
Animal diseases	546	717	1263
Pathological conditions, signs and symptoms	92	91	183
Total	10,443	12,733	23,176

We need more considerations in specializing the text categorization system to the medical domain. Additionally, we may need the external knowledge about the medicine for building the classification capacity through the learning process. The medical terms should be characterized in terms of their semantic ranges; some medical terms may span over very wide categories and others may be specific to a particular category. Because each medical term consists of more than one word, in the tokenization process, bi-grams or n-grams should be generated rather than only uni-grams. The words which appear frequently in the medical domains such as patient, nurse, and doctor should be treated as stop words.

8.3   F1 Measure

In this section, we describe the process of evaluating the performances of text categorization systems by computing the F1 measures. In Sect. 8.3.1, we mention the process of building the contingency table from a binary classification and

Table 8.5 Contingency table for binary classification

	#True positives	#True negatives
#Classified positives	A	B
#Classified negatives	C	D

computing the F1 measure. In Sect. 8.3.2, we explain the micro-averaged F1 measure which is built from merging the contingency tables. In Sect. 8.3.3, we introduce the macro-averaged F1 as another evaluation measure. In Sect. 8.3.4, we demonstrate the process of computing both, the micro-averaged F1 and the macro-averaged F1, by a simple example.

8.3.1 Contingency Table

This section is concerned with the contingency table which is the results from a binary classification. It was originally used for evaluating the information retrieval systems, focusing on the number of retrieved relevant items; the columns stand for true relevant and true irrelevant and the rows stand for retrieved and non-retrieved. When the contingency table is introduced into the binary classification task, its columns stand for true positive and true negative and its rows stand for classified positive and classified negative, as shown in Table 8.5. The modified contingency table is used for evaluating the binary classification performance, focusing on the number of correctly classified positive items. In this section, we explain how to use the contingency table for doing so.

Table 8.5 illustrates the contingency table which records the results from the binary classification. It is assumed that the binary classification is to classify each item in the test set into one of the two categories: positive class and negative class. True positive indicates the number of test data items which are originally labeled with positive class and classified positive indicates the number of test data items which are classified into the positive class by the given classifier. The summation of the entries in the contingency table which is shown in Table 8.5 is the total number of test data items. In order to evaluate the binary classification performance, we need to construct the contingency table after classifying the test data items.

We need to compute the evaluation metrics from the contingency table which is presented in Table 8.5. The accuracy which is simple and popular is computed by Eq. (8.1),

$$\text{Accuracy} = \frac{A + D}{A + B + C + D} \tag{8.1}$$

We need to focus on the positive examples, if the given classification task is decomposed into binary classifications. The precision and the recall are computed by Eqs. (8.2) and (8.3), respectively,

$$\text{Precision} = \frac{A}{A + B} \tag{8.2}$$

$$\text{Recall} = \frac{A}{A + C} \tag{8.3}$$

The precision is the rate of the number of correctly classified positive examples to the number of examples which are classified into the positive class, and the recall is the rate to number of examples which are initially labeled with the positive class.

Let us mention the F-alpha measure as the evaluation metric which integrated the precision and the recall. They are anti-correlated with each other in evaluating classification systems. The F-alpha measure is computed by Eq. (8.4),

$$F_\alpha = (1 + \alpha^2) \frac{\text{precision} \cdot \text{recall}}{(\alpha^2 \cdot \text{precision}) + \text{recall}} \tag{8.4}$$

If $\alpha > 1$, the precision is weighted higher, whereas if $\alpha < 1$, the recall is weighted higher. If $\alpha = 1$, both metrics are balanced; the F1 measure is computed by Eq. (8.5),

$$F_1 = \frac{2 \cdot \text{precision} \cdot \text{recall}}{\text{precision} + \text{recall}} \tag{8.5}$$

As the alternative metric to the F1 measure, we consider the G measure which is computed by Eq. (8.6),

$$G = \sqrt{\text{precision} \cdot \text{recall}} \tag{8.6}$$

The text categorization and information retrieval systems are usually evaluated with the balanced portion of the recall and the precision, but depending on application area, we need to weight the two metrics differently. If we want to retrieve technical documents from a limited source, we need to collect relevant technical references as many as possible; in this case, the recall should be weighted higher than the precision. In the case of spam mail filtering, we need to put more weight to the precision, in order to avoid misclassifying ham mails into spam ones, as much as possible. In the case of text summarization, the recall is more important than the precision, because we avoid missing essential part as much as possible.

8.3.2 Micro-Averaged F1

This section is concerned with the scheme of averaging the F1 measures from contingency tables corresponding to binary classification; it is called micro-averaged F1. It is assumed that the classification task is decomposed into binary classification tasks as many as categories. A single contingency table is constructed by merging ones which correspond to the binary classifications. By the process which is described in Sect. 8.3.2, the precision, the recall, and the F1 measure are computed from the integrated contingency table. In this section, we explain the micro-averaged F1 and point out its characteristics.

It is assumed that the multiple classification is decomposed into the binary classifications. The decomposition process was explained in Chap. 5. The categories in the multiple classification are given as $C_1, \ldots, C_{|C|}$ and it is decomposed into the binary classification tasks as $BT_1, \ldots, BT_{|C|}$. There are the two classes in the binary classification task, BT_i: the positive class indicates that it belongs to the category, C_i, and the negative class indicates that it does not. Relabeling and reallocating the training examples become overhead for doing the decomposition.

After doing the decomposition, contingency tables are constructed as many as categories. A_i, B_i, C_i, and D_i are the number of correctly classified positive examples, the number of negative examples which are misclassified into the positive class, the number of positive examples which are misclassified into the negative class, and the number of correctly classified negative examples, in the contingency table which corresponds to the binary classification task, BT_i. The integrated contingency table is constructed by summing elements, category by category, as follows: $A = \sum_{i=1}^{|C|} A_i$, $B = \sum_{i=1}^{|C|} B_i$, $C = \sum_{i=1}^{|C|} C_i$, and $D = \sum_{i=1}^{|C|} D_i$. The precision and recall are computed from the integrated contingency table by Eqs. (8.2) and (8.3), and the micro-averaged F1 measure is computed by Eq. (8.5). The micro-averaged F1 reflects the fact that the predefined categories are discriminated by their sizes.

Let us demonstrate the process of finding the micro-averaged F1 measure through a simple example which is presented in Fig. 8.6. In the example, the total number of correctly classified positive examples is counted by summing as follows:

$$A = 100 + 100 + 200 + 10 = 410$$

We count the total number of negative examples which are misclassified into the positive class and the total number of positive examples which are misclassified into the negative class as follows:

$$B = 150 + 150 + 200 + 10 = 510$$

$$C = 100 + 100 + 100 + 40 = 340$$

The precision and the recall are computed as follows:

$$\text{recall} = \frac{410}{410 + 340} = 0.5466$$

$$\text{precision} = \frac{410}{410 + 510} = 0.4456$$

		True Positive	True Negative
Politics	Classified Negative	100	150
	Classified Positive	100	50
		True Positive	True Negative
Economics	Classified Negative	100	150
	Classified Positive	100	50
		True Positive	True Negative
IT	Classified Negative	200	200
	Classified Positive	100	100
		True Positive	True Negative
Sports	Classified Negative	10	10
	Classified Positive	40	40

Fig. 8.3 Example of binary classifications

The F1 measure is computed by using the precision and the recall which are computed above as follows:

$$F1 = \frac{2 \cdot 0.5466 \cdot 0.4456}{0.5466 + 0.4456} = 0.4909$$

Let us mention the scheme of averaging the F1 measures from the decomposed binary classifications. In the example which is presented in Fig. 8.3, the category, IT, is most sparse. The recall, the precision, and F1 measures are computed as 0.2, 0.5, and 0.2857, respectively. The category influences very weakly on computing the micro-averaged F1. In the next section, we observe the differences between the two averaging schemes.

8.3.3 Macro-Averaged F1

This section is concerned with another scheme of averaging F1 measures, called macro-averaged F1 measure. Both the averaging schemes are used for evaluating text categorization systems in previous literatures, altogether. F1 measures, which correspond to binary classifications into which the text categorization is decomposed, are averaged regardless of category sizes in the macro-averaged F1. In this averaging scheme, sparse categories have relatively higher influence, compared with the micro-averaged F1. In this section, we explain the characteristics of this averaging scheme and describe the process of computing it from F1 measures.

Let us consider the distributions over categories, before describing the process of computing the macro-averaged F1. If the distribution over categories is completely balanced, both the micro-averaged F1 and the macro-averaged F1 are the same as each other. A variable distribution over categories causes the gap between both the averaging schemes. In the micro-averaged F1 which was covered in Sect. 8.3.2, categories are weighted implicitly, proportionally to their sizes. However, in the macro-averaged F1, categories are weighted identically regardless of their size.

The contingency tables corresponding to the predefined categories are constructed under the assumption that the given classification task is decomposed into the binary classifications. The recall, the precision, and the F1 measure are computed from each contingency table instead of merging them into the integrated version. The macro-averaged recall and the macro-averaged precision are computed by averaging those of contingency tables by Eqs. (8.7) and (8.8),

$$\text{macro-recall} = \frac{1}{|C|} \sum_{i=1}^{|C|} \text{recall}_i \tag{8.7}$$

$$\text{macro-precision} = \frac{1}{|C|} \sum_{i=1}^{|C|} \text{precision}_i \tag{8.8}$$

The macro-averaged F1 is computed by averaging over the F1 measures, by Eq. (8.9),

$$\text{macro-F1} = \frac{1}{|C|} \sum_{i=1}^{|C|} \text{F1}_i \tag{8.9}$$

As shown in Eq. (8.9), identical weights are assigned to the categories, regardless of their sizes.

Let us compute the macro-averaged F1 from the example which is illustrated in Fig. 8.3. In Sect. 8.3.4, we already computed the micro-averaged F1 as 0.4909. We may compute the four F1 measures which correspond to the contingency tables: 0.4444, 0.4444, 0.5714, and 0.2857. By averaging them, we obtain the macro-averaged F1, 0.4364. The fact that the macro-averaged F1 is less than the micro-averaged F1 means weak tolerance to the sparse category, sports, in the example which is presented in Fig. 8.3.

The micro-averaged F1 measure is used more frequently for evaluating the performances of text categorization and information retrieval systems. The macro-averaged F1 is usually used for evaluating the tolerances to sparse categories of both kinds of systems, rather than performance. If the distribution over categories completely balanced, the values of both F1 measures are the same as each other. The macro-averaged F1 is less than the micro-averaged as shown in the example in Sect. 8.3.3, it means the weak tolerance to the sparse category, compared to its performance. In conclusion, we recommend that both kinds of F1 measures should be used for evaluating text categorization systems.

Fig. 8.4 Process of
decomposing into four binary
classifications

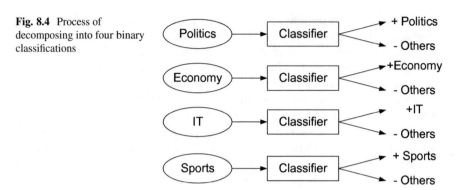

8.3.4   Example

This section is concerned with the demonstration of evaluating a text categorization
system by a simple example. The example was previously mentioned in Sects. 8.3.2
and 8.3.3, for explaining the computation of the micro-averaged F1 and the macro-
averaged F1. The task which is mentioned in this section is to classify each text
into one of the four categories: politics, economy, IT, and sports. The results are
illustrated from classifying texts by a particular algorithm in Fig. 8.3. In this section,
we present the process of computing both the kinds of F1 measure by an example.

Figure 8.4 illustrates the process of decomposing the task of classifying a text
into one of the four categories into the four binary classifications. The binary
classifiers as many as the categories are allocated; each text is classified into the
positive class or the negative class. The sample texts which belong to their own
categories are given as positive examples, and some sample texts which do not
belong to the category are given as negative examples. The positive class means that
the text belongs to the corresponding category, whereas the negative class means
that it does not belong to the category. The decomposition process is explained in
Sect. 5.2.3.

The results from the four binary classifications are presented as the four
contingency tables in Fig. 8.3. The contingency tables were mentioned in
Sects. 8.3.2 and 8.3.3, as the example of computing the micro-averaged F1 and
the macro-averaged F1. The explanation about the contingency table was provided
in Sect. 8.3.1. The category, IT, is the majority category which contains the largest
number of texts, and the category, sports, is the most sparse category which contains
the smallest number of texts. The F1 measure of a contingency table was already
mentioned in Sect. 8.3.3.

Table 8.4 presents the values from evaluating the classification results which are
illustrated in Fig. 8.3. The micro-averaged recall, precision, and F1 are computed by
merging the contingency tables into one table. The macro-averaged ones are computed by averaging recalls, precisions, and F1 measures of the contingency tables.
The macro-averaged values are less than the micro-averaged ones in evaluating a
text classification system; that means that its tolerance to sparse categories is weak
compared with its own performance. In evaluating text categorization systems on

the collection, 20NewsGroups, there is little difference between the two averaging schemes, because of all most identical categorical sizes in the collection, whereas in doing so on the collection, Reuter21578, there is outstanding difference between them, because of very variable categorical sizes.

Let us interpret the micro-averaged F1 and the macro-averaged one in comparing the two text categorization systems. They may be evaluated by either of the two averaged schemes on the test collection with almost identical categorical sizes, such as 20NewsGroups. When evaluating the systems on the collection, such as Reuter21578, both of them are needed. It is assumed that as the results from evaluating the two systems, text categorization system A has its better micro-averaged F1 and text categorization system B has its better macro-averaged F1. From the evaluation, it is concluded that system A has its better general performance and system B has its better tolerance to the sparse category.

8.4   Statistical *t*-Test

This section is concerned with the Student's *t*-test for comparing two approaches to the text categorization with each other. In Sect. 8.4.1, we introduce and study the Student's *t*- distribution. In Sect. 8.4.2, we discuss the process of comparing the two samples from their own populations based on the *t*-distribution. In Sect. 8.4.3, we mention the process of comparing pairs each of which is measured in its unique condition with each other. In Sect. 8.4.4, we demonstrate the process of comparing the two approaches with each other and making the hypothesis by a simple example.

8.4.1   Student's *t*-Distribution

This section is concerned with the Student's *t*-distribution which is a continuous probability distribution. The Gaussian distribution is also a continuous probability distribution which is assumed to be the most popular distribution over data. The Student's *t*-distribution is the variant of the Gaussian distribution which allows more spreading, depending on variable sample sizes. If the sample size reaches more than 30, the two distributions will be the same as each other. In this section, we describe the Student's *t*-distribution which is used for comparing approaches to the text categorization.

Both the Gaussian distribution and the *t* distribution are illustrated in Fig. 8.5, which is taken from the website , http://tstudent.altervista.org/site/tstudentnormal. Both the distributions form bell shapes; the *t* distribution is broader than the Gaussian distribution. The means of both the distributions are zeros but the *t*-distributions have their standard deviation which is greater than that of the Gaussian distribution. The *t*-distribution was intended to reflect the higher fluctuation around the population mean by the small-sized sample one. Therefore, the *t*-distribution was derived from the Gaussian distribution by widening its shape.

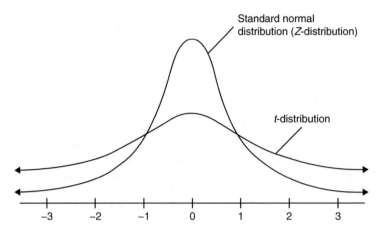

Fig. 8.5 Gaussian and Student's t-distribution from http://tstudent.altervista.org/site/tstudentnormal

The table of the t-distribution for finding the z values is illustrated in Fig. 8.6, which is taken from the website, http://albertsblog.blogspot.kr/2010/08/student-t-distribution-table.html. In Fig. 8.6, the columns indicate the error probability and rows indicate the degree of freedom which is sample size minus one. Each entry in Fig. 8.6 which is determined by the error probability, α, and the degree of freedom indicates the critical Z value. For example, if the sample size is ten and $\alpha = 0.05$, the z value becomes 1.833 which is crossed by 0.05 and 9 which is the degree of freedom. The Z value which is shown as an entry in the table is the scaled difference between two experiments.

The two purposes of using the Student's t-distribution is to estimate population values and to make the hypothesis test of sample means. Samples are gathered from a population and sample mean and variance are computed from them. The interval of the population mean is defined by the sample mean and variance. There are two hypotheses in the test: one is the null hypothesis that the sample mean and the population mean are little different and the other is the alternative hypothesis that they are very different. The null hypothesis is initially assumed, the z value is computed, and it is compared with the z-threshold which is gained from the table which is shown in Fig. 8.6. If the z value is less than the z-threshold, the null hypothesis is accepted, otherwise it is rejected.

Figure 8.7, which is from http://www.statsdirect.co.uk/help/content/distributions/t.htm, presents the changes of Student's t-distributions by the degree of freedom. The three degrees of freedom, 5, 15, and 25, are given in the Student's t-distributions. The Student's t-distribution with the degree of freedom 25 is most concentrated, but the one with the degree of freedom 5 is most spread, as shown

TABLE of CRITICAL VALUES for STUDENT'S t DISTRIBUTIONS

Column headings denote probabilities (α) **above** tabulated values.

d.f.	0.40	0.25	0.10	0.05	0.04	0.025	0.02	0.01	0.005	0.0025	0.001	0.0005
1	0.325	1.000	3.078	6.314	7.916	12.706	15.894	31.821	63.656	127.321	318.289	636.578
2	0.289	0.816	1.886	2.920	3.320	4.303	4.849	6.965	9.925	14.089	22.328	31.600
3	0.277	0.765	1.638	2.353	2.605	3.182	3.482	4.541	5.841	7.453	10.214	12.924
4	0.271	0.741	1.533	2.132	2.333	2.776	2.999	3.747	4.604	5.598	7.173	8.610
5	0.267	0.727	1.476	2.015	2.191	2.571	2.757	3.365	4.032	4.773	5.894	6.869
6	0.265	0.718	1.440	1.943	2.104	2.447	2.612	3.143	3.707	4.317	5.208	5.959
7	0.263	0.711	1.415	1.895	2.046	2.365	2.517	2.998	3.499	4.029	4.785	5.408
8	0.262	0.706	1.397	1.860	2.004	2.306	2.449	2.896	3.355	3.833	4.501	5.041
9	0.261	0.703	1.383	1.833	1.973	2.262	2.398	2.821	3.250	3.690	4.297	4.781
10	0.260	0.700	1.372	1.812	1.948	2.228	2.359	2.764	3.169	3.581	4.144	4.587
11	0.260	0.697	1.363	1.796	1.928	2.201	2.328	2.718	3.106	3.497	4.025	4.437
12	0.259	0.695	1.356	1.782	1.912	2.179	2.303	2.681	3.055	3.428	3.930	4.318
13	0.259	0.694	1.350	1.771	1.899	2.160	2.282	2.650	3.012	3.372	3.852	4.221
14	0.258	0.692	1.345	1.761	1.887	2.145	2.264	2.624	2.977	3.326	3.787	4.140
15	0.258	0.691	1.341	1.753	1.878	2.131	2.249	2.602	2.947	3.286	3.733	4.073
16	0.258	0.690	1.337	1.746	1.869	2.120	2.235	2.583	2.921	3.252	3.686	4.015
17	0.257	0.689	1.333	1.740	1.862	2.110	2.224	2.567	2.898	3.222	3.646	3.965
18	0.257	0.688	1.330	1.734	1.855	2.101	2.214	2.552	2.878	3.197	3.610	3.922
19	0.257	0.688	1.328	1.729	1.850	2.093	2.205	2.539	2.861	3.174	3.579	3.883
20	0.257	0.687	1.325	1.725	1.844	2.086	2.197	2.528	2.845	3.153	3.552	3.850
21	0.257	0.686	1.323	1.721	1.840	2.080	2.189	2.518	2.831	3.135	3.527	3.819
22	0.256	0.686	1.321	1.717	1.835	2.074	2.183	2.508	2.819	3.119	3.505	3.792
23	0.256	0.685	1.319	1.714	1.832	2.069	2.177	2.500	2.807	3.104	3.485	3.768
24	0.256	0.685	1.318	1.711	1.828	2.064	2.172	2.492	2.797	3.091	3.467	3.745
25	0.256	0.684	1.316	1.708	1.825	2.060	2.167	2.485	2.787	3.078	3.450	3.725
26	0.256	0.684	1.315	1.706	1.822	2.056	2.162	2.479	2.779	3.067	3.435	3.707
27	0.256	0.684	1.314	1.703	1.819	2.052	2.158	2.473	2.771	3.057	3.421	3.689
28	0.256	0.683	1.313	1.701	1.817	2.048	2.154	2.467	2.763	3.047	3.408	3.674
29	0.256	0.683	1.311	1.699	1.814	2.045	2.150	2.462	2.756	3.038	3.396	3.660
30	0.256	0.683	1.310	1.697	1.812	2.042	2.147	2.457	2.750	3.030	3.385	3.646
31	0.256	0.682	1.309	1.696	1.810	2.040	2.144	2.453	2.744	3.022	3.375	3.633
32	0.255	0.682	1.309	1.694	1.808	2.037	2.141	2.449	2.738	3.015	3.365	3.622
33	0.255	0.682	1.308	1.692	1.806	2.035	2.138	2.445	2.733	3.008	3.356	3.611
34	0.255	0.682	1.307	1.691	1.805	2.032	2.136	2.441	2.728	3.002	3.348	3.601
35	0.255	0.682	1.306	1.690	1.803	2.030	2.133	2.438	2.724	2.996	3.340	3.591
36	0.255	0.681	1.306	1.688	1.802	2.028	2.131	2.434	2.719	2.990	3.333	3.582
37	0.255	0.681	1.305	1.687	1.800	2.026	2.129	2.431	2.715	2.985	3.326	3.574
38	0.255	0.681	1.304	1.686	1.799	2.024	2.127	2.429	2.712	2.980	3.319	3.566
39	0.255	0.681	1.304	1.685	1.798	2.023	2.125	2.426	2.708	2.976	3.313	3.558
40	0.255	0.681	1.303	1.684	1.796	2.021	2.123	2.423	2.704	2.971	3.307	3.551
60	0.254	0.679	1.296	1.671	1.781	2.000	2.099	2.390	2.660	2.915	3.232	3.460
80	0.254	0.678	1.292	1.664	1.773	1.990	2.088	2.374	2.639	2.887	3.195	3.416
100	0.254	0.677	1.290	1.660	1.769	1.984	2.081	2.364	2.626	2.871	3.174	3.390
120	0.254	0.677	1.289	1.658	1.766	1.980	2.076	2.358	2.617	2.860	3.160	3.373
140	0.254	0.676	1.288	1.656	1.763	1.977	2.073	2.353	2.611	2.852	3.149	3.361
160	0.254	0.676	1.287	1.654	1.762	1.975	2.071	2.350	2.607	2.847	3.142	3.352
180	0.254	0.676	1.286	1.653	1.761	1.973	2.069	2.347	2.603	2.842	3.136	3.345
200	0.254	0.676	1.286	1.653	1.760	1.972	2.067	2.345	2.601	2.838	3.131	3.340
250	0.254	0.675	1.285	1.651	1.758	1.969	2.065	2.341	2.596	2.832	3.123	3.330
inf	0.253	0.674	1.282	1.645	1.751	1.960	2.054	2.326	2.576	2.807	3.090	3.290

Fig. 8.6 Z values in Student's t-distribution from http://albertsblog.blogspot.kr/2010/08/student-t-distribution-table.html

in Fig. 8.7. In the small-sized sample which corresponds to df (degree of freedom) = 5, the probability of mean is decreased but probabilities of extreme values are increased. Therefore, the Student's t- distribution is viewed as the Gaussian distribution which is adjusted by the degree of freedom which indicates the sample size.

Fig. 8.7 Changes of
Student's *t*-distributions by df
from http://www.statsdirect.
co.uk/help/content/
distributions/t.htm

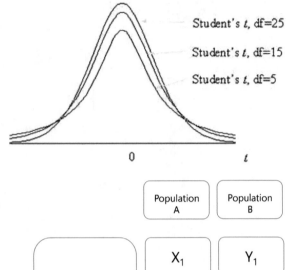

Student's *t*, df=25

Student's *t*, df=15

Student's *t*, df=5

Fig. 8.8 Values from two
populations for unpaired
difference inference

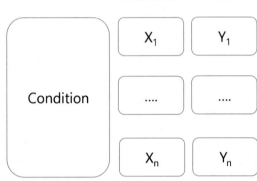

8.4.2   *Unpaired Difference Inference*

This section is concerned with the process of making the inferences from two
populations. The two groups of samples are made by selecting them at random from
their own populations. The means and the variances are computed from the two
groups of samples. By the Student's *t*-distribution, based on the mean difference, it
is decided whether the two populations are different from each other, or not. In this
section, we describe the process of making the inferences about the two populations.

The values are derived from the two populations as samples, as illustrated in
Fig. 8.8. The values are selected from the two populations at random, to make the
two groups of samples: $X_1, X_2, \ldots, X_n$ and $Y_1, Y_2 \ldots, Y_n$. It is assumed that the
condition under which the sample values are generated applied identically to the
both populations. The sample means are computed in Eq. (8.10),

$$\bar{X} = \sum_{i=1}^{n} X_i, \bar{Y} = \sum_{i=1}^{n} Y_i \tag{8.10}$$

Generating the two groups of samples is tended to test the difference between the two populations by observing the difference between the means.

The t value is computed as the important factor for making the hypothesis test based on the t distribution. It is assumed that two sample sizes are identically set to n. The standard deviation which spans the two populations is computed by Eq. (8.11),

$$s = \sqrt{\frac{\sum_{i=1}^{n}(X_i - \bar{X})^2 + \sum_{i=1}^{n}(Y_i - \bar{Y})^2}{2(n - 1)}} \qquad (8.11)$$

The t value is computed by Eq. (8.12),

$$t = \frac{|\bar{X} - \bar{Y}|}{s\sqrt{\frac{1}{2n}}} \qquad (8.12)$$

The t value is proportional to the difference between the two means, but anti-proportional to the standard deviation as shown in Eq. (8.12).

After computing the t-value by the process which is mentioned above, we need to make the hypothesis test, in order to decide whether the null hypothesis is accepted or rejected. If the statement that there is no difference between the two populations is set as the null hypothesis, the hypothesis test is made, based on the two-tailed test. The t-value threshold is retrieved under the confidence 95%, $\alpha = 0.05$ based on the degree of freedom, $2(n - 1)$, from the table which is illustrated in Fig. 8.6. The t-value which is computed by Eq. (8.12) is compared with the threshold which is from the table; if the t value is greater than the threshold, the null hypothesis is rejected. The rejection indicates that the two populations should be distinguished from each other with the confidence 95%.

Let us mention several cases of applying the unpaired difference inferences. It is used for testing the difference of average incomes of labors in two cities. We may consider the difference of academic achievements of students between two universities as a case of using the unpaired difference inference. We may test the difference of averaging living cost of an urban life and a rural life, using what is described in this section. The difference of grain harvest amounts between two regions is considered for using it.

8.4.3 Paired Difference Inference

This section is concerned with the hypothesis test for paired differences. The samples from two populations are measured under an identical condition in the unpaired case. In this case, difference conditions are applied to sample pairs, one by one. What is covered in this section will be applied for evaluating two text categorization systems on different test collections. In this section, we explain the process of making the hypothesis test to paired differences, in detail.

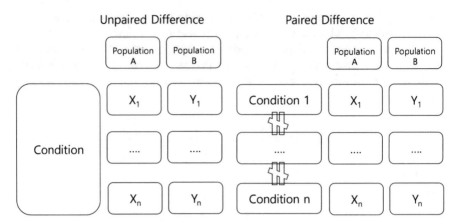

Fig. 8.9 Values from two populations for unpaired vs paired difference inference

Figure 8.9 illustrates the differences between the unpaired difference inference which was covered in Sect. 8.4.2 and the paired one. In both inferences, the two groups of samples are generated from their own populations at random. In the unpaired difference inference, the same condition of generating samples is applied, whereas in the paired one, the different conditions are applied. In each group, in the unpaired difference inference, samples in each group are unordered, while in the paired one, samples are ordered. The goal of both kinds of inference is to make the hypothesis test with the two samples based on their difference.

Let us describe the process of computing the t-value from the two sample sets, $X_1, X_2, \ldots, X_n$ and $Y_1, Y_2 \cdots, Y_n$. It is assumed that different conditions are applied for each pair of sample values, X_i and Y_i. From the two sample sets, we generate the difference samples, $d_1, d_2, \ldots, d_n$ where $d_i = X_i - Y_i$, and the mean and the standard deviation are computed by Eqs. (8.13) and (8.14), respectively,

$$\bar{d} = \frac{1}{n} \sum_{i=1}^{n} d_i \tag{8.13}$$

$$s_d = \sqrt{\frac{1}{n} \sum_{i=1}^{n} (d_i - \bar{d})^2} \tag{8.14}$$

The t value is computed by Eq. (8.15),

$$t = \frac{\bar{d}}{s_d \sqrt{\frac{1}{n}}} \tag{8.15}$$

By the t value, it is decided whether the two populations are different from each other, or not.

We need to make the hypothesis about the difference between the two groups of samples. The null hypothesis is set as the statement that the two populations are not different from each other. The t value threshold is retrieved from the table which is shown in Fig. 8.6, with the degree of freedom, $n - 1$ and $\alpha = 0.05$. The t value which is computed by the above process is compared with the t value threshold which is taken from the table; if the computed t value is greater than or equal to the t value threshold, the null hypothesis is rejected. It means that the two populations are different from each other, with the confidence 95%.

Let us mention some cases to which the paired difference inference is applicable. The hypothesis test is used for comparing the effectiveness of two medicines to different patients. The performances of two vehicles in different kinds of roads are compared using this kind of the hypothesis test. The better performance of the proposed text categorization approach is confirmed through the hypothesis test. It may be used for comparing the populations of two political candidates, distinct by distinct.

8.4.4 Example

This section is concerned with the practice of evaluating two classifiers based on the t-distribution. It is assumed that the two classifiers are evaluated on the different test collections: different conditions. So, we adopt the paired different inference, and make the inference about the results of the classifiers by the process which was described in Sect. 8.4.3. The confidence level is set as 95%, the degree of freedom is given as the number of conditions minus one, and the threshold value is taken from the table in Fig. 8.3. In this section, we demonstrate the process of making the inference on the comparison of two classifiers by the example which is shown in Table 8.6.

The results from evaluating the two classifiers are presented in Table 8.6. The second and third columns indicate the results of the two classifiers, and the last column shows the differences between them. Each row indicates its own test collection on which the two classifiers are evaluated. The values in Table 8.6 in the second and third columns are F1 measures of classifiers. In Table 8.6, classifier II has its better performance in three of the four test collections.

The t-value is computed from the example of paired difference inference. We discover that the difference mean is 0.0837 by averaging values in the last column in Table 8.6. The value, 0.0611, is obtained as the standard deviation of the differences.

Table 8.6 Accuracies of two classifiers

Case	Classifier I	Classifier II	Difference
Case 1	0.732	0.843	0.111
Case 2	0.766	0.923	0.157
Case 3	0.822	0.812	−0.01
Case 4	0.834	0.911	0.077

The t value is computed by Eq. (8.15) as 0.2847. Because the two classifiers are evaluated on their different conditions, the unpaired difference inference which was covered in Sect. 8.4.2 is not applicable to this case.

We need to make the hypothesis about the difference between the two groups of samples. The null hypothesis is set as the statement that the two populations are not different from each other. The t value threshold is retrieved from the table which is shown in Fig. 8.6, with the degree of freedom, $n - 1$ and $\alpha = 0.05$. The t value which is computed by the above process is compared with the t value threshold which is taken from the table; if the computed t value is greater than or equal to the t value threshold, the null hypothesis is rejected. It means that the two populations are different from each other, with the confidence 95%.

The hypothesis test is made by the t distribution about the experimental results which are presented in Table 8.6. Classifier II shows actually better performances than classifier I in three of the four test collections. In spite of that, by the hypothesis test, it was concluded that the classifiers have almost the same performances on the four tests. In order to reject the hypothesis, we must make experiments on the 100 test collections from Eq. (8.14). With only about 50%, we support that Classifier II has its better performances, from the results in Table 8.6.

8.5   Summary and Further Discussions

This chapter is summarized as evaluating text categorization systems by the F1 measures and comparing two systems based on the t-distribution. We introduced some text collections, such as 20NewsGroups and Reuter21578, for evaluating the text categorization systems. The text categorization system was evaluated by the F1 measure, assuming that the text categorization task is decomposed into binary classification tasks as many as categories. The two systems are compared with each other, based on the hypothesis test of the paired differences between them, by the t distribution. In this section, we make the further discussion about what we studied in this chapter.

Other text collections are available other than ones which were covered in Sect. 8.1. Spambase data set which is the collection of emails was used for evaluating the spam mail filtering systems. Amazon book reviews, which is the collection of positive and negative reviews on books, are used for evaluating sentimental analysis systems. The collection of sentences rather than articles is used for classifying sentences by authors' intentions. The test collections for evaluating the machine learning algorithms as well as text collections are provided from the website , https://archive.ics.uci.edu/ml/index.html.

The performance of text categorization may be dependent on the domain. Even if a classifier has very excellent performance on a particular domain, it is not guaranteed that it has such performance on another domain. The collections of news articles, such as 20NewsGroup and Reuter21578, are used popularly for validating empirically performances of text classifiers. Recently, Jo suggested the empirical validation on test collections specific domains such as medicine, law, and engineering, after validating on ones on broad domains. Prior domain knowledge as well as training examples are utilized for improving text classification performances, especially in specific domains.

The ROC (Receiver Operating Characteristic) curve is considered as an alternative evaluation metric. It involves the true positive rate which is the rate of the positive examples which are classified correctly to the total positive examples and the false positive rate which is the rate of the positive examples which are classified into negative class to the total positive examples. The curve is drawn in the two dimensional coordination where the x-axis is the false positive rate and the y-axis is the true positive rate. If the curve is drawn in the top-left direction, it indicates the desirable performance. The top-left indicates the high true positive rate and the low false positive rate.

The ANOVA (Analysis of Variance) is the statistical analysis for comparing more than two groups with their means. The t-test of the paired difference inference is used for comparing the two approaches with each other in different conditions. The goal of ANOVA is to decide whether more than two groups are different from each other, or almost sample. In other words, the ANOVA is used for comparing at least three approaches with one another. However, it is not used for deciding which approach is best.

Part III
Text Clustering

Part III is concerned with the concepts, the approaches, the implementation, and the evaluation of the text clustering systems. We explore the text clustering types in the functional view as the start of this part. We mention the typical text clustering approaches and present implementations in Java. We describe the schemes of evaluating the text clustering system. We cover the text categorization and the text clustering in Parts II and III, respectively.

Chapter 9 is concerned with the text clustering in the functional and conceptual view. We will mention the four types of text clustering by results: hard vs soft text clustering, flat vs hierarchical text clustering, and single vs multiple viewed text clustering. We present some directions of clustering texts, similarity-based clustering, prototype-based clustering, and growth- based clustering. We also mention fusion tasks of clustering and classification. In the subsequent chapters, we specify the text clustering with respect to the approach and the implementation.

Chapter 10 is concerned with some representative approaches to the text clustering which are the unsupervised learning algorithms. This chapter begins with the AHC algorithm which is a simple and popular approach. Afterward, we describe in detail the k means algorithms and its variants. We present the SOM (Self-Organizing Map), together with the WOBSOM which is the typical case of applying it to the text clustering. In this book, the scope of text clustering approaches is restricted to only machine learning algorithms.

Chapter 11 is concerned with implementing the text clustering system in Java. In the beginning of this chapter, we present a class list, together with their roles in implementing the system. We explain in detail the methods in the classes which are involved in implementing it, together with the method implementations. We show the demonstration of the implemented system, through a simple example. Note that the AHC algorithm is adopted in implementing the system, in the chapter.

Chapter 12 is concerned with the schemes of evaluating results from clustering texts. We mention schemes of evaluating clustering results and point out their limits. Afterward, we describe the evaluation schemes which are specialized for the clustering task, including the clustering index. We mention the modified version of the existing clustering algorithm where the parameter tuning based on the clustering index is installed. Therefore, the text clustering is covered from Chaps. 9–12, in this book.

Chapter 9
Text Clustering: Conceptual View

This chapter is concerned with the conceptual view of text clustering tasks. We make the general discussions concerned with text clustering in Sect. 9.1, and compare the task of data clustering with other data mining tasks in Sect. 9.2. In Sect. 9.3, we explore the mutually opposite types of text clustering depending on dichotomies. We mention the real tasks which are derived from text clustering in Sect. 9.4, and make the summary and the further discussions on what we study in this chapter in Sect. 9.5. In this section, we describe the text clustering with its functional view.

9.1 Definition of Text Clustering

Text clustering refers to the process of segmenting a text group into subgroups of content-based similar ones. It is possible to cluster texts by their character combinations, called lexical clustering, but the kind of clustering is covered in this chapter. It is called semantic clustering to cluster texts based on their contents or meaning, and it will be covered in this study. It is assumed that texts are not labeled at all and unsupervised learning algorithm which are mentioned in Chap. 10 are adopted for this task. In this section, we describe briefly the text clustering before discussing it in detail, together with the review of text categorization which was covered in the previous part.

Let us review briefly the text categorization which was covered in the previous part, in order to help to characterize the text clustering. The text categorization was mentioned as the process of classifying texts into one or some of the predefined categories. The supervised machine learning algorithms, such as K Nearest Neighbor, Naive Bayes, or Support Vector Machine, are used as the main approaches. The performances of text categorization systems and approaches are evaluated by the accuracy or the F1 measure which were mentioned in Chap. 8. Both the text

© Springer International Publishing AG, part of Springer Nature 2019
T. Jo, *Text Mining*, Studies in Big Data 45,
https://doi.org/10.1007/978-3-319-91815-0_9

categorization and the text clustering may be integrated into a particular task, rather than two independent tasks, as mentioned in Chap. 16.

The text clustering is characterized with several agenda differently from other text mining tasks, such as the text categorization and the text association. A group of texts is given as the input and anonymous groups of them are the results from clustering the texts. The text clustering proceeds depending strongly on the similarities among the texts; it is very important task how to define similarity metrics. Cluster prototypes are initialized randomly or arbitrarily, and optimized into to maximize item similarities within each cluster and minimize similarities between clusters. The set of texts is never partitioned into the training set and the test set, in evaluating text clustering systems.

The text clustering is coupled with the text categorization for automating the preliminary tasks. Because manual labeling is very tedious job, it is not easy to gather labeled examples, but it is easy to gather unlabeled ones. The labeled text collection is able to be constructed by clustering unlabeled texts and naming the clusters as the sample texts for implementing text categorization systems. The list of cluster names becomes that of predefined categories and texts within each cluster become the sample texts which are labeled with its cluster name. Coupling of both the tasks is proposed and implemented by Jo in 2006 [25].

The text clustering is compared with the two tasks, the text categorization and the text association in the previous parts. In comparing the text association and the text clustering with each other, the former defines unidirectional relations among texts, while the latter defines the bidirectional relations among them. The preliminary tasks, such as category predefinition and sample text preparation, are required for the text categorization. The text clustering tends to be coupled with the text categorization into a single task, as mentioned above. Association rules among texts which are generated from the text association become useful contexts for doing the text clustering and the text categorization.

9.2   Data Clustering

This section is concerned with the generic data clustering in its functional and conceptual view and consists of the four sections. The clustering itself is described in its functional view in Sect. 9.2.1. In Sect. 9.2.2, the clustering is compared with the association, in order to distinguish them from each other. In Sect. 9.2.3, the clustering and the classification are compared with each other. In Sect. 9.2.4, the constraint clustering is mentioned as the special clustering where some labeled examples are provided.

Fig. 9.1 Data clustering

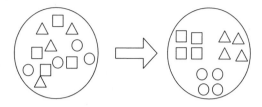

9.2.1   SubSubsectionTitle

The data clustering is illustrated in Fig. 9.1. A single group of various data items is given as the initial clustering status. The single group is segmented into the several subgroups which are called clusters, depending on the similarities among data items. The data items within their own subgroup should be similar as each other, ones in different subgroups should be discriminated among each other. In this section, we describe the data clustering, in its functional view.

Computing the similarities among examples and clusters is very important task for clustering data items. The similarity metrics between two individual items which are represented into numerical vectors are previously expressed in Eqs (6.1) and (6.4). We set the similarity between an individual item and a cluster as the similarity between its represented vector and the mean vector of the cluster. The similarity between two clusters is computed by the similarity between mean vectors which correspond to them. We may use the maximum or the minimum of similarities of all possible pairs of items as the similarity between the two clusters.

Even if the clustering proceeds differently depending on clustering algorithm, the general frame of doing it exists. We mentioned above the computation of similarities among items or clusters as the core operation which is necessary for clustering data items. The clusters are characterized by their prototypes or their representative examples, and the data items are arranged by the likelihoods with the clusters. The two steps, cluster characterization and data item arrangement, are iterated until the cluster characteristics become stable; they change very little. The stable cluster characteristics and the clusters of data items are the final results from the data clustering.

Let us mention some clustering algorithms which are approaches to the text clustering. The AHC algorithm is mentioned as a typical clustering algorithm, and it is simplest but has highly time complexity. We may consider the k means algorithm as the most popular clustering algorithm. The k means algorithm may be the fuzzy version where memberships of data items in clusters are computed. It belongs to the EM (Estimation and Maximization) algorithm where data items are clustered with the two steps: the estimation step and the maximization step.

Let us compare the clustering task with other data mining tasks: association, classification, regression, and outlier detection. The data association is referred to the process of extracting association rules which are given as if–then ones from a collection of item sets. The data classification is defined as the process of assigning

Table 9.1 Association vs
clustering

	Association	Clustering
Input	Item sets	Item group
Output	If then rules	Clusters
Relation	Causal	Similar
Direction	One direction	Both directions

one or some of the predefined categories to each data item by analyzing the labeled
sample ones. The regression is the process of estimating a continuous output value or
values by analyzing input factors and previous output values. The outlier detection
which is an additional data mining task is the process of detecting and removing
alien data items and noisy ones for cleaning data items.

9.2.2   Association vs Clustering

This section is concerned with the comparison of the text clustering with the text
association. The text association means the process of extracting association rules
as if–then forms between texts; it was studied in Chap. 4. As the distinguished
characteristics of the two tasks, the text association pursues for casual relations
among texts, whereas the text clustering does for similarities among them. They
are distinguished from each other, but they may be combined with each other. In
this section, we investigate the differences between tasks and connect them with
each other.

We described the process of associating data items in Chap. 4. In the association,
item sets are given as the input, initially. All association rules each of which consists
of a single item which is the conditional part, and items which are its causal parts
are extracted. For each association rule, we compute its support and confidence and
select some with their higher supports and confidences than the given threshold
among them. The difference between the two tasks is that in the association, the
item sets are given as the input whereas in the clustering a single group of items is
given as the input.

Table 9.1 illustrates the comparison of the two tasks: clustering and association.
In the association, item sets are given as the input, whereas a single group of
items is given as the input in the clustering. In the association, if–then rules are
generated as the output, whereas in the clustering, clusters of similar items are
generated. In the association, its results indicate causal relations among items,
whereas in the clustering, its results indicate similarities. The causal relations are
unidirectional which does not guarantee its reverse equivalence, but the similarities
are bidirectional which does it.

The results from clustering data items may become the input for doing the data
association. The clusters which are generated as the output are viewed as item
sets. The association rules are extracted from the clusters, using the association

algorithm, apriori algorithm. It becomes possible to find the association rules from individual items by means of data clustering. However, note that this type of data association is sensitive to the results from data clustering.

The word association rules are generated from texts by the above process. The list of words which are given the conditional parts of association rules becomes the list of taxonomies. In each association rule, texts which are relevant to the words in the causal part are retrieved in each association rule. In generating taxonomies from the association rules, each association rule is transformed into a topic and texts which belong to it. Texts should be encoded into lists of words through text indexing for doing the word association.

9.2.3 Classification vs Clustering

The classification was already covered in the previous part, by studying the text categorization. We mention the classification in this section for distinguishing the clustering from the classification. The classification task is referred to the process of assigning one or some of the predefined categorization to each data item. The clustering task looks similar as the classification in that each item is arranged into one or some clusters. In this section, we explain briefly the classification task in the functional view and compare it with the clustering task.

Let us review the process of classifying data items which was covered in Part III. We predefined categories as a list or a tree, and gather training examples which are labeled with one or some of the predefined ones. The classification capacity is constructed by the process called supervised learning with the training examples. Novice examples are classified with the constructed classification capacity. The supervised learning is performed by minimizing differences between the true labels and the classified ones of training examples.

The comparisons between the classification and the clustering are illustrated in Table 9.2. The classification categories are defined automatically through clustering, whereas they are defined manually in the classification. The supervised learning algorithms are applied to the classification, whereas the unsupervised learning algorithms are done to the clustering. In the clustering, a group of data items is given as the input, whereas in the classification, a single data item is given as the input. In the clustering, subgroups each of which contains similar data items are generated as the output, whereas one or some among the predefined categories are assigned to each item in the classification.

Table 9.2 Clustering vs classification

	Clustering	Classification
Classification categories	Automatic definition	Manual definition
Machine learning type	Unsupervised learning	Supervised learning
Input	Group of data items	Single data item
Output	Subgroups of similar data items	Category or some categories

   The results from clustering the data items are significant to doing the data classification. The clustering results consist of the predefined categories and the sample texts which belong to each of them. A list of clusters or a cluster hierarchy becomes the predefined category structure and the texts which are contained in each cluster are sample ones which are labeled with its corresponding cluster identifier. The data clustering automates the preliminary tasks for the text categorization, here. The results from clustering data items are used as sample texts for learning a supervised learning algorithm for performing the text categorization.

   The event detection and tracking was mentioned as a typical case of combining the text categorization with the text clustering [2]. The event detection means the process of deciding whether the current news article deals with a particular event, or not, as the special instance of text clustering [8]. The event tracking is the process of tracing events of other news articles using the detected ones as the special case of text categorization. News articles about a particular event are grabbed by the event detection and related news articles are retrieved by the event tracking. Detection and tracking of topics as well as events exist.

9.2.4   Constraint Clustering

This section is concerned with the constraint clustering where some labeled examples are provided. Figure 9.2 illustrates the constraint clustering as a diagram. Both labeled examples and unlabeled ones are used for clustering both kinds of them. Only small number of labeled examples is usually given as references for clustering unlabeled ones. In this section, we explain the constraint clustering in detail, and compare it with other tasks.

Fig. 9.2 Constraint clustering

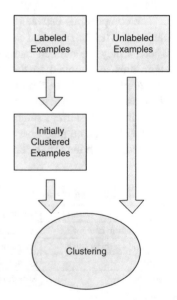

Let us compare the constraint clustering with the clustering which was mainly covered in this part. In the clustering, it is assumed that all examples are unlabeled, whereas in the constraint clustering some examples may be labeled by the prior knowledge or accidently. In the clustering, it proceeds depending on similarities or distances among examples purely, and in the constraint clustering, it does depending on similarities or distance and labels of some examples. In both the clustering and the constraint clustering, their ultimate goal is to segment a group of items into subgroups of similar ones. In the clustering, clustering prototypes are initialized at random, and in the constraint clustering, they are initialized based on some labeled examples.

The constraint clustering is similar as the pure clustering except the fact that some labeled examples are given, additionally. In the constraint clustering, a small number of labeled examples and a large number of unlabeled ones are given as the input; unlabeled examples are given as the majority of data items. The labeled examples are arranged into their clusters which correspond to their labels, in advance, and the cluster prototypes are initialized based on the labeled ones. The unlabeled examples are clustered by a clustering algorithm, afterward. The both kinds of examples are used for learning like the semi-supervised learning.

Let us compare the constraint clustering with the semi-supervised learning. In both the tasks, both labeled examples and unlabeled ones are used, as mentioned above. However, they have their different goals; the semi-supervised learning is intended for the classification and the regression, whereas the constraint clustering is done for the clustering. The semi-supervised learning involves both the supervised learning algorithm and the unsupervised ones, while the constraint clustering involves only unsupervised one [73]. In the semi-supervised learning, unlabeled examples are used as the additional ones, whereas in the constraint clustering, the labeled ones are used as constraints for clustering data items.

Let us mention the alien cluster which consists of very sparse number of data items. When plotting data points in the two- dimensional space, it may be discovered that some points are far from main clusters. The groups of the points are called alien clusters or outliers [91]. The sparse clusters are generated as some among a large number of small sized clusters, or they may be generated as noises in the case of a small number of large clusters. This kind of clusters should be removed by the process which is called outlier detection [91].

9.3 Clustering Types

This section is concerned with the clustering types by various views. In Sect. 9.3.1, we mention the offline vs the online clustering. In Sect. 9.3.2, we explain the two clustering types, the crisp clustering and the fuzzy clustering. In Sect. 9.3.3, we describe the flat clustering and the hierarchical clustering, depending on whether any nested cluster is allowed, or not. In Sect. 9.3.4, we introduce the single viewed and the multiple viewed clustering.

9.3.1 Static vs Dynamic Clustering

This section is concerned with the two types of data clustering: static clustering and dynamic clustering. The dichotomy for dividing the clustering task into the two types is whether data clusters as the clustering results are fixed or continually variable. In the static clustering, results from clustering data items are fixed, whereas in the dynamic clustering, results from doing so are updated, continually. In the previous works on data clustering, the static clustering has been assumed for explaining the clustering algorithm. In this section, we explain the two types of clustering, individually, and compare them with each other.

The static clustering is one where it is assumed that the clustering results are fixed, once data items are clustered. Until now, the clustering algorithms have been explained under the assumption of the static clustering. A group of various data items is given as input and clusters of similar ones which are results are fixed. However, there is possibility of adding more data items and deleting some from the results, so the data items should be reorganized, accordingly. In the static clustering, the real situation is overlooked.

Let us consider the dynamic clustering which is opposite to the static clustering. The dynamic clustering is one where as more data items are added or some items are deleted, continually, the data items are reorganized, accordingly. There are two kinds of reorganizations: the hard organization which organizes all of data items again by a clustering algorithm and the soft organization which merges some clusters into one or divides one into several clusters for reorganizing data items. We need to measure the current organization quality, in order to decide whether items should be reorganized, or not. The dynamic clustering will be explained in detail in Chap. 16.

The static clustering and the dynamic clustering are compared with each other. The assumption in the static clustering is that all of data items are given at a time, whereas in the dynamic clustering is that data items are added and deleted, continually. In the static clustering, a single group of data items is clustered into subgroups only one time, whereas in the dynamic clustering, data items are clustered continually after clustering so. In the static clustering, the organization of data items is fixed, whereas in the dynamic clustering, it is variable. In the static clustering, the operations such as division and merging of existing clustering are not considered, whereas in the dynamic clustering, they are considered as the soft organization.

The organization management in the dynamic clustering is to decide one of the maintenances, the soft organization and the hard organization. The two metrics which indicate the quality of clustering results are computed: the intra-cluster similarity and the inter-cluster similarity. The direction of clustering data items is to maximize the intra-cluster similarity and minimize the inter-cluster similarity, at same time. The decision of one of the three actions is made by comparing values of the two metrics after adding or deleting items with those after doing that. In Chap. 12, we mention in detail the two measures for evaluating the results from clustering data items.

	Cluster1	Cluster2	Cluster3	Cluster4
Item 1	O	X	X	X
Item 2	X	O	X	X
Item 3	X	X	X	O
Item 4	X	O	X	X
Item 5	X	O	X	X
Item 6	O	X	X	X
Item 7	X	X	O	X
Item 8	X	O	X	X

Fig. 9.3 Example of crisp clustering

9.3.2   Crisp vs Fuzzy Clustering

This section is concerned with the two types of clustering: crisp and fuzzy clustering. The dichotomy criteria is whether each item is allowed to belong to more than one cluster, or not. The crisp clustering is one where every item belongs to only one cluster, whereas the fuzzy clustering is one where at least one item belongs to more than one cluster. In the fuzzy clustering, rather than belonging to more than one cluster, to each item, membership values of clusters are assigned. In this section, we explain the two types of clustering and their differences.

We present an example of the crisp clustering in Fig. 9.3. The table in Fig. 9.3 is given as a matrix of the eight items and the four clusters. All of eight items belong to only one of the four clusters; no item belongs to more than one cluster. As results from the exclusive clustering, cluster 1 contains item 1 and 6, cluster 2 does item 2, 4, 5, and 8, cluster 3 contains only item 7, and cluster 4 contains item 3. No overlapping between clusters exists in the crisp clustering.

The simple example of the fuzzy clustering is illustrated in Fig. 9.4. The table in Fig. 9.4 has the same frame to that in Fig. 9.3. The five of the eight items are labeled with more than one category. Item 2 is labeled with the three categories and the four items are labeled with two categories. If at least one item is labeled with more than one category, it becomes the results from the fuzzy clustering.

Table 9.3 illustrates the differences between the crisp clustering and the fuzzy clustering which are covered in this section. The crisp clustering does not allow the overlapping between clusters at all, whereas the fuzzy one does it. The membership values are given binary values in the crisp clustering whereas the membership values are given as continuous values between zero and one. In the crisp clustering, each data item is arranged into its most similar cluster, whereas in the fuzzy clustering, it is arranged into more than one cluster whose similarity is more than the given threshold. The crisp clustering is usually applied to the opinion clustering where each data item is arranged into one of the positive cluster, the neutral one, and the negative one, whereas the fuzzy clustering is applied to topic-based text clustering where each data item is arranged into its relevant topic clusters.

	Cluster1	Cluster2	Cluster3	Cluster4
Item 1	O	X	X	X
Item 2	X	O	X	X
Item 3	X	X	X	O
Item 4	X	O	X	X
Item 5	X	O	X	X
Item 6	O	X	X	X
Item 7	X	X	O	X
Item 8	X	O	X	X

Fig. 9.4 Example of fuzzy clustering

Table 9.3 Crisp vs fuzzy clustering

	Crisp clustering	Fuzzy clustering
Overlapping	No	Yes
Membership	0 or 1	Between 0 and 1
Arrangement	Highest similarity	Higher than threshold
Case	Opinion mining	Topic-based clustering

	Cluster1	Cluster2	Cluster3	Cluster4
Item 1	0.71	0.34	0.45	0.63
Item 2	0.82	0.68	0.45	0.31
Item 3	0.34	0.35	0.61	0.59
Item 4	0.11	0.87	0.76	0.43
Item 5	0.23	0.74	0.62	0.19
Item 6	0.53	0.57	0.59	0.34
Item 7	0.43	0.86	0.83	0.48
Item 8	0.32	0.59	0.73	0.26

Fig. 9.5 Item–cluster matrix

In Fig. 9.5, the item–cluster matrix is illustrated as the results from the fuzzy clustering. There are the two types of fuzzy clustering results: one are the clusters of items in which some items belong to more than one cluster and the other is the item–cluster matrix which is shown in Fig. 9.5. In the matrix frame, each column corresponds to a cluster, and each row does to an item. The entry which is crossed of a column and a row is a membership value of the item to the cluster. The item–cluster matrix will be mentioned again for explaining the fuzzy k means algorithm in Chap. 10.

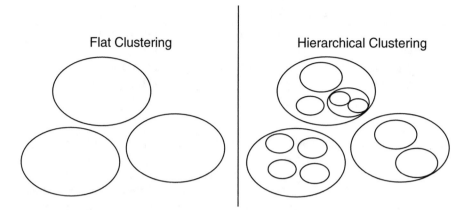

Fig. 9.6 Flat vs hierarchical clustering

9.3.3   Flat vs Hierarchical Clustering

This section is concerned with the two opposite types of clustering: flat clustering and hierarchical clustering. The dichotomy criteria of the two types is whether to allow nested clusters in a particular cluster. The flat clustering is one where no nested cluster is available, while the hierarchical one is one where at least, one nested cluster is available. In the flat clustering, a list of clusters is given as its results, while in the hierarchical clustering, a tree of clusters is given as ones. In this section, we explain the two types of clustering and their comparisons.

Figure 9.6 illustrates the flat clustering and the hierarchical clustering. The flat clustering is shown in the left side of Fig. 9.6. In the flat clustering, no cluster is nested in any cluster and a list of clusters is given as the results. The k means algorithm which were mentioned in Sect. 10.2.1 is usually used as the approach to this kind of clustering. The flat clustering is used for dividing a group of data items into several subgroups based on their similarities.

The hierarchical clustering is one where at least a nested cluster is allowed in a particular cluster as shown in the right side of Fig. 9.6. The three clusters are given in both the sides in Fig. 9.6; there is no nested cluster in the left side, while there are nested clusters in the right side. A tree of clusters, where its root node is a group of entire items, the lead nodes are individual data items, and the intermediate nodes are subgroups, is generated as the results. The AHC algorithm which was covered in Sect. 10.2.1, and the divisive algorithm which was covered in Sect. 10.2.2 are typical approaches to this type of clustering. The hierarchical clustering is used for constructing a hierarchical organization of data items for browsing them.

Table 9.4 illustrates the differences from the two types of clustering. As mentioned above, in the flat clustering, no nested cluster is allowed whereas in the hierarchical clustering, any nested cluster is allowed. Even if various clustering algorithms are used for both the types of clustering, in the flat clustering, the k

Table 9.4 Flat vs hierarchical clustering

	Flat clustering	Hierarchical clustering
Nested cluster	No	Yes
Typical algorithm	K means algorithm	AHC algorithm
Clustering proceed	Arrangement	Merging or division
Results	List	Tree

means algorithm is usually used, whereas in the hierarchical one, the AHC algorithm is usually used. The flat clustering is characterized as the process of arranging data items into one or some among clusters, whereas the hierarchical one is characterized as the process of dividing in the top–down direction or merging in the bottom–up direction. The flat clustering results are given as a list of clusters, whereas the hierarchical clustering results are given as a tree of clusters.

Let us consider the process of evaluating the hierarchical clustering results. By computing the intra-cluster similarity and the inter-cluster similarity, the flat clustering results are evaluated, easily. The results from the hierarchical clustering tend to be underestimated by the lower intra-cluster similarities of higher clusters and the higher inter-cluster similarities among nested ones. Even if the adopted clustering algorithm was really to the hierarchical clustering task, it tends to be evaluated in the flat clustering tasks before applying it. In this study, the clustering index is proposed as the metric of evaluating the clustering results, and the evaluation process will be mentioned in Sect. 12.3.4.

9.3.4   Single vs Multiple Viewed Clustering

This section is concerned with the two types of clustering, single viewed clustering and multiple viewed one. The dichotomy for dividing the clustering task into the two types is whether multiple results are allowed to be accommodated, or not. Even by the same clustering algorithm, different results may be produced depending on its external parameter values. As the difference between the hierarchical clustering and the multiple viewed clustering, a single tree of clusters is generated from the former, whereas forests which consist of multiple trees are produced from the latter. In this section, we explain the two types of clustering tasks and their comparisons.

The single viewed clustering is conceptually illustrated in Fig. 9.7. From a group of items, several clusters are generated by a clustering algorithm. The clusters are regarded as the entire results whether they are overlapping or exclusive. In the area of machine learning and data mining, until now, it has been assumed that results from clustering data items by a single algorithm are everything. However, afterward, we need to accommodate other clustering results which are generated in other views.

As shown in Fig. 9.8, in the multiple viewed clusters, various results from clustering data items are accommodated. The fact that data items are organized

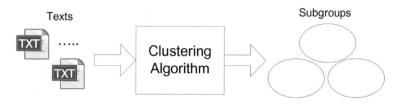

Fig. 9.7 Single viewed clustering

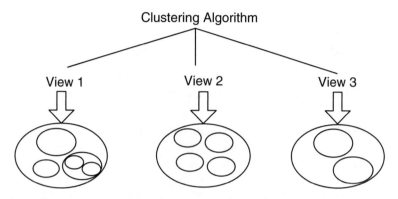

Fig. 9.8 Multiple viewed clustering

differently by different subjective even in the manual clustering is the motivation for considering this type of data clustering. The results from clustering data items are different depending on values of its external parameters, even by the same algorithm. For examples, in the case of the k means algorithm, different results are caused by changing the number of clusters. In this type of clustering, different multiple results from clustering data items are allowed as their organizations.

We need to integrate multiple results from clustering data items into a single one. Multiple different results which are called multiple organizations were mentioned above and one of them may be selected by a user. By merging similar clusters, they are integrated into one as the union of multiple organizations. The scheme of integrating multiple organizations which are clustering results is described in detail in [93].

The hierarchical clustering and the multiple viewed clustering are illustrated in Fig. 9.9. More than one list of clusters exists in both clustering types, so they look confusing for distinguishing them from each other. A single organization of data items is given as the results in the hierarchical clustering. However, the multiple organizations are given independently of each other in the multiple viewed clustering. In other words, a single tree of clusters is given as the results in the hierarchical clustering, whereas the forests are given as ones in the multiple viewed clustering.

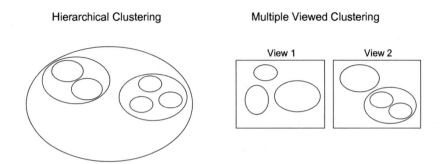

Fig. 9.9 Hierarchical clustering and multiple viewed clustering

9.4   Derived Tasks from Text Clustering

This section is concerned with the specific tasks of data clustering and other tasks which derive from it and it consists of the four subsections. In Sect. 9.4.1, we mention the cluster naming which is the subsequent task after the text clustering. In Sect. 9.4.2, we describe the subtext clustering as the process of generating virtual texts. In Sect. 9.4.3, we explain the process of using the text clustering results as sample texts for performing the text categorization. In Sect. 9.4.4, we cover the process of applying the text clustering for detecting redundant national projects.

9.4.1   Cluster Naming

Figure 9.10 illustrates the process of assigning names to clusters which are results from clustering data items. The cluster naming is the process of identifying each cluster, relevantly to its contents with the symbolic names. We will mention some principles of naming clusters; identifying each cluster with a primary key value or a numerical value, irrelevantly to its content, is out of cluster naming. Reflecting text contents in each cluster in its name is the requirement of cluster naming. In this section, we describe the cluster naming for enabling browsing of text clusters.

The principles of naming clusters symbolically were defined in 2006 in his PhD dissertation [25]. The first principle of naming clusters is to reflect cluster contents in their symbolic names. The second principle is that each cluster name should not be too long; the cluster name consists of only couple of words. The third principle is that each cluster name should be unique; there is no redundant name among clusters. The cluster names are helpful for accessing texts by browsing.

In 2006, Jo described the process of naming clusters based on the above principles, in his PhD dissertation [25]. In each cluster, its texts are indexed into a list of words with the basic three steps which were mentioned in Chap. 2. For each cluster, weights of indexed words are computed and the words with their highest weights are selected as the cluster name candidates. If more than two clusters have

Fig. 9.10 Cluster naming

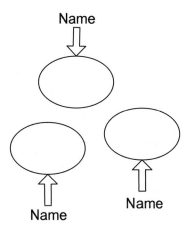

the same name, the action is taken against it, by adopting one among schemes which will be mentioned in the next paragraph. Other approaches are available than this way and cluster naming is regarded as a text mining task which is separated from the text clustering.

More than two clusters with the same name violates against the principle, so let us mention the schemes of avoiding the redundancy. In this case, the word with its highest weight is replaced by one with its second highest weight in naming the cluster. The clusters with their redundant name are merged into one cluster, as the alternative scheme. We use multiple schemes of weighting words and the word is replaced by one which is weighted by another scheme. In spite of avoiding the redundant cluster name, it is possible that the clusters which are named with their different words have their identical meanings.

The taxonomy generation is achieved by combining the text clustering and the cluster naming with each other. The taxonomy generation is referred to the process of generating topics, their relations, and their associated texts, from a corpus. The taxonomy generation process consists of clustering texts by a clustering algorithm and naming clusters by the above process. The taxonomy generation may be expanded to the process of constructing manually or semiautomatically ontologies. The taxonomy generation will be mentioned in detail, in Chap. 15.

9.4.2   Subtext Clustering

This section is concerned with the subtext clustering which is illustrated in Fig. 9.11. The subtext means a part of full text, such as paragraphs, sentences, and words, in the broad meaning. The texts in the corpus are partitioned into paragraphs and they are clustered into subgroups of content-based similar ones. The results from clustering subtexts become the source of synthesizing subtexts into an artificial text. In this section, we describe the process of clustering subtexts.

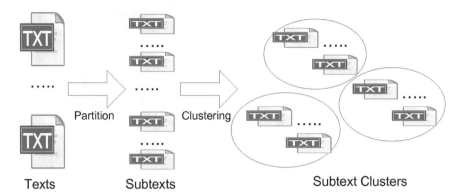

Fig. 9.11 Subtext clustering

A text may be viewed as a hierarchical form of its components. The texts in the corpus are partitioned into paragraphs by the carriage return. Each paragraph is partitioned into sentences by punctuation mark. Each sentence is partitioned into words; paragraphs, sentences, and words belong to subtexts. Each full text is expressed in a hierarchical form whose root node is the full text, intermediate nodes are paragraphs or sentences, and the terminal nodes are words.

Even if a subtext is a part of the given full text, it belongs to textual data; it could become a text clustering target. Subtexts are extracted from texts in the given collection, and they are encoded into numerical vectors. They are clustered into subgroups based on their similarities by a clustering algorithm. There is the possibility that subtexts within a given text are scattered over different clusters. Note that numerical vectors which represent subtexts may be sparser than those which represent full texts.

Let us mention some similar tasks with subtext clustering which is covered in this section. The summary-based clustering which is mentioned in Chap. 13 is the process of clustering texts by encoding their summaries, instead of full texts. If clustering paragraphs in a long text into subgroups, the paragraph subgroups are also called subtexts. A full text is segmented into subtexts based on their contents through the text segmentation, and each subtext is called topic-based subtext.

Let us point out some issues from subtext clustering. The numerical vectors which represent subtexts tend to be sparser than ones representing full texts. The summary-based clustering which is mentioned in Chap. 13 belongs to the kind of subtext clustering for clustering texts fast. Clustering paragraphs within a text is called unordered text segmentation which is mentioned in Chap. 14. It is the critical factor how to select subtexts from a full text in this task.

9.4.3   Automatic Sampling for Text Categorization

This section is concerned with the process of generating sample texts automatically for the text categorization. The preliminary task for the text categorization is to predefine categories as a list or a tree and to allocate texts to each category as samples. The preliminary tasks is automated by the two tasks: the text clustering and the cluster naming. A list or a tree of cluster names is given as the predefined categories and texts which belong to the clusters become sample labeled texts. In this section, we describe the text clustering which is viewed as the process of automating the preliminary tasks.

The preliminary tasks for the text categorization, such as the category predefinition and sample text allocation, become the tedious jobs. Topics or categories are predefined as a list or a tree. Texts are collected from the external source and labeled manually. The sample texts are encoded into numerical vectors by the process which was described in Chap. 3. However, more reliable quality of sample texts is expected for training text classifiers.

Let us explain the process of automating the above preliminary tasks for the text categorization. In this case, unlabeled texts are gathered from external sources. The collection of the gathered ones is segmented into subgroups of similar ones, and each cluster is named symbolically. The cluster names are given as categories and texts in each cluster are given as labeled ones. However, as the payment for automating the preliminary tasks, we expect less reliable quality of sample texts which are automatically generated by this process.

We need to customize the system in using the text clustering for automating the preliminary tasks for the text categorization. We decide the classification types such as crisp vs fuzzy classification and flat vs hierarchical classification. The clustering should be executed following the classification type which we decide. The identifiers are assigned to the clusters and they are named symbolically through the cluster naming which was mentioned in Sect. 9.4.1. As an optional one, we may add the process of redistributing cluster examples as sample examples for the binary classifications, which are decomposed from the multiple classification.

We should remind the poor reliability of sample examples which are generated automatically by the above process. So, we need to consider the learning type where both labeled and unlabeled examples are used. The learning which combines the supervised learning with the unsupervised one is called semi-supervised learning. In Chap. 10, we mention the LVQ (Learning Vector Quantization) as the semi-supervised learning algorithm. In clustering items by the k means algorithm or the Kohonen Networks, it is required to decide the number of clusters in advance.

9.4.4   Redundant Project Detection

This section is concerned with the process of detecting redundant projects as a derived task from the text clustering. The task which is discovered in this section is defined as the process of building groups of content-based very similar proposals of research projects. The research project proposals which are given as texts are entered and project clusters are constructed by the text clustering. In 2003, Jo implemented the system of doing that, using the single pass algorithm whose similarity threshold is set closely to 1.0 [24]. In this section, we describe the process of detecting the redundant research projects by clustering them.

The research project proposal consists of the three parts: project scope, project contents, and project goals. The first part, project scope, is concerned with area or coverage of the project which researcher tries to propose. The second part, project contents, describes what is proposed, implemented, or researched in the project [24]. The last part, project goals, indicates what the project is intended. It is assumed that the project proposal is given as an XML documents with the three components which are given tags.

Let us mention the process of detecting the redundancy of research project proposals through the text clustering. The research project proposals each of which consists of the three parts are collected in the given system. The collection is clustered into subgroups by a clustering algorithm; the single pass algorithm was used as the clustering algorithm in [24]. The research project proposals within each cluster are regarded as potential candidates of redundant proposals or associated ones. Either of two actions should be taken to each cluster: selection of only one proposal or integration of projects into a single project.

We need to customize the text clustering to be more suitable for detecting redundant project proposals. The direction is to generate large number of clusters each of which contains very small number of items. In this case, the similarity threshold is set closely to one in using the single pass algorithm. Discriminant weights are assigned to the three components of proposal; the higher weight is assigned to the research content and research goal, and the lower weight is assigned to the research scope. More than half of clusters in the results are skeletons each of which has only one item.

The text association becomes the alternative way of detecting redundant research projects to the text clustering. Words are represented as text sets from the corpus. Association text rules are extracted from the text sets by the Apriori algorithm. Texts are associated as redundant project candidates or associated ones. In this case, the support threshold and the confidence threshold are set closely to 1.0.

9.5   Summary and Further Discussions

In this chapter, we described the text clustering in the functional view. We defined the clustering and compared it with other tasks. We surveyed the dichotomies of two opposite text clustering types. We mentioned the tasks which are similar as the text clustering and derived from it. In this section, we make some further discussions from what we studied in this chapter.

Let us refer to the association mining to the process of extracting implicit knowledge from a collection of association rules. They are extracted from a collection or clusters of sets and become a kind of data, in this case. They are classified or clustered; they become instances of association mining. We need to define operations on association rules for implementing association mining system. The association rules which are concerned with relation among association rules are extracted from them; the task is called meta association rules.

Let us consider the online clustering which proceeds the clustering to the data items which arrive continually. The assumption underlying in the offline clustering which is opposite to the online clustering is that all clustering targets are given at a time. No data is available initially and whenever data items arrive continually, the data clustering proceeds by updating current clusters. The traditional clustering algorithms which were mentioned in Chap. 10 were inherently designed for the offline clustering, so we need to modify them into online clustering version. The online clustering will be mentioned in Chap. 10.

The taxonomy generation is for creating a frame of organizing texts from a corpus. It is defined as the process of generating a list, a tree, or a network of topics which are called taxonomies. The word categorization, the word clustering, and the keyword extraction are involved in executing the taxonomy generation. Each taxonomy is linked to its own relevant texts and is associated to its scripts. It will be described in detail in Chap. 15.

The organization evolution may be considered as the process of changing the text organization, gradually to the better one. Reorganizing entire texts is the hard action, whereas modifying some organization parts is the soft action. Typical operations for maintaining the organization are to merge some clusters into a single cluster and to divide a cluster into several ones. Outlier detection is considered as the process of detecting whether a text is an alien or not, for creating one more cluster. The schemes for evolving the organization will be described in detail in Chap. 16.

Chapter 10
Text Clustering: Approaches

This chapter is concerned with the unsupervised learning algorithms which are approaches to text clustering. We explain briefly the unsupervised learning in Sect. 10.1, and describe some simple clustering algorithms in Sect. 10.2. In Sect. 10.3, we describe the k means algorithm and its variants which are the most popular approaches to text clustering in detail. We cover the competitive neural networks as advanced approaches to text clustering, in Sect. 10.4, and make summarization and further discussions on this chapter in Sect. 10.5. In this chapter, we describe unsupervised learning algorithms and neural networks as the approaches to text clustering.

10.1 Unsupervised Learning

The learning process which depends on similarities of input data for the clustering task is called unsupervised learning. In this chapter, we mention unsupervised learning algorithms such as the k means algorithm and the Kohonen Networks, as the approaches to text clustering. The clustering prototypes are initialized at random, and the similarities between the cluster prototypes and the input vectors are computed. The unsupervised learning process is the process of optimizing the cluster prototypes in order maximize the similarities between examples and the prototypes of their belonging clustering. In this section, we describe briefly the unsupervised learning as the basis for performing the text clustering, together with review of the supervised learning.

Let us review briefly the supervised learning which was described in Chap. 6 as the approaches to text categorization, before mentioning the unsupervised learning. It is assumed that labels of training examples are always available in the supervised learning. In the learning process, the differences between target labels and classified ones should be minimized. The supervised learning is applied to the classification

© Springer International Publishing AG, part of Springer Nature 2019
T. Jo, *Text Mining*, Studies in Big Data 45,
https://doi.org/10.1007/978-3-319-91815-0_10

where the outputs are given as discrete values and the regression where they are given as continuous values. K Nearest Neighbor, Naive Bayes, Perceptron, and Support Vector Machine are typical supervised learning algorithms, but they are able to be modified into their unsupervised ones.

In spite of various learning processes depending on types of learning algorithms, let us examine the frame of the unsupervised learning algorithms. It is assumed that unlabeled examples are given as training examples; the collection is not partitioned into the training set and the test set for evaluating unsupervised learning algorithms. Each cluster is initially characterized by its own prototypes and given examples are arranged, depending on their similarities with the cluster prototypes; each example belongs to the cluster whose similarity is highest. Cluster prototypes are optimized by iterating arranging examples into clusters and updating the cluster prototypes. The optimized cluster prototypes and unnamed clusters of similar examples are results from the unsupervised learning.

Even if it is possible to modify machine learning algorithms between the supervised and the unsupervised, we need to mention some unsupervised learning algorithms. The k means algorithm is the most representative unsupervised learning algorithm and will be described in detail in Sect. 10.3.1. The k means algorithm may be generalized into EM algorithm which is the frame of clustering algorithm consisting of E(Estimation)-step and M(Maximization)-step, and its particular version will be described in Sect. 10.3.3. The Kohonen Networks proposed by T. Kohonen in 1972 are the typical neural networks of unsupervised learning, and are described in detail in Sect. 10.4.1 [51]. The neural gas is the unsupervised learning algorithm derived from the Kohonen Networks, and is described in detail in Sect. 10.4.4.

The supervised and unsupervised learning are mentioned above. Now, let us mention the semi-supervised learning as a mixture of both of them. The semi-supervised learning is motivated by the fact that labeled training examples are expensive but unlabeled ones are cheap in gathering them. Not only labeled examples, but also labeled examples are used in the semi-supervised learning. Because the semi-supervised is intended for the tasks, classification and regression; it is closer to the supervised learning. If both kinds of examples are intended for clustering, it is called constraint clustering.

10.2   Simple Clustering Algorithms

This section is concerned with some simple clustering algorithms as starting to study the text clustering approaches. In Sect. 10.2.1, we study the AHC (Agglomerative Hierarchical Clustering) algorithm which clusters data items in the bottom-up direction. In Sect. 10.2.2, we will cover the alternative simple clustering algorithm called divisive clustering algorithm which clusters data items in the top-down direction. In Sect. 10.2.3, we mention the single pass algorithm which is known as a fast algorithm. In Sect. 10.2.4, we mention the growing algorithm where the radii around data items grow and merge with adjacent ones.

Fig. 10.1 Initial stage in
using AHC algorithm

Fig. 10.2 Merging two
clusters into a cluster

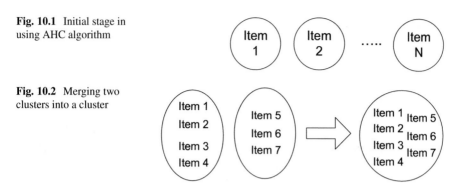

10.2.1 AHC Algorithm

This section is concerned with the simple clustering algorithm, AHC algorithm.
It is required to define a similarity measure between data items and a scheme
of computing a similarity between clusters. Clusters are constructed by merging
smaller clusters by their similarities in the bottom-up direction. The number of
clusters or the critical similarity is set as the stopping criterion of this algorithm. In
this section, we describe the AHC algorithm with respect to its clustering process.

In Fig. 10.1, we illustrate the initial stage of clustering N data items with the
AHC algorithm. The N data items are given as the input. The N clusters are
created, and each cluster contains single item. The N clusters which are called single
skeletons are generated as the results from the initialization. After the initialization,
all possible pairs are generated from the list of clusters and similarity between
clusters in each pair is computed.

In the AHC algorithm, clustering data items proceeds by merging initial clusters.
All possible pairs of clusters are generated and their similarities are computed. The
cluster pair with its maximum similarity is selected and the clusters in the pair are
merged into a single cluster, as shown in Fig. 10.2. When merging the pair of clusters
into a cluster, the total number of clusters is decreased by one; the above process is
repeated until the number of clusters reaches the desired one. Instead of the number
of clusters, the similarity threshold may be given as the alternative parameter; they
are iterated until the maximum similarity of a pair is less than the threshold.

Various versions may be derived from the above AHC algorithm as its variants.
The similarities among clusters or individual data items may be computed, discriminating the given attributes. As a fast version, the variant where the initial item of
each cluster is fixed as its representative one may be derived. More than one pair
whose similarities are higher than the threshold are merged in each iteration, rather
than merging only one pair with its maximum similarity. There are two directions
for deriving the variants: more reliable schemes of computing the similarities and
fast execution of data clustering.

Let us consider various criteria for stopping merging clusters to avoid reaching
a single cluster of all data items. The number of clusters may be considered as the

Fig. 10.3 Initial stage in
using divisive clustering
algorithm

stopping criterion, but there is no way of knowing the desired number of clusters, in advance. If the maximum similarity is less than the threshold in an iteration, the merge of clusters is stopped. When the maximum size of cluster is greater than the given threshold, it may terminate. The intra-cluster similarity and the intercluster similarity are considered as the stopping criterion, in addition.

10.2.2   Divisive Clustering Algorithm

This section is concerned with the clustering algorithm which proceeds the data clustering in the top-down direction. The clustering algorithm is called divisive clustering algorithm and is opposite to the AHC algorithm which was mentioned in Sect. 10.2. It starts with a single cluster of all items and proceeds the clustering by dividing one cluster into several clusters. Some variants may be derived, depending on the scheme of partitioning clusters. So, in this section, we describe the divisive clustering algorithm and its variants in detail.

Figure 10.3 illustrates the initial status of using the divisive clustering algorithm. The individual items are given as the input, and they are initially grouped into a single cluster. The single cluster in the initial state characterizes this clustering algorithm, oppositely to the case in using the AHC algorithm. The main operation for proceeding the clustering is to partition a single clustering into two or more clusters. In the divisive clustering, a hierarchical structure is constructed in the top-down direction.

Figure 10.4 illustrates the direction of proceeding the data clustering by the divisive clustering algorithm. The clusters of similar data items are generated by dividing the clustering in the top-down direction. From the cluster, the pair with its least similarity is selected, the cluster is divided into two clusters by arranging the other toward higher similarity to one in the pair. The two items in the selected pair become the representative of the two divided clusters. Proceeding the clustering by the divisive clustering algorithm is visualized like a binary tree.

Some variants may be derived from the divisive clustering algorithm which is mentioned above. Its various versions are considered depending on how to define the similarity metric between data items and the scheme of computing the similarity between clusters. Multiple pairs with their smallest similarities are selected and a cluster is partitioned into more than two clusters, in proceeding the data clustering. Some variants are built by nominating medoids from clusters or computing cluster

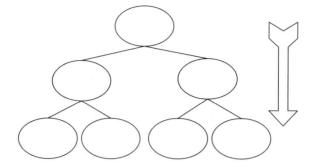

Fig. 10.4 Direction of proceeding clustering in divisive clustering algorithm

mean vectors as cluster representatives, and excluding some examples which are weakly similar to the cluster representatives from given clusters. This type of data clustering proceeds by applying a particular clustering algorithm such as k means algorithm within each cluster.

Let us consider various criteria for stopping the cluster division, in order to avoid reaching the results which are given as singletons. The number of clusters may be set as the stopping criterion; when the current number of clusters reaches the threshold, the division stops. When the least similarity of all possible pairs is greater than the threshold, the division proceeds no more. When the minimum size of cluster is less than the threshold, the execution is stopped. We may also consider the intra-cluster similarity and the intercluster similarity as the additional criteria.

10.2.3 Single Pass Algorithm

This section is concerned with the clustering algorithm which is simple and fast, called single pass algorithm. The clustering algorithm was adopted in 2003 by Jo for detecting redundancy of national research and development projects proposals [23]. It starts with a single cluster which has a single item, and the data clustering proceeds by arranging next data items into one of existing clusters or creating a new cluster. This algorithm costs almost linear complexity for clustering data items, in the case of a small number of clusters compared with the number of data items. In this section, we describe the single pass algorithm as one of simple clustering algorithms.

The process of clustering the data items by the clustering algorithm is explained through Figs. 10.5, 10.6, and 10.7. The initial status consists of a single cluster and an item as shown in Fig. 10.5; the item becomes the initial prototype of the cluster. The similarity between the first item and the second item is computed and one of the two cases is selected depending on it. If the similarity is greater than the threshold, the second item joins to the exiting cluster, and otherwise, one more cluster is created and it is the initial item of the cluster. The threshold is given as the external parameter in the single pass algorithm.

Fig. 10.5 First text in using single pass algorithm

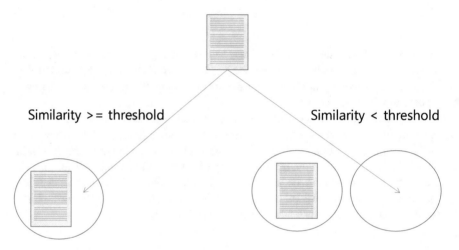

Fig. 10.6 Second text in using single pass algorithm

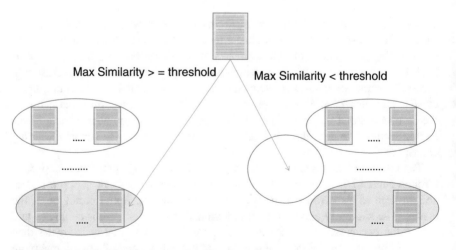

Fig. 10.7 nth text in using single pass algorithm

Let us consider the case to the nth data item after constructing some clusters as shown in Fig. 10.7. Currently, it is assumed that the $n - 1$ data items were already presented, and k clusters are constructed less than $n - 1$. When the nth data item is presented, its similarities with the k clusters are computed and the maximum similarity is selected among them. If the maximum similarity is greater than or equal to the threshold, it joins to the corresponding cluster. Otherwise, one more cluster is created and the data item joints to the cluster as its initial one. If the number of clusters is much smaller than the number of data items, it takes almost linear complexity for clustering data items.

Some variants are derived from the version which is illustrated in Figs. 10.5, 10.6, and 10.7. In the initial version, the initial item is given as the representative one in each cluster, whereas the mean vector or the medoid is set as the representative one in the variant. Instead of the maximum similarity, the intra-cluster similarity and the intercluster similarity are used for deciding whether the item joins to one of existing clusters or creates one more cluster. The distribution over similarities of the item with existing clusters may be used for building the fuzzy version of single pass algorithm where at least one item is allowed to join into more than one cluster. The variants are characterized as the fact that clustering starts with a single skeleton, altogether with the initial version.

The single pass algorithm is characterized as the fast clustering algorithm with almost linear complexity [23]. The clustering algorithm was adopted in implementing the Korean national R&D project system before inventing the evaluation metric of clustering results, called clustering index. In 2006, when the clustering index was invented, it was discovered that the quality of clustering data items by the single pass algorithm is very poor compared with that of doing so by other clustering algorithms. In the clustering algorithm, subsequent items are arranged depending on the initial items of clusters without optimizing the cluster prototypes. There is trade-off between the clustering speed and the clustering quality in adopting clustering algorithms [25].

10.2.4 Growing Algorithm

This section is concerned with another simple clustering algorithm which is called growing algorithm. Like the AHC algorithm, it starts with the skeletons as many as individual items. Each cluster increases its radius from its own example from zero, gradually. When the two clusters face with each other by their growths, they are merged with each other into a cluster, and the center is updated into their facing point. In this section, we describe the growing clustering algorithm with respect to the initial stage, the clustering stage, and the variants.

A group of data items is given as the input like any other clustering algorithms. The skeletons are created as many as data items and they match with data items, one to one. Each of the initial clusters has its zero radius and its center is its own data item. The radius and the center of the cluster, $\mathbf{C_k}$, is notated by r_k and $\mathbf{C_k} = \begin{bmatrix} C_{k1} & C_{k2} & \cdots & C_{kd} \end{bmatrix}$. After initializing the clusters so, they grow parallel and independently.

As the clustering proceeds, the clusters grow continually and two clusters are merged into one cluster; each cluster is modeled as a hypersphere. The clusters are initialized as Eqs. (10.1) and (10.2) which initialize the center and the radius, respectively,

$$\mathbf{C_k} = \begin{bmatrix} x_1 \ x_2 \ \cdots \ x_d \end{bmatrix} \tag{10.1}$$

$$r_k = 0 \tag{10.2}$$

and the cluster grows by updating the radius by Eq. (10.3),

$$r_k(t + 1) = r_k(t) + \theta \tag{10.3}$$

where $\theta \geq 0$ is a constant. The two clusters, $\mathbf{C_k}$ and $\mathbf{C_m}$, with the parameters, $\mathbf{C_k} = \begin{bmatrix} C_{k1} \ C_{k2} \ \cdots \ C_{kd} \end{bmatrix}$, r_k, and $\mathbf{C_m} = \begin{bmatrix} C_{m1} \ C_{m2} \ \cdots \ C_{md} \end{bmatrix}$, r_m, respectively, start to overlap with each other, when the distance is less than the addition of their radii shown in Eq. (10.4),

$$\text{dist}(\mathbf{C_k}, \mathbf{C_m}) = \sqrt{\sum_{i=1}^{d}(C_{ki} - C_{mi})^2} \leq r_k + r_m \tag{10.4}$$

The two clusters, $\mathbf{C_k}$ and $\mathbf{C_m}$, are merged into a new cluster, $\mathbf{C_n}$ by updating the involved parameters by Eqs. (10.5) and (10.6),

$$\mathbf{C_n} = \begin{bmatrix} \frac{1}{2}(C_{k1} + C_{m1}) \ \frac{1}{2}(C_{k2} + C_{m2}) \ \cdots \ \frac{1}{2}(C_{kd} + C_{md}) \end{bmatrix} \tag{10.5}$$

$$r_n = \frac{1}{2}(r_k + r_m) \tag{10.6}$$

This process is iterated until reaching the desired number of clusters.

Some variants are considered, depending on the scheme of modeling clusters, the condition of merging two clusters, and the similarity metric between clusters. Each cluster may be modeled as a Gaussian distribution whose parameters are the mean vector and the covariance matrix, instead of the hypersphere. The intercluster similarity and the intra-cluster similarity are defined as the condition for merging two clusters. Instead of the Euclidean distance, we use the similarity metrics, such as the cosine similarity or other variants as the condition of merging two clusters. Instead of a number of clusters, the similarity threshold is set as the termination condition.

Let us make some remarks on the growing algorithm as its characterization. In our reality, it is not feasible to cluster a very large number of data items by a single algorithm. However, because each cluster grows independently of others, it is possible to implement the clustering by the parallel and distributed computing. It starts with the distributed agents as many as data items and it merges with others. The complexity is reduced from the quadratic one to linear one by implementing it as the parallel and distributed system.

10.3 *K* Means Algorithm

This section is concerned with the k means algorithm which is the most popular clustering algorithm in any application domain. In Sect. 10.3.1, we explain the crisp k means algorithm which is the initial version. In Sect. 10.3.2, we expand the initial version into the fuzzy version. In Sect. 10.3.3, we cover the complicated version of the k means algorithm where each cluster is interpreted as a Gaussian distribution. In Sect. 10.3.4, we mention the variant which is called k medoid algorithm.

10.3.1 Crisp K Means Algorithm

This section is concerned with the initial version of k means algorithm. The version is called crisp k means algorithm, in order to distinguish it from the version which is mentioned in Sect. 10.3.2. The clustering task which is performed by the initial version corresponds to the crisp clustering where each item is allowed to belong to only one cluster. The clustering proceeds by iterating arranging data items into their own clusters and updating mean vectors of clusters. In this section, we describe in detail the initial version of k means algorithm as an approach to the text clustering.

A group of data items is initially given as the input. The number of clusters is decided in advance, and data items as many as clusters are selected from the input, at random. The selected data items are the initial mean vectors of clusters, and the others are arranged into clusters, depending on their similarities with the initially selected ones. The mean vectors are updated by averaging data items in each cluster, and the data items are arranged again by the updated ones. Different results are expected in the k means algorithm, depending on the selected data items and the number of clusters which is given as an external parameter.

The k means clustering algorithm is illustrated as the pseudocode in Fig. 10.8. The number of clusters is given as the external parameter, data items as many as clusters are selected at random, and the cluster mean vectors are initialized by the examples. The following two steps are iterated until the cluster mean vectors converge. The first step is to arrange examples based on their similarities with the cluster mean vectors and the second step is to update the cluster mean vectors by computing mean vectors in each cluster. When the cluster mean vectors change very little, the iteration may be terminated.

It is important to decide the termination condition, in using the k means algorithm. The simplest scheme is to give the iteration number as the external parameter. Another termination condition is whether the difference between previous members and current members in each cluster is less than the threshold, or not. The difference between the previous mean vector and the current mean vector decides whether the iteration terminates, or not. We may consider that different termination conditions are assigned to the clusters.

Input: Unlabeled Examples (Numerical Vectors); #Clusters
Select Examples at Random as many as Clusters
Initialize the Means Vectors of Clusters with the Selected Ones
Iterate until Convergence of Mean Vectors
    For each example
        Arrange each example into one of cluster where the mean vector is most similar with itself
    For each cluster
        Compute the mean vector in each cluster

Fig. 10.8 K means algorithm

Other schemes of initializing the cluster mean vectors are available as the alternatives to one where they are initialized by randomly selected ones. The cluster mean vectors may be initialized by the prior knowledge about the application domain. They may be initialized by selecting data items at random, avoiding similar items as previous ones. In the process of initializing them, we may define constraints about similarities among the cluster mean vectors. After selecting data items at random, the cluster mean vectors may be adjusted.

10.3.2   Fuzzy K Means Algorithm

This section is concerned with the fuzzy version of k means algorithm. It is assumed that the given task is the fuzzy clustering where each item has its own membership values of clusters. Instead of arranging an item into its most similar cluster, its membership values of clusters are computed. By the linear combination of data items weighted by their membership values, the cluster mean vectors are computed. In this section, we describe the fuzzy k means algorithm as the approach to the fuzzy clustering.

Memberships of items to the clusters are viewed as the item by cluster matrix in Fig. 10.9. The rows correspond to items and the columns do to clusters in the matrix. The entry of the j row and the i column, which is notated as $\mu_{C_j}(\mathbf{x_i})$, is the membership of the item, $\mathbf{x_i}$ to the cluster, C_j. If it is the crisp clustering, only one membership becomes one and the others are zero in each row. The membership is given as a continuous value between zero and one, in the fuzzy clustering.

Figure 10.10 illustrates the update of rows and columns of item–cluster matrix. The mean vectors are initialized by selecting items as many as clusters at random. By dividing their similarities by the maximum one, for each item, its membership values of clusters are computed. For each cluster, the mean vectors are updated by the linear combination of items which are weighted by their membership values. In Fig. 10.10, the rows are updated, corresponding to membership values, and the columns are updated, corresponding to the mean vectors.

Let us consider the relation between the two versions of k means algorithm: the crisp one and the fuzzy one. In the crisp version, 1.0 is assigned to only one cluster and zeros are assigned to the others in the matrix, whereas in the fuzzy k

Clusters

$$
\begin{bmatrix}
\mu_{C_1}(x_1) & \mu_{C_2}(x_1) & \cdots & \mu_{C_M}(x_1) \\
\mu_{C_1}(x_2) & \mu_{C_2}(x_2) & \cdots & \mu_{C_M}(x_2) \\
\cdots & \cdots & \cdots & \cdots \\
\mu_{C_1}(x_N) & \mu_{C_2}(x_N) & \cdots & \mu_{C_M}(x_N)
\end{bmatrix}
$$

Items

Fig. 10.9 Item by cluster matrix for fuzzy clustering

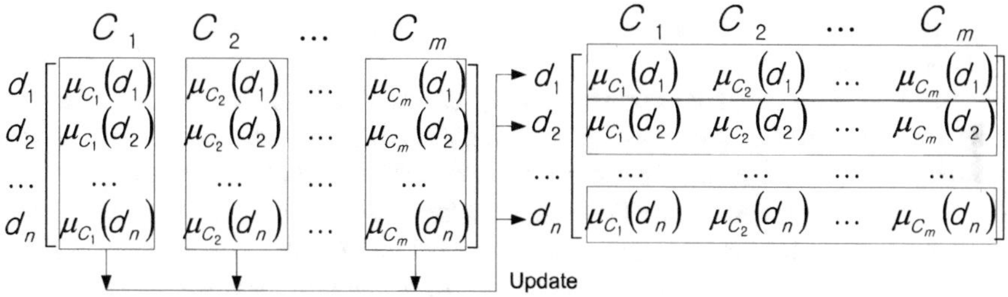

Fig. 10.10 Update of rows and columns of the item by cluster matrix

means clustering, various continuous values are given any cluster. In both versions, columns are scanned for computing mean vectors of clusters and rows are scanned for arranging items into clusters. In the crisp one, each data item is arranged into only one cluster whose membership is 1.0, whereas it is arranged into more than one cluster by the alpha-cut. From both versions, we derive the variant, the *k* medoid algorithm, which is mentioned in Sect. 10.4.

10.3.3 Gaussian Mixture

This section is concerned with the state-of-the-art version of *k* means algorithm. The assumption underlying in this version is that each cluster is represented into a Gaussian distribution which is called the normal distribution in Chap. 6. Each cluster is expressed as its own mean vector and its own covariance matrix which are the Gaussian distribution parameters. The initial version of the *k* means algorithm is expanded by considering additionally the covariance matrix as well as the mean vector. In this section, we describe the version of the *k* means algorithm which is called Gaussian mixture.

Let us explain the Gaussian distribution on vectors before describing the process of clustering data items by this version. The Gaussian distribution is expressed as Eq. (6.25) which involves the two parameters. The mean vector is computed by averaging individual vectors in a given group. The entry of covariance matrix is computed averaging over products of absolute differences corresponding to elements. Considering the covariance matrix additionally is the difference from the initial version.

The process of clustering data items by the Gaussian mixture is identical to that of doing them by the k means algorithm except considering the covariance matrix. In the initial stage, some data items are selected at random as the initial mean vectors and the determinant of covariance matrix is initialized to one. The mean vectors and the covariance matrices of clusters are computed by the process which was mentioned above. The likelihoods of data items to clusters are computed by Eq. (6.25), and each data item is arranged to cluster whose likelihood is maximum. This process is iterated until the mean vectors and covariance matrices of clusters are converged.

The Gaussian mixtures algorithm may be modified into the version for the fuzzy clustering. The process of initializing the mean vectors and the covariance matrices is identical to that in the crisp version. The likelihoods are computed by Eq. (6.25) as the membership of each item to the clusters. The mean vectors and the covariance matrices are updated by summing data items and differences which are weighted by memberships. The versions of k means algorithm which were covered in Sects. 10.3.1 and 10.3.2 are mixtures of Gaussian in case of fixing the covariance matrix determinants to one.

In using the k means algorithm, it is assumed that the clusters may be given as various distributions other than the Gaussian distribution. The clusters may be defined as the uniform distribution whose parameters are given as the start vector and the end vector; the uniform distribution was not adopted. It is able to define the clusters as fuzzy distributions such as triangle distribution and trapezoid distribution. We may consider other statistical distributions such as Poisson distribution, hypergeometric distribution, and exponential distribution, and the hybrid distribution which is the mixture of multiple distributions. However, the Gaussian distribution is usually used for representing the clusters because of the central limit theorem which asserts that any distribution is converged to the Gaussian distribution as trials are made, infinitely.

10.3.4   K Medoid Algorithm

This section is concerned with a variant of k means algorithm, which is called k medoid algorithm. The mean vector has been representative to each cluster in applying the k means algorithm. In the k medoid algorithm, the representative one which is called medoid is one among cluster members. The process of clustering data items by the k medoid algorithm is to iterate arranging data items into clusters and nominating a medoid for each cluster, until the cluster medoids are converged. In this section, we omit the detailed clustering process, because it is redundant with the k means algorithm, and explain only the scheme of nominating the cluster medoids.

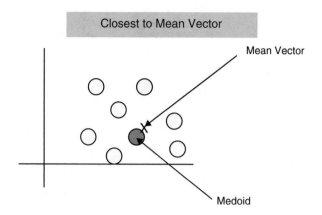

Fig. 10.11 Medoid: closet to mean vector

Let us mention the scheme of selecting a medoid among vectors, called closet to mean vectors, as shown in Fig. 10.11. The cluster, C_i is expressed as a set of vectors, $C_i = \{\mathbf{x}_{i1}, \mathbf{x}_{i2}, \ldots, \mathbf{x}_{im}\}$, and the mean vector of cluster, C_i is computed by Eq. (10.7),

$$\bar{\mathbf{x}}_i = \frac{1}{m} \sum_{i=1}^{m} \mathbf{x}_{ik} \tag{10.7}$$

The vector which is least distant from the mean vector is selected as the method of this cluster, by Eq. (10.8),

$$\mathbf{x}_{med} = \operatorname{argmin}_k ||\mathbf{x}_{ik} - \bar{\mathbf{x}}_i|| \tag{10.8}$$

The vector with maximum cosine similarity with the mean vector, which is expressed by Eq. (10.9), is selected as the medoid,

$$\mathbf{x}_{med} = \operatorname{argmax}_k \frac{\mathbf{x}_{ik} \cdot \bar{\mathbf{x}}_i}{||\mathbf{x}_{ik}|| \cdot ||\bar{\mathbf{x}}_i} \tag{10.9}$$

This scheme may be used for selecting representative members from the final cluster rather for proceeding the clustering.

Let us mention another scheme which is called minimum distance variance, as shown in Fig. 10.12. For each vector, $\mathbf{x}_{ik}$, its distances from others are computed and averaged by Eq. (10.10),

$$\text{mean\_dist}_{ik} = \frac{1}{m-1} \sum_{\substack{j \neq k}}^{m} ||\mathbf{x}_{ij} - \mathbf{x}_{ik}|| \tag{10.10}$$

Fig. 10.12 Minimum
variance of distances
from rest

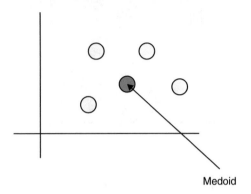

Medoid

The distance variances are computed by Eq. (10.11),

$$\text{dist\_var}_{ik} = \frac{1}{m-1} \sum_{\substack{j \neq k}}^{m} (\text{mean\_dist}_{ik} - ||\mathbf{x}_{ij} - \mathbf{x}_{ik}||)^2 \tag{10.11}$$

The vector with its minimum distance variance is selected as the medoid as expressed in Eq. (10.12),

$$\mathbf{x}_{\text{med}} = \text{argmin}_k (\text{dist\_var}_k) \tag{10.12}$$

The cluster medoids which are selected by this scheme look like the cluster medians.

Let us consider another scheme of selecting methods which is called the smallest maximum distance, as shown in Fig. 10.13. For each vector, its maximum distance with other vectors is computed by Eq. (10.13),

$$\text{max\_dist}_{ik} = \text{argmax}_{j \neq k} ||\mathbf{x}_{ij} - \mathbf{x}_{ik}|| \tag{10.13}$$

The vector with its smallest maximum distance is selected by Eq. (10.14),

$$\mathbf{x}_{\text{med}} = \text{argmin}_k (\text{max\_dist}_k) \tag{10.14}$$

If the cosine similarity is applied for selecting the medoids, this scheme is called the largest minimum similarity. To the number of data items in the cluster, m, it takes the linear complexity, $O(m)$, in the closet to the mean vector, while it takes the quadratic complexity, $O(m^2)$, in the others.

It makes the remarks on the k medoid algorithms which are variants of k means algorithm. One which represents a cluster is selected among cluster members. In the k medoid algorithm, earlier convergence is expected than in the k means algorithm.

Fig. 10.13 Smallest
maximum distance

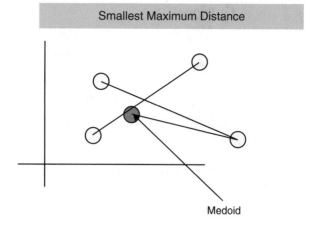

Fig. 10.13 Smallest
maximum distance

If the minimum distance variance or the smallest maximum distance is adopted as
the scheme of selecting the medoid, it takes much cost for updating it. The k medoid
algorithm is applicable to other types of data where it is possible to measure their
similarities, as well as numerical vectors.

10.4 Competitive Learning

This section is concerned with the Kohonen Networks and its variants and consists
of the four subsections. In Sect. 10.4.1, we explain the initial version of Kohonen
Networks and its competitive learning process. In Sect. 10.4.2, we describe the
LVQ (Learning Vector Quantization) which is the supervised version of Kohonen
Networks. In Sect. 10.4.3, we mention the expansion of Kohonen Networks into the
SOM (Self-Organizing Map). In Sect. 10.4.4, we cover the Neural Gas which is a
Kohonen Networks variant.

10.4.1 Kohonen Networks

This section is concerned with the initial version of Kohonen Networks which
is an unsupervised neural network. The neural networks were initially invented
by Kohonen in 1972 [51]. Since then, the neural networks have been used for
clustering data items as the unsupervised learning algorithm. It was modified into
the supervised version, called LVQ, and was expanded into the biological version
called SOM. In this section, we describe the unsupervised learning algorithm of
Kohonen Networks.

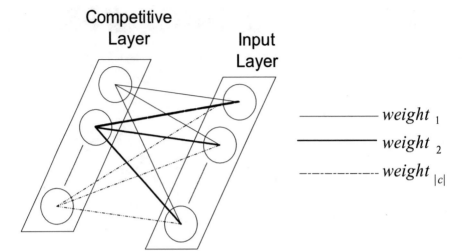

Fig. 10.14 Kohonen networks architecture

The Kohonen Networks architecture is illustrated in Fig. 10.14. There are two layers in the neural networks: one is the input layer and the other is the output layer which is called the competitive layer. The weight vectors between the two layers play their roles of prototype vectors which characterize clusters. The number of input nodes indicates the dimension of input vectors and the number of competitive nodes does the number of clusters. The dimension of input vectors and the number of clusters should be decided in advance, in order to use the Kohonen Networks for real problems.

Let us explain the learning process of Kohonen Networks which is the process of optimizing weight vectors. Here, we notate the number of input nodes, the number of competitive nodes, the weight vector, and the input vector, by d, C, $\mathbf{w}_i = \begin{bmatrix} w_{i1} & w_{i2} & \cdots & w_{id} \end{bmatrix}$, $1 \leq i \leq |C|$, and $\mathbf{x} = \begin{bmatrix} x_1 & x_2 & \cdots & x_d \end{bmatrix}$, respectively. The net input is computed by Eq. (10.15), as the inner product of the weight vector which corresponds to the cluster, i, and the input vector, $\mathbf{x}$,

$$\text{net}_i = \mathbf{x} \cdot \mathbf{w}_i \qquad (10.15)$$

The index of weight vector whose inner product with the input vector is maximum is selected by Eq. (10.16),

$$\text{max} = \text{argmax}_i \text{net}_i \qquad (10.16)$$

The corresponding weight vector is updated by Eq. (10.17),

$$\mathbf{w}_{\text{max}}(t+1) = \mathbf{w}_{\text{max}}(t) + \eta(\mathbf{x} - \mathbf{w}_{\text{max}}(t)) \qquad (10.17)$$

The competitive node from which the weight vector was updated is called winner, and the process of optimizing weights by this algorithm is called competitive learning.

Let us consider the design of Kohonen Networks for applying it to the text clustering task. The number of nodes in the input layer is the dimension of numerical vectors and the number of nodes in the competitive layer is the number of clusters which is decided in advance. The weights between the two layers are initialized at random, and they are optimized by the above process. The optimized weight vectors become the final prototype vectors which characterize their own clusters and texts are arranged into clusters, depending on their similarities with the prototype vectors, subsequently. The Kohonen Networks are expanded into SOM (Self-Organizing Map) which is mentioned in Sect. 10.4.3, and it is applied to the text clustering, by Kohonen et al. in 2000 [52].

The two clustering algorithms, the Kohonen Networks and the k means algorithms, are compared with each other. In advance, the number of clusters is decided in both the algorithms. The mean vector of each cluster is its representative one in the k means algorithm and the prototype vectors which are given as the weight vectors are updated by the learning algorithm, gradually as their representative ones in the Kohonen Networks. In the k means algorithm, the mean vectors of clusters are initialized by selecting data items as many as clusters at random, and the weight vectors are selected at random in the Kohonen Networks. In the k means algorithm, the learning process is to update the mean vectors by averaging items in the clusters and in the Kohonen Networks, the weight vectors whose inner product with the input vector is maximum are updated.

10.4.2 Learning Vector Quantization

This section is concerned with the supervised version of Kohonen Networks which is called Learning Vector Quantization. The initial version of Kohonen Networks is intended for the unsupervised learning which is used for clustering data items. In the LVQ, the training examples are labeled with their targets, and the weights which are connected to the competitive nodes which correspond to the target labels are updated. This version of Kohonen Networks is usually used for the classification tasks. In this section, formally, we describe the supervised version of Kohonen Networks with respect to its learning process.

The LVQ architecture is the same to that of the initial version of Kohonen Networks. There are two layers in LVQ like the initial version of Kohonen Networks: the input layer and the competitive layer. The number of input nodes is the dimension of the input vector and the number of competitive nodes is the number of categories. Each competitive node corresponds to a cluster in the initial version and it does to a category in the LVQ. The architectures of both versions are the same to each other and the learning process proceeds in the competitive style.

Let us explain briefly the learning process of LVQ as a supervised learning algorithm. The training examples which are labeled with their own categories are given, and the input vector dimension and the number of predefined categories are notated by d and C, respectively. The weight vector is notated by $\mathbf{w}_i = \begin{bmatrix} w_{i1} & w_{i2} & \cdots & w_{id} \end{bmatrix}$, $1 \leq i \leq |C|$ and the input vector is notated by $\mathbf{x} = \begin{bmatrix} x_1 & x_2 & \cdots & x_d \end{bmatrix}$. If the input vector is labeled with the category, C_k the weight vector is updated by Eq. (10.18),

$$\mathbf{w}_j(t+1) = \mathbf{w}_j(t) + \eta(\mathbf{x} - \mathbf{w}_j(t)) \tag{10.18}$$

The competitive nodes indicate clusters in the Kohonen Networks, whereas they do predefined categories in the LVQ.

We use the LVQ for the text categorization, rather than the text clustering, as a supervised learning algorithm. The sample labeled texts are given and encoded into numerical vectors. By the above process, the weights are optimized between the two layers. A novice text is encoded and its inner products with the weight vectors are computed. It is classified into the category which corresponds to the competitive node from which the inner product is maximal.

The LVQ may be modified into the semi-supervised version and the constraint clustering version by the mixture of the supervised and the unsupervised. In both kinds of learning, we use the labeled and unlabeled examples. The semi-supervised learning is intended for the classification and the regression by adding unlabeled examples, and the constraint clustering is intended for the clustering, by adding labeled examples as the constraints. To the labeled examples, the weights which are connected to the competitive node corresponding to the target label, and to the unlabeled ones, the weights which are connected to one corresponding to the winner, in the modified version. The number of competitive nodes is decided by the number of the predefined categories in this case.

10.4.3   Two-Dimensional Self-Organizing Map

This section is concerned with the expanded version of the Kohonen Networks which is called SOM (Self-Organizing Map). In Sect. 10.4.2, we studied the Kohonen Networks, where the competitive nodes correspond to the clusters. The competitive nodes are conceptually arranged as a two-dimensional grid in the SOM. The weights which are connected from the winner and its neighbors are updated in this version. In this section, we describe the SOM in detail with respect to its architecture and its learning process.

The SOM architecture is illustrated in Fig. 10.15. The input layer is identical to that in the Kohonen Networks; the number of nodes is given as the input vector dimension. The difference from the previous version is that the competitive layer is given as a two- dimensional grid; the number of nodes is given by multiplying the number of rows, $SizeX$, by the number of columns, $SizeY$. The number of weights

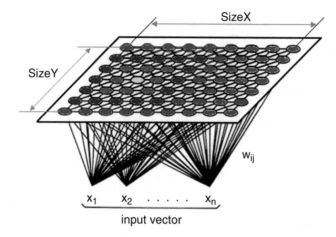

Fig. 10.15 Self Organizing Map Architecture from https://codesachin.wordpress.com/2015/11/28/self-organizing-maps-with-googles-tensorflow/

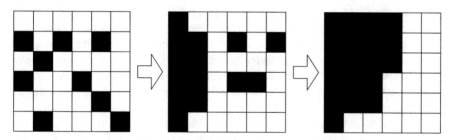

Fig. 10.16 Visualizing learning process of SOM

is computed by Eq. (10.19),

$$SizeX \times SizeY \times d \tag{10.19}$$

where d is the number of input nodes, and d dimensional weight vectors characterizes the individual nodes in the competitive layer. Another difference from the initial version is to update weights which connect with the winner and its neighbors.

In Fig. 10.16, the learning process of SOM is visualized and the learning algorithm is presented in the pseudocode, in Fig. 10.17. The two kinds of nodes are scattered over the grid in the initial stage, as shown in the left part of Fig. 10.16, and they are separated from each other in the final stage as shown in the right part. In the learning process, for each input vector, its cosine similarity with the weight vectors is computed, the weight vector with its maximal similarity nominated, and the competitive node which corresponds to the weight vector is selected as the winner. The weight vectors of the winner and its neighborhoods are updated with the degree which is defined by the theta function. The total number of competitive nodes is given arbitrary.

```
For each weight vector, wⱼ(0)
     wⱼ(0) ← random values
Iterate the following process from t=0 to t = M -1
     for each input vector, xᵢ
          maxIndex ← 0, maxSimilarity ← 0
          for each weight vector, wⱼ(t)
               if(maxSimilarity < cosineSimilarity(xᵢ , wⱼ(t)))
                    maxIndex ← j
                    maxSimilarity ← cosineSimilarity(xᵢ , wⱼ(t))
          for each weight vector, wⱼ(t)
               wⱼ(t + 1)← wⱼ(t) + θ(j, maxIndex)η(xᵢ - wⱼ(t))
```

Fig. 10.17 Learning process of SOM in Pseudocode

The text clustering system which is called WEBSOM was implemented by adopting the SOM as the approach by Kaski et al. in 1998. It used the 315 nodes in the input layer as 315 dimensional input vector which represents a text. The two versions of competitive nodes, called word category map, which correspond to words defined: 18×45 in the smaller version and 204×512 in the larger version. The values which are close to one are generated from competitive nodes which correspond to relevant words. The WEBSOM was intended for displaying relevant words as the fingerprint of the given text.

Let us compare the initial version of Kohonen Networks and the SOM with each other. The number of clusters in the initial version should be decided in advance, whereas it is not necessary in the SOM. The competitive nodes are given as clusters in the initial version, whereas they are given as a two-dimensional grid which represents a map in the SOM. In the initial version, the weights which are connected to the winner are updated, whereas in the SOM, the weights which are done to its neighbors as well as the winner, itself. The initial version of Kohonen Networks is intended for clustering data items, whereas the SOM is intended for visualizing data items by a map.

10.4.4   Neural Gas

This section is concerned with the Kohonen Networks variant, which is called Neural Gas. It was invented by Matinetz and Schulten in 1991 [67], based on the SOM which was covered in Section. In this version, the all weight vectors are updated, proportionally to the similarity with the input vector; the Neural Gas becomes a very smooth version of Kohonen Networks. As learning proceeds, the degree of updating weight vectors is decreased. In this section, we describe the Neural Gas as another variant of Kohonen Networks, with respect to its characteristics and learning process.

Let us explore some characteristics of Neural Gas as the differences from the Kohonen Networks. The weight vectors are adapted like the gas which is distributed from the input vector; that is the reason of calling the neural networks Neural Gas. In the initial version of Kohonen Networks, the only weight vector which is connected to the winner is updated, whereas in the Neural Gas, all weight vectors are updated anti-proportionally to the distance from the input vector. In the initial version of Kohonen Networks, as learning proceeds, the weight vectors are updated constantly, whereas in the Neural Gas, the degree of updating the weight vectors is decreased. The initial version of Kohohen Networks is characterized as a crisp clustering tool, whereas the neural gas is done as a fuzzy clustering tool.

Let us explain briefly the unsupervised learning process in the neural gas. We notate the number of input nodes, the number of competitive nodes, the weight vector, and the input vectors, by d, c, $\mathbf{w}_i = \begin{bmatrix} w_{i1} & w_{i2} & \cdots & w_{id} \end{bmatrix}$ $1 \leq i \leq c$, and $\mathbf{x} = \begin{bmatrix} x_1 & x_2 & \cdots & x_d \end{bmatrix}$, respectively. All the weight vectors are updated by Eq. (10.20),

$$\mathbf{w}_i = \mathbf{w}_i + \eta e^{\frac{\kappa}{\lambda}} (x - \mathbf{w}_i), 1 \leq i \leq c \tag{10.20}$$

where κ is the range or distance from the winner, and λ is the temperature. The degree of updating the weight vectors is decreased exponentially by the stance from the input vector by the term, $e^{\frac{\kappa}{\lambda}}$ in Eq. (10.20). As the learning proceeds, the degree of updating weight vectors is reduced as the temperature decreases, λ.

Let us mention the ART (Adaptive Resonance Theory) as the unsupervised neural networks which are alternative to the Kohonen Networks and their variants which were covered in this section. In 1987, Grossberg invented the ART which was intended to simulate the information process in the human brain [19]. Its architecture consists of the two layers, the input layer and the competitive layer, and the vigilance parameter is added between the layers. In the competitive layer, the winner is selected by the learning process which is identical to that of the Kohonen Networks, but it is validated by the vigilance parameter, unlike the Kohonen Networks. When the winner is refused by the validation, it is not participated in the next competition, and another node is selected as the winner.

10.5 Summary and Further Discussions

In this chapter, we described the clustering algorithms and the unsupervised neural networks such as the Kohonen Networks as the text clustering approaches. We mentioned the AHC algorithm, the single pass algorithm, the divisive algorithm, and the growing algorithm as the simple clustering algorithms. The k means algorithm and its variants are known as the most popular clustering algorithms based on the EM algorithm. The unsupervised neural networks, the Kohonen Networks and their variants were described as state-of-the-art approaches to the text clustering. In this section, we make some further discussions from what we study in this chapter.

The reinforced learning is mentioned as another type of learning which is separated from the supervised and the unsupervised. We used the supervised learning and the unsupervised learning for the text categorization and the text clustering as the approaches to them, respectively. The reinforced learning is the learning type where the reward or the penalty is given as the output and the action or the environment is given as the input. The action is decided to the reward in this type of learning. It is used for implementing an autonomous robot or a game agent.

We need to discriminate attributes in computing the similarity between data items. The KNN variant where different weights are assigned to attributes was mentioned in Chap. 6. When using the k means algorithm for clustering data items, the similarity between the two input vectors is computed by the weighted attributes. Another k means algorithm variant may be given as the discriminated weighted version in this case. The discriminations among attributes are applicable to other machine learning algorithms.

The k means algorithm and its variants which were mentioned in Sect. 10.3 belong to the EM algorithm class. The EM algorithm consists of the two steps: E-step and M-step, and each cluster is viewed as a popular statistical distribution. The E-step is the process of estimating memberships of each item to the clusters, and the M-step is the process of estimating parameters which are involved in the distribution. The above two steps are iterated until the parameters of distributions which represent clusters converge to fixed values. A specific version of EM algorithm is decided by the statistical distribution and the scheme of estimating memberships of each example.

Recently, deep learning becomes hot trends in the area of machine learning. The learning algorithms which were mentioned as the approaches to the text categorization and the text clustering in Chaps. 6 and 10 belong to the swallow learning. Deep learning is the machine learning paradigm where several steps are required for doing the classification, the regression, and the clustering; each step gets the input from its previous step and generates its output which becomes the input of the next step. Because each example is associated with only its final output in the supervised learning, the unsupervised learning is mainly applied in intermediate steps. In order to get the detailed explanation about deep learning, refer to the literature.

Chapter 11
Text Clustering: Implementation

This chapter is concerned with the implementation of text clustering system which is its prototype version. We present the architecture of text clustering system which we implement in this chapter, in Sect. 11.1, and define the classes which are involved in the system, in Sect. 11.2. In Sect. 11.3, we illustrate and explain the implementations of methods in the involved classes. We demonstrate the process of executing text clustering system in Sect. 11.4, and make the summarization and further discussions on this chapter, in Sect. 11.5. In this chapter, we implement and demonstrate the prototype version of text clustering system in order to provide the guide for implementing its real version.

11.1 System Architecture

In Fig. 11.1, we illustrate the text clustering system architecture which we implement in this chapter. The first step is to gather texts as clustering targets. The features are generated from the texts and they are encoded into numerical vectors in the next step. The numerical vectors which represent texts are clustered into subgroups by the AHC algorithm which is adopted for implementing the system. In this section, we describe modules which are involved in implementing the text clustering system in the functional view.

The text encoding module plays its role of encoding given texts into numerical vectors. The first step of executing the text clustering system is to gather texts to be clustered. The gathered texts are indexed into a list of words which are feature candidates and some among them are selected as features. Each text is encoded into a numerical vector whose attributes correspond to the selected features. The text encoding module was already developed in Chap. 7, and it is attached to the text clustering system in this chapter.

© Springer International Publishing AG, part of Springer Nature 2019
T. Jo, *Text Mining*, Studies in Big Data 45,
https://doi.org/10.1007/978-3-319-91815-0_11

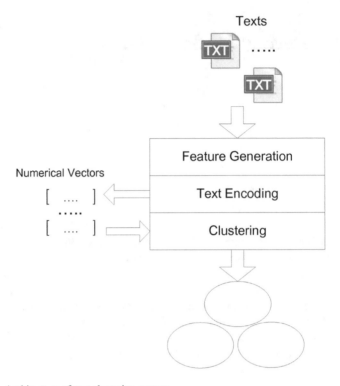

Fig. 11.1 Architecture of text clustering system

In text clustering module, the similarities between numerical vectors are computed after encoding texts into numerical vectors. A cluster pair is given as the input in the similarity computation module, a normalized value between zero and one is generated as the output. In this system, we adopted the cosine similarity as the similarity metric between examples, in implementing the text clustering system. Even if the process of computing the similarities is not displayed in the system architecture which is illustrated in Fig. 11.1, it exists in the text clustering module as the core operation. More similarity metrics will be added in implementing the subsequent version of this system.

The text clustering is executed in the text clustering module, depending on its tested similarity computation module. The AHC algorithm is adopted as the approach in implementing this text clustering system. It starts with skeletons with one item and builds clusters by merging pairs of their highest similarity. The similarity between clusters is computed by averaging over similarities of all possible pairs of elements. Merging of these clusters from individual items is continued until reaching number of clusters which is set as an external parameter.

Let us explain the process of executing the text clustering system which is developed in this chapter. Texts are collected as text files which are clustering targets. The collected texts are clustered into subgroups of content based similar ones. Clustered texts are displayed as a list of file names together with cluster boundary lines. The number of clusters is fixed to two in this system as the prototype and demo version.

11.2 Class Definitions

In this section, we present and explain the class definitions which are involved in implementing the text clustering system in Java. In Sect. 11.2.1, we review the classes which were defined in Chap. 7, and reused for implementing the system. In Sect. 11.2.2, we describe the class, "Cluster," with respect to its properties and methods. In Sect. 11.2.3, the interface, "ClusterAnalyzer," which is the frame for implementing various clustering algorithms. In Sect. 11.2.4, we explain the class, "AHCAlgorithm," which is the class specific to the interface, "ClusterAnalyzer."

11.2.1 Classes in Text Categorization System

This section is concerned with the four classes which are involved implementing the text clustering system as well as the text categorization system. In the previous chapters, we already defined and explained the classes. The four classes, "FileString," "Word," "Text," and "PlainText," are used for implementing the basic operations on texts which are necessary for the both systems. The class, "Text," was defined as an abstract class which inherits its properties and methods to its subclass, "PlainText." In this section, we review the four cases which were previously covered.

The class, "FileString," is defined in Fig. 11.2 for loading and saving a string from and to file. The properties are "fileName" which is the name of file which we try to access, and "fileString" which is the content of file which is given as a string. A file is created for loading a string with only its file name but it is created for saving it with not only its name but also the string which is the saving target. The methods are "saveFileString" which saves the string into the file, and "loadFileString" which loads the string from the file. This class is needed for implementing any application program, because any program requires file accesses.

The class, Word, is defined in Fig. 11.3. The properties of this class are "wordName" which indicates a word itself, and "wordFrequency" which indicates its frequency in a given text. An object of the class, Word, is created by initializing the property, "wordName," which is the symbolic identifier. The methods, "setWord-Frequency,' "getWordName," and "getWordFrequency," are ones for accessing and mutating the properties. The class, Word, is needed for implementing individual words as objects. In Fig. 11.4, the class, Text, is defined as an abstract class. The abstract class was defined for implementing the text categorization system in Chap. 7, and is reused for implementing the text clustering system. The properties of this class are "textFileName," "wordList," "fullText," "textLabel," and "featureVec-

```
package TextClusterAnalysis;
public class FileString {
    String fileName;
    String fileString;
    public FileString(String fileName){
        this.fileName = fileName;
        this.fileString = "";
    }

    public FileString(String fileName, String fileString){
        this.fileName = fileName;
        this.fileString = fileString;
    }
    public String getFileString(){
        return this.fileString;
    }
    public void setFileString(String fileString){
        this.fileString = fileString;
    }
    public void loadFileString(){}
    public void saveFileString(){}
}
```

Fig. 11.2 The class: FileClass

Fig. 11.3 The class: word

```
package TextClusterAnalysis;

public class Word {
    String wordName;
    int wordFrequency;
    public Word(String wordName){
        this.wordName = wordName;
        this.wordFrequency = 0;
    }

    void setWordFrequency(int wordFrequency){
        this.wordFrequency = wordFrequency;
    }

    String getWordName(){
        return this.wordName;
    }

    int getWordFrequency(){
        return this.wordFrequency;
    }
}
```

```
package TextClusterAnalysis;
import java.util.*;
public abstract class Text {
    String textFileName;
    Vector<Word> wordList;
    String fullText;
    String textLabel;
    double featureVector[];

    abstract void setTextLabel(String textLabel);
    abstract void setFullText(String fullText);
    abstract String getTextFileName();
    abstract String getTextLabel();
    abstract String getFullText();
    abstract void loadFullText();
    abstract void indexFullText();
    abstract void encodeFullText(int dimension);
    abstract double computeSimilarity(Text another);
}
```

Fig. 11.4 The class: text

tor." The methods are "loadFullText," "indexFullText," "encodeFullText," and "computeSimilarity." The differences of the abstract class from the interface are that the properties may be included and the methods may have their implementations.

In Fig. 11.4, the class, Text, is defined as an abstract class. The abstract class was defined for implementing the text categorization system in Chap. 7, and is reused for implementing the text clustering system. The properties of this class are "textFile-Name," "wordList," "fullText," "textLabel," and "featureVector." The methods are "loadFullText," "indexFullText," "encodeFullText," and "computeSimilarity." The differences of the abstract class from the interface are that the properties may be included and the methods may have their implementations.

The class, "PlainText," is defined in Fig. 11.5. The two classes, Text and "PlainText," have the inheritance relation. The class, Text, becomes the super class which provides the inheritance, and the class, "PlainText," becomes the subclass which receives it. By defining more classes which receive the inheritance from the abstract class, Text, it is possible to expand the version which supports more types of text, such as XML document, HTML one, and other types of documents. Implementations of method which are defined in the class, Text, are provided in its subclasses.

Fig. 11.5 The class:
PlainText

```
package TextClusterAnalysis;

public class PlainText extends Text {
    public PlainText(String textFileName){
        this.textFileName = textFileName;
    }
    void setTextLabel(String textLabel) {
        this.textLabel = textLabel;
    }
    void setFullText(String fullText){
        this.fullText = fullText;
    }
    String getTextFileName() {
        return this.textFileName;
    }
    String getTextLabel() {
        return this.textLabel;
    }
    String getFullText() {
        return this.fullText;
    }

    void loadFullText() {}
    void indexFullText() {}
    void encodeFullText(int dimension) {}
    double computeSimilarity(Text another) {
        return 0.0;
    }
}
```

11.2.2   Class: Cluster

This section is concerned with defining the class, Cluster, which is illustrated in
Fig. 11.6. The class is intended to define a group of texts as an object. An object of
this class is created with its cluster identifier as an initially empty cluster. As texts
are added, the cluster grows, and an individual text is accessed by its own index. So,
in this section, we explain the class, Cluster, which is defined in Fig. 11.6.

The properties of this class are "clusterID" and "textList." The former stands
for the cluster identifier which is given as a primary key value. The latter which
is declared as a dynamic array of objects of the class, "Text," stands for the list of
texts which belong to the current cluster. The class, "Text," is the super class of the
current one. In the next version, the property, "clusterName," which stands for its
symbolic cluster name which reflects its contents, will be added.

The methods which are involved in the class, Cluster, were already displayed in
Fig. 11.6. An object of class, Cluster is viewed as a list of texts and an individual text
is added to the property, "textList," by the method, "addTextItem." The information
about the cluster is accessed by the methods, "getClusterID," "getTextItem" and

```
public class Cluster {
    int clusterID;
    Vector<Text> textList;

    public Cluster(int clusterID){
        this.clusterID = clusterID;
        this.textList = new Vector<Text>();
    }

    public int getClusterID(){
        return clusterID;
    }

    void addTextItem(Text textItem){
        this.textList.addElement(textItem);
    }

    void addTextItem(String textFileName){
        Text textItem = new PlainText(textFileName);
        this.textList.addElement(textItem);
    }

    Text getTextItem(int index){
        return this.textList.elementAt(index);
    }

    Vector getTextList(){
        return this.textList;
    }

    int getSize(){
        return this.textList.size();
    }

    void mergeCluster(Cluster another){}

    double computeIntraClusterSimilarity(){return 0.0;}

    double computeInterClusterSimilarity(Cluster another){return 0.0;}

    double computeSimilarity(Cluster another){return 0.0;}
}
```

Fig. 11.6 The class: cluster

"getSize." The method, "mergeCluster," which takes an object of class, Cluster, as the argument, and adds the texts in the cluster to the property, "textList." The method which deletes a text item from the property, "textList," is not included because it is not used in this program.

Let us consider some potential methods which we add for upgrading the text clustering system in future. We add the method, "splitCluster" for splitting a cluster into several clusters. The method, "computeIntraClusterSimialrity," may be considered for computing the intra-cluster similarity which measures how much data items are coupled within a cluster. Accordingly we may add the method, "computeInterClusterSimilarity," for computing the inter-cluster similarity which

Fig. 11.7 The interface:
ClusterAnalyzer

```
public interface ClusterAnalyzer {
    void setTextList(Vector<Text> textList);
    Vector<Cluster> getClusterList();
    void clusterTextList();
}
```

measure how much data items between different clusters are discriminated. In upgrading the system, we consider the property, "clusterPrototype," and the method, "optimizeClusterPrototype," for optimizing the cluster characteristics.

Let us consider integrating the text clustering system with the text categorization system which was previously implemented in Chap. 7. It is the requirement for using the text categorization system to predefine the categories and allocate the sample texts, manually. The text clusters which are automatically generated by the text clustering system are given as a list of categories and allocated sample texts. A list of the cluster identifiers becomes a list of categories and the texts in each cluster become sample texts which are labeled with the cluster identifier. The integration of the two systems with each other is left as the subsequent task in future.

11.2.3   Interface: ClusterAnalyzer

The class, "ClusterAnalyzer" is defined as an interface, as shown in Fig. 11.7. The interface in Java programming is defined as list of methods without their implementations. The three methods, "setTextList," "getClusterList," and "clusterTextList," are defined in the interface. Their implementations are given in its specific class, "AHCAlgorithm," and more clustering algorithm may be added as specific classes in upgrading the system in future. In this section, we describe the interface as the programming concept, explain the methods which are defined in the interface, and compare the interface and the abstract class with each other.

Defining the class "ClusterAnalyzer" as an interface aims to add more clustering algorithm easily. The methods which are defined in the interface are implemented differently according to clustering algorithms. In the current version, we use only the AHC algorithm as the clustering algorithm. In subsequent versions, more clustering algorithms such as k means algorithm, k medoid algorithm, single pass algorithm, and Kohonen Networks will be added as the classes which are specific to the interface. The clustering algorithms will be treated as objects which are typed identically by the class, "ClusterAnalyzer," for implementing the combined model of multiple clustering algorithms.

The methods which are contained in the interface which is shown in Fig. 11.7 are defined as operations of individual objects of the class specific to the interface. The method, "setTextList," assigns a list of texts as clustering targets. The method, "getClusterList," gets a list of clusters of content based similar texts. The method, "clusterTextList," clusters a list of texts which is assigned by the method, "setTextList," by a clustering algorithm. The methods will be implemented in the classes which are specific to the interface.

Let us mention the abstract class as the alternative to the interface, with respect to the specification. In the relation between the interface and its specific classes, the methods are listed in the interface and their implementations are provided in the specific classes. The abstract class may have the both roles of the interface for making a list of methods and the super class for inheriting its properties and method to its subclasses. It is possible to define both properties and methods with and without their implementations in the abstract class. The implementations of the methods are given in defining a super class.

Let us mention some classes which we will add in upgrading the text clustering system in figure as ones specific to the interface, "ClusterAnalyzer." In the current version, we defined and implemented the class, "AHCAlgorithm." We add the class "KMeansAlgorithm" which indicates the k means algorithm which is covered in Sect. 10.3, as a specific class to the interface. It is possible to add the class, "KohonenNetworks," which corresponds to the unsupervised neural networks, Kohonen Networks, as such a kind of class. We upgrade the system continually by adding more different clustering algorithms as the classes which are specific to the interface.

11.2.4 Class: AHCAlgorithm

The class, "AHCAlgorithm," is defined as one which is specific to the interface, "ClusterAnalyzer," in Fig. 11.8. The methods which are defined in the interface are implemented in this class. Objects of this class are treated as ones which belong to the interface, "ClusterAnalyzer." The implementations of the involved methods are explained in Sect. 3.3. In this section, we explain the defined class with respect to its object creation and properties.

Let us consider the process of creating the AHC Algorithm as an object. The declaration of creating the object is stated as follows:

ClusterAnalyzer clusteringAlgorithm = new AHCAlgorithm(2);

The object which is created by the above statement is given as an object of the interface, "ClusterAnalyzer," rather than the specific class, "AHCAlgorithm." Its methods are executed following the implementations in the specific class. The above statement shows that the method implementations are separated from the object type.

In the class which is shown in Fig. 11.8, the properties, "clusterListSize," "textList," and "clusterList," are declared. The property, "clusterListSize," indicates the desired number of clusters as the condition for terminating the iterations. The property, "textList," is the list of texts which is a group of data items. The property, "clusterList," is the list of clusters which is resulted from clustering data items. The two properties, "clusterListSize" and 'textList, are initialized before proceeding the data clustering.

```
public class AHCAlgorithm implements ClusterAnalyzer {
    int clusterListSize;
    Vector<Text> textList;
    Vector<Cluster> clusterList;

    public AHCAlgorithm(int clusterListSize){
        this.clusterListSize = clusterListSize;
        this.clusterList = new Vector<Cluster>();
    }

    public void setTextList(Vector<Text> textList) {
        this.textList = textList;
    }

    public Vector<Cluster> getClusterList() {
        return this.clusterList;
    }

    public void initializeClusterList(){

    }

    public void clusterTextList() {

    }
}
```

Fig. 11.8 The class: AHCAlgorithm

The methods which are defined in the interface, "TextClusterAnalyzer" in Fig. 11.7, are defined also in the class, "AHCAlgorithm" in Fig. 11.8. The implementation of the method, "setTextList," is identical to any of specific class. The implementation of the method, "initializeClusterList," is different, depending on the clustering algorithm; clusters as many as data items are constructed in the AHC algorithm. The implementation of the method, "clusterTextList," is also different, depending on the clustering algorithm; it is explained in detail in Sect. 11.3. The three methods which are defined in the interface are used externally as the operations for proceeding the clustering, and the method, "initialClusterList," should be invoked in the method, "clusterTextList."

Let us consider the relations of the class, "AHCAlgorithm," with others. In this class, the methods which are defined in the interface, "TextClusterAnalyzer," are implemented. An object is created with the class, "TextClusterAnalyzer," and methods which are implemented in the class, "AHCAlgorithm," are invoked. The class, "AHCAlgorithm," provides the specific implementations of the methods for the interface, "TextClusterAnalyzer." We add more clustering algorithms as alternative clustering approaches to the AHC algorithm, by adding more classes which are specific to the interface, "TextClusterAnalyzer."

11.3   Method Implementations

This section is concerned with the method implementations which are involved in the classes. In Sect. 11.3.1, we review the methods in the previous classes. In Sect. 11.3.2, we explain the implementations of the methods which are defined in the class, Cluster. In Sect. 11.3.3, we describe the method implementations in the class, AHCAlgorithm. In this section, we explain the method implementations which are involved in implementing the text clustering system.

11.3.1   Methods in Previous Classes

This section is concerned with method implementation in the classes which were mentioned in the previous parts. The methods which are defined in the class, "FileString," are reviewed for accessing to secondary storages. The methods which are defined in the classes, "Text" and "Word," are mentioned. We explain the process of generating features by the method, "generateFeatureList." Therefore, in this section, we review the methods which are defined in the previous classes.

The implementations of the methods, "loadFileString" and "saveFileString," are illustrated in Fig. 11.9. The method, "loadFileString," loads contents from a file as a string, whereas the method, "saveFileString," saves a string into a file as its contents. For doing the both tasks, an object of the class, "RandomAccessFile," should be created in the both method implementations. The exceptional handling should be included for dealing with the failure in opening the file. One of the two methods is executed by creating an object of the class, "FileString," and invoking either of them.

Figure 11.10 illustrates the implementation of the method, "computeWordFrequency." If the argument is given as an empty string, the method does nothing. The first occurrence of the word which is given as the property of this class is positioned by invoking the method, "indexOf." If there is no occurrence of the argument, the method returns 0, and otherwise, it adds one to the value which is returned by the recursive call of this method with the updated one, the substring which starts after the first occurrence. The style of implementing the method by the recursive call is called recursive programming.

Figure 11.11 illustrates the implementation of the method, "encodeFullText." The list of features which are attributes of numerical vectors and the dimension are given as the arguments, and if the dimension and the number of features do not match, it terminates. The features are given as words and the frequencies of words are computed by the method, "computeWordFrequency." The numerical vector whose elements are frequencies of words is generated from this method. It represents the full text and is given as the property of this class.

```
public void loadFileString(){
    try{
        RandomAccessFile stream = new RandomAccessFile(this.fileName, "r");
        long length = stream.length();
        byte[] byArray = new byte[(int)length];
        stream.readFully(byArray);
        this.fileString = new String(byArray);
        stream.close();
    }catch (IOException e){
        System.out.println("There is the error in file processing!!" + e);
    }
}

public void saveFileString(){
    try{
        RandomAccessFile stream = new RandomAccessFile(this.fileName, "rw");
        stream.writeBytes(this.fileString);
        stream.close();
    }catch (IOException e){
        System.out.println("There is the error in file processing!!" + e);
    }
}
```

Fig. 11.9 The methods: saveFileString and loadFileString

```
int computeWordFrequency(String fullText){
    if(fullText.length() == 0){
        return 0;
    }
    int offset = fullText.indexOf(this.wordName);
    if(offset == -1){
        return 0;
    }
    return 1 + computeWordFrequency(fullText.substring(offset + 1));
}
```

Fig. 11.10 The methods: computeWordWeight

```
void encodeFullText(int dimension, Vector<String> featureList){
    int featureSize = featureList.size();
    if(dimension != featureSize){
        System.out.println("Mismatch between dimension and feature size!!");
        return;
    }
    this.featureVector = new int[dimension];
    for(int i = 0;i < dimension;i++){
        String featureName = featureList.elementAt(i);
        this.featureVector[i] = this.computeWordFrequency(featureName, this.fullText);
    }
}
```

Fig. 11.11 The methods: encodeFullText

```
Vector<String> generateFeatureList(int dimension){
    this.indexFullText();
    if(this.wordList == null)
        return null;
    int size = this.wordList.size();
    Vector<String> sortedWordNameList = new Vector<String>();
    for(int i = 0;i < dimension;i++){
        int maxFrequency = 0;
        int maxIndex = 0;
        for(int j = 0;j < size;j++){
            String wordItem = this.wordList.elementAt(j);
            int wordFrequency = this.computeWordFrequency(wordItem,this.fullText);
            if(maxFrequency < wordFrequency){
                maxFrequency = wordFrequency;
                maxIndex = j;
            }
        }
        String maxWordItem = this.wordList.elementAt(maxIndex);
        sortedWordNameList.addElement(maxWordItem);
        this.wordList.remove(maxIndex);
        size = this.wordList.size();
    }
    return sortedWordNameList;
}
```

Fig. 11.12 The methods: generateFeatureList

Figure 11.12 illustrates the implementation of the method, generateFeatureList. Externally, an object of class, Text, is created with the full text which is concatenated from texts in the corpus. The concatenated text is indexed into a list of words, and some words among them are selected as features. The criteria for selecting them is total frequency of each word in the concatenated text. This method returns the features which are attributes of numerical vector which represents a text.

11.3.2   Class: Cluster

This section is concerned with the implementations of methods which are defined in the class, "Cluster." The class, "Cluster," was defined in Fig. 11.6 with its properties, constructor, and methods. This section covers the methods, "computeSimilarity", "computeIntraClusterSimilarity," "computeInterClusterSimilarity," and "mergeCluster." The properties and the constructor were already explained in Sect. 11.2.2. In this section, we explain the method implementations, in detail.

In Fig. 11.13, we illustrate the implementation of the method, "computeWordSimilarity" which computes the similarity of the current cluster with another. Another object of the class, Cluster, is given as an argument. The similarities of all possible pairs between two clusters are computed as shown in the tested for loop. The average over the similarities is the similarity between the two clusters and returned from this method. Because the similarity between individual items is given as a normalized value, the similarity between the two clusters is also given as normalized value.

```
double computeSimilarity(Cluster another){
    int size1 = this.getSize();
    int size2 = another.getSize();
    double totalSimilarity = 0.0;
    for(int i = 0;i < size1; i++){
        Text textItem1 = this.getTextItem(i);
        for(int j = 0;j < size2;j++){
            Text textItem2 = another.getTextItem(j);
            double similarity = textItem1.computeSimilarity(textItem2);
            totalSimilarity = totalSimilarity + similarity;
        }
    }
    return totalSimilarity/(double)(size1 * size2);
}
```

Fig. 11.13 The methods: computeSimilarity

```
double computeIntraClusterSimilarity(){
    int size = this.getSize();
    double totalSimilarity = 0.0;
    for(int i = 0;i < size; i++){
        Text textItem1 = this.getTextItem(i);
        for(int j = i;j < size;j++){
            Text textItem2 = this.getTextItem(j);
            double similarity = textItem1.computeSimilarity(textItem2);
            totalSimilarity = totalSimilarity + similarity;
        }
    }
    return (2 * totalSimilarity)/(size * (size -1));
}
```

Fig. 11.14 The methods: computeClusterSimilarity

Figure 11.14 presents the method implementation, computeClusterSimilarity. The similarities of all possible pairs of item are computed. They are summed through the for-loop with the nested one. The average over the similarities computed by dividing the sum by half of product of the cluster size and cluster size minus one, and the average is returned as the intra-cluster similarity of the clustering results. If individual similarities are given as normalized values, the intra-cluster similarity is also given as a normalized value.

Figure 11.15 illustrates the implementation of the method, "computeInterClusterSimilarity." Another object of the class, Cluster, is the opposite one which is given as the method argument. The similarities between the current cluster elements and the opposite cluster ones are computed and summed. It is averaged by dividing the sum by the product of the two cluster sizes and returned as the inter cluster similarity between the two clusters. The inter-cluster similarity becomes the reverse of discriminations between them.

Figure 11.16 illustrates the implementation of the method, "mergeCluster." An object of class, Cluster, is given as an argument. Elements in the opposite cluster are added to this cluster. The results from executing the method is the cluster which consists of its exiting elements and the added ones from the opposite one. The method is needed for proceeding clustering by the AHC algorithm.

```
double computeInterClusterSimilarity(Cluster another){
    int size1 = this.getSize();
    int size2 = another.getSize();
    double totalSimilarity = 0.0;
    for(int i = 0;i < size1; i++){
        Text textItem1 = this.getTextItem(i);
        for(int j = 0;j < size2;j++){
            Text textItem2 = another.getTextItem(j);
            double similarity = textItem1.computeSimilarity(textItem2);
            totalSimilarity = totalSimilarity + similarity;
        }
    }
    return totalSimilarity/(size1 * size2);
}
```

Fig. 11.15 The methods: computeInterClusterSimilarity

Fig. 11.16 The methods:
mergeCluster

```
void mergeCluster(Cluster another){
    int size = another.getSize();
    for(int i = 0;i < size; i++){
        Text textItem = another.getTextItem(i);
        this.textList.addElement(textItem);
    }
}
```

```
public void initializeClusterList(){
    int size = this.textList.size();
    for(int i = 0;i < size;i++){
        Text textItem = (Text)this.textList.elementAt(i);
        Cluster clusterItem = new Cluster(i);
        clusterItem.addTextItem(textItem);
        this.clusterList.addElement(clusterItem);
    }
}
```

Fig. 11.17 The methods: mergeCluster

11.3.3   Class: AHC Algorithm

This section is concerned with method implementations which are involved in the class, "AHCAlgorithm." In implementing the text clustering system, we adopted the AHC algorithm which was covered in Sect. 10.2.1, as the approach. As shown in Fig. 11.7, the class, "ClusterAnalyzer," is defined as an interface, and the class, "AHCAlgorithm" is defined as a specific class to the interface. The methods in the interface, "ClusterAnalyzer," are implemented in this class. In this section, we explain the method implementations.

Figure 11.17 presents the implementation of the method, "initializeClusterList." This method initializes a list of clusters for executing the AHC algorithm. Objects of class, Cluster, as many as data items are created as skeletons. The list of clusters each of which has only one data item is the results from executing the method. This method is invoked in implementing the method, "clusterTextList."

```
public void clusterTextList() {
    this.initializeClusterList();
    while(this.clusterList.size() != 2){
        int clusterSize = this.clusterList.size();
        int maxIndex1 = 0;
        int maxIndex2 = 0;
        double maxSimilarity = 0.0;
        for(int i = 0;i < clusterSize;i++){
            Cluster clusterItem1 = (Cluster)this.clusterList.elementAt(i);
            for(int j = i+1 ;j < clusterSize;j++){
                Cluster clusterItem2 = (Cluster)this.clusterList.elementAt(j);
                double similarity = clusterItem1.computeSimilarity(clusterItem2);
                if(maxSimilarity < similarity){
                    maxSimilarity = similarity;
                    maxIndex1 = i;
                    maxIndex2 = j;
                }
            }
        }
        Cluster selectedCluster1 = (Cluster)this.clusterList.elementAt(maxIndex1);
        Cluster selectedCluster2 = (Cluster)this.clusterList.elementAt(maxIndex2);
        selectedCluster1.mergeCluster(selectedCluster2);
        this.clusterList.setElementAt(selectedCluster1,maxIndex1);
        this.clusterList.remove(maxIndex2);
    }
}
```

Fig. 11.18 The methods: clusterTextList

Figure 11.18 shows the implementation of the method, "clusterTextList." The clusters are initialized by invoking the method, "initializeClusterList." The similarities of all possible pairs of clusters are computed within the nested for-loop in the implementation. The two clusters whose similarity is maximum are merged into one cluster. The process is iterated until the number of clusters is decremented to two; the binary clustering is implemented as the demonstration version.

11.4   Class: ClusterAnalysisAPI

This section is concerned with the two final classes and the demonstrations and consists of four subsections. In Sect. 11.4.1, we explain the definition of class, "ClusterAnalysisAPI" and the involved method implementations. In Sect. 11.4.2, we explore the class, "ClusterAnalysisGUI," which is involved directly in implementing the system interface. In Sect. 11.4.3, we demonstrate the process of clustering texts by the developed system. In Sect. 11.4.4, we mention some points of upgrading the text clustering system.

```
public class TextClusterAnalysisAPI {
    Vector<Text> textList;
    Vector<Cluster> clusterList;
    Vector<String> featureList;

    int dimension;

    public TextClusterAnalysisAPI(int dimension){
        this.textList = new Vector<Text>();
        this.dimension = dimension;
    }

    public void setTextList(Vector<Text> textList){
        this.textList = textList;
    }

    public Vector<Cluster> getClusterList(){
        return this.clusterList;
    }

    public void addTextFileName(String textFileName){
        Text textItem = new PlainText(textFileName);
        this.textList.addElement(textItem);
    }

    public void loadTextList(){
        if(this.textList == null){
            System.out.println("Null Pointer!!");
            return;
        }
        int size = this.textList.size();
        for(int i = 0;i < size; i++){
            Text textItem = this.textList.elementAt(i);
            textItem.loadFullText();
        }
    }
}
```

```
    public void encodeTextList(){
        if(this.textList == null){
            System.out.println("Null Pointer!!");
            return;
        }
        int size = this.textList.size();
        for(int i = 0;i < size; i++){
            Text textItem = this.textList.elementAt(i);
            textItem.encodeFullText(this.dimension,this.featureList);
        }
    }

    public void clusterTextList(){
        System.out.println("TextClusterAnalysisAPI::clusterTextList();");
        ClusterAnalyzer textClassifier = new AgglerativeHierarhicalClustering(
        textClassifier.setTextList(this.textList);
        textClassifier.clusterTextList();
        this.clusterList = textClassifier.getClusterList();
        int size = this.clusterList.size();
        for(int i = 0; i < size; i++){
            Cluster clusterItem = this.clusterList.elementAt(i);
            int clusterSize = clusterItem.getSize();
            for(int j = 0;j < clusterSize;j++){
                Text textItem = clusterItem.getTextItem(j);
                String textFileName = textItem.getTextFileName();
                System.out.println(textFileName);
            }
            System.out.println();
        }
    }

    public void generateFeatureList(){
        if(this.textList == null){
            System.out.println("Null Pointer!!");
            return;
        }
        int size = this.textList.size();
        String integratedFullText = "";
        for(int i = 0;i < size; i++){
            Text textItem = this.textList.elementAt(i);
            integratedFullText = integratedFullText + textItem.getFullText();
        }
        Text integratedText = new PlainText("dummyFileName");
        integratedText.setFullText(integratedFullText);
        this.featureList = integratedText.generateFeatureList(this.dimension);
    }
}
```

Fig. 11.19 Class: ClusterAnalysisAPI

11.4.1   Class: ClusterAnalysisAPI

Figure 11.19 illustrates the class definition, "ClusterAnalysisAPI," in Fig. 11.19. A text list, a cluster list, and a feature list are given as the properties. An object of this class is created by taking dimension of numerical vectors which represent texts as the argument. The method implementations which are involved in executing the clustering are presented in Fig. 11.19. In this section, we explain the properties and the methods of this class.

Let us mention the properties of defining in the class, "textList," "clusterList," and "featureList." The property, "textList," is the list of texts which are given as the input. The property, "clusterList," means the list of clusters which are the results from clustering texts in the property, "textList." The property, "featureList," indicates the list of words which are selected as the features from the texts in the property, "textList." We use the property, "featureList," for encoding texts into numerical vectors.

Let us explain methods which are involved in executing text clustering. The method, "loadTextList," loads full texts from text files, and stores them as a list in the property, "textList." The method, "encodeTextList," encodes texts which are stored in the property, "textList," into numerical vectors. The method, "clusterTextList,"

```
public class TextClusterAnalysisGUI extends JFrame{
    Vector<Cluster> clusterList;
    Vector<String> textFileNameList;

    JLabel textLabel;
    JList textList;
    JButton addTextListButton;
    JButton clusterTextListButton;
    TextClusterAnalysisAPI textClusterAnalyzer;

    public TextClusterAnalysisGUI(){

        super("Testing Buttons");
        setLayout(new FlowLayout());
        this.clusterList = new Vector<Cluster>();
        this.textFileNameList = new Vector<String>();

        textClusterAnalyzer = new TextClusterAnalysisAPI(50);

        this.textLabel = new JLabel("Texts:");
        this.add(textLabel);
        this.textList = new JList();
        textList.setFixedCellHeight(30);
        textList.setFixedCellWidth(300);
        this.add(new JScrollPane(textList));

        this.addTextListButton = new JButton("Add Texts");
        this.add(addTextListButton);
        this.clusterTextListButton = new JButton("Cluster Texts");
        this.add(clusterTextListButton);

        ButtonHandler handler = new ButtonHandler();
        this.addTextListButton.addActionListener(handler);
        this.clusterTextListButton.addActionListener(handler);
    }
```

```
    private class ButtonHandler implements ActionListener{
        public void actionPerformed(ActionEvent event){
            String eventName = event.getActionCommand();
            System.out.println(eventName);

            if(eventName == "Add Texts"){
                JFileChooser fileChooser = new JFileChooser();
                fileChooser.setMultiSelectionEnabled(true);
                int result = fileChooser.showOpenDialog(TextClusterAnalysisGUI.this);
                File[] fileList = fileChooser.getSelectedFiles();
                String currentDirectory = fileChooser.getCurrentDirectory().getAbsolutePath();
                for(int i = 0; i < fileList.length; i++){
                    String fileName = currentDirectory + fileList[i].getName();
                    TextClusterAnalysisGUI.this.textFileNameList.addElement(fileName);
                    TextClusterAnalysisGUI.this.textClusterAnalyzer.addTextFileName(fileName);
                }
                TextClusterAnalysisGUI.this.textList.setListData(TextClusterAnalysisGUI.this.textFileNameList);
            }

            if(eventName == "Cluster Texts"){
                String resultStream = "";
                TextClusterAnalysisGUI.this.textClusterAnalyzer.loadTextList();
                TextClusterAnalysisGUI.this.textClusterAnalyzer.generateFeatureList();
                TextClusterAnalysisGUI.this.textClusterAnalyzer.encodeTextList();
                TextClusterAnalysisGUI.this.textClusterAnalyzer.clusterTextList();
                Vector<Cluster> clusterList = TextClusterAnalysisGUI.this.textClusterAnalyzer.getClusterList();
                int clusterSize = clusterList.size();

                for(int i = 0;i < clusterSize;i++){
                    Cluster clusterItem = clusterList.elementAt(i);
                    int textSize = clusterItem.getSize();
                    for(int j = 0; j < textSize;j++){
                        Text textItem = clusterItem.getTextItem(j);
                        String textFileName = textItem.getTextFileName();
                        resultStream = resultStream + textFileName + "\n";
                    }
                    resultStream = resultStream + "--------------------\n";
                }

                JOptionPane.showMessageDialog(TextClusterAnalysisGUI.this, resultStream);
            }
        }
    }
}
```

Fig. 11.20   Class: ClusterAnalysisGUI

clusters texts in the property, "textList," into subgroups, and stores them in the property, "clusterList." The method, "generateFeatureList," indexes the list of texts in the property, "textList," into the list of words and selects some among them as features.

Let us consider relations of this class, "ClusterAnalysisAPI," with other classes. The API class whose objects are created and methods are invoked in the main program is final to a console program. Objects of other classes are created and methods of other classes are invoked in this class. In the GUI (Graphic User Interface) programming, the objects of this class are created and methods are involved in the class, "ClusterAnalysisGUI." In Sect. 11.4.2, we mention the definition, properties, and methods of the class, "ClusterAnalysisGUI."

11.4.2   Class: ClusterAnalyzerGUI

This section is concerned with the class, "ClusterAnalysisGUI," for displaying and manipulating the graphic user interface, as defined in Fig. 11.20. The properties correspond to the components in the interface which is illustrated in Fig. 11.21. The constructor in the class, "ClusterAnalysisGUI," displays the graphic user interface. The class which is nested in "ClusterAnalysisGUI," "ButtonHandler," is for manipulating components in the interface. In this section, we explain the class which is defined in Fig. 11.21, with respect to the properties, methods, and the nested class.

The four components of interface and the object of the class, "TextCluster-AnalysisAPI," are given in the class, as the properties. In Fig. 11.21, the property,

Fig. 11.21 Class: interface
of text clustering system

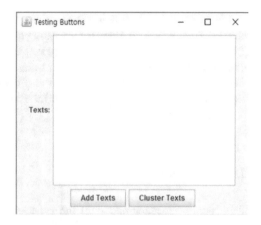

"textList," is the area for displaying a list of texts which is given as the input, and the property, "textLabel," is the label which is written following, "Texts:". The property, "addTextListButton," is given for initiating adding texts to the list as clustering targets. The property, "clusterTextListButton," is for initiating clustering texts into subgroups. The property, "textClusterAnalyser," is an object of the class, "TextClusterAnalysisAPI," for executing the text clustering.

The method in the class, "TextClusterAnalysisGUI," is the only constructor, so we explain its implementation. The memory is allocated to objects of the classes, "clusterList" and "textFileNameList." The object of the class, 'TextClusterAnalyzerAPI, is created for executing the text clustering. The components which are shown in Fig. 11.21, are configured. In order to implement the event driven programming, the two buttons of adding texts and executing the text clustering are added to the action listener, by invoking the method, "addActionListener."

The class, "ButtonHandler," is nested in the class, "TextClusterAnalyzerGUI," and it has the implementation of the method, actionPerformed'. The objects of the class, "JButton," are created with the component names in the constructor of class, "TextClusterAnalyzerGUI," as follows:

$$\text{this.addTextListButton} = \text{new JButton("Add Texts");}$$

$$\text{this.clusterTextListButton} = \text{new JButton("Cluster Texts");}$$

The two conditional statements in the implementation exist for responding to pushing one of the two buttons. If the button, "AddTexts," is pushed, the window is opened for choosing files and the files which we select are stored as clustering targets. If the button, "Cluser" Texts, is pushed, the selected texts are clustered and the text clusters are displayed as a message box. We need to register the two buttons as a listener by invoking "addActionListener," in order to execute the corresponding actions in pushing the button.

Let us consider the relation of the class, "TextClusterAnalyzerGUI," with other classes. This class is intended for activating the graphic user interface for performing the text clustering. In the main program, the object of this class is created and methods which are implemented in this class. In this class, the object of the class, "TextClusterAnalyzerAPI," is created and the methods in it are invoked. In this class, as a GUI class, the graphic user interface which is related with the API class is implemented.

11.4.3   Demonstration

This section is concerned with the process of demonstrating the text clustering system which is developed in this chapter. The left in the interface is intended for adding texts through the file chooser window. The file names which are selected by a user are displayed in the area whose label is "Text:". Texts are encoded into numerical vectors and the text clustering is executed by pressing the right button. In this section, we demonstrate the text clustering system, explaining the detail process.

In Fig. 11.22, the process of gathering texts is demonstrated. The system has no text in the initial stage. The button, "add-text" is pressed and the file chooser is popped up. The files are selected through the file chooser and the left button is pressed in it. The files are displayed in the text area, as shown in Fig. 11.23.

The text clustering is executed by pressing the button, "Cluster-Texts." The features are generated from texts which are selected by the above process. Texts are encoded into numerical vectors. They are clustered into subgroups by the AHC algorithm. The clusters which are most similar with each other are merged until the number of clusters reaches two.

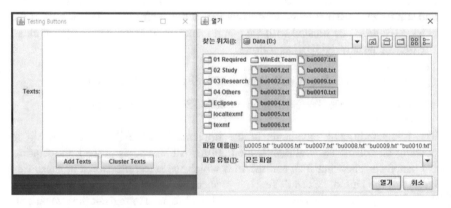

Fig. 11.22 Gathering texts

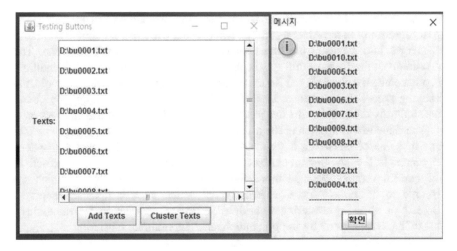

Fig. 11.23 Selecting files

The results from demonstrating the text clustering system are illustrated in Fig. 11.23. In the text area, texts which are selected in Fig. 11.22 are displayed. The results from clustering the selected texts are displayed in a message box in the right part in Fig. 11.23. The dashed line which is shown in the message box between text identifiers means the boundary between clusters. The ten texts are selected and they are clustered into two clusters: one has 8 texts and the other has 2 texts in this demonstration.

Let us point out some limits in the current version of text clustering system as the directions for upgrading it. The system deals with only plain texts which are given as text files whose extension is "txt." Only AHC algorithm is used as the text clustering approach, so we need to install more approaches in the system. The results from clustering texts are displayed in the message box as a textual form as shown in Fig. 11.23. The scope of clustering types is restricted to crisp and flat clustering in this system.

11.4.4   System Upgrading

This section is concerned with the directions for upgrading the current version of the text clustering system. In Sect. 11.4.3, we demonstrated the text clustering system which is a prototype but not a real version. In the current version, only AHC algorithm is adopted as the approach, in spite that other approaches are available, and it supports only the plain texts. We need to upgrade the text clustering system by adding more clustering algorithms and supporting more kinds of texts such as XML documents. In this section, we point out limits of the current version and present directions for upgrading the system.

The number of clusters is set as the termination condition of executing the AHC algorithm. In the AHC algorithm, two clusters with their maximum similarity is merges into a cluster and the number of clusters is decreased by one. The maximum similarity threshold may be used as the alternative termination condition, in proceeding the clustering. If the maximum similarity is less than the threshold, the merging process is terminated. It allows the variable number of clusters by giving the similarity threshold as the hyper parameter, instead of the number of clusters.

The direction of upgrading the text clustering system is to allow users to select one or some of clustering algorithms by the menu. The AHC algorithm whose argument is the number of clusters is used as the text clustering approach in the current version. We add more cluster algorithms as the classes which are specific to the class, "TextClusterAnalyzer," which is defined as an interface. Objects of different clustering algorithms are treated as ones in the same class, "TextClusterAnalyzer," by the polymorphism. The multiple clustering algorithms may be combined for clustering texts and clustering results are integrated with each other; this may be considered as the direction of upgrading the system.

Another direction of upgrading the system is to allow it to process various formats of texts. In the current version, only plain texts are processed for clustering them. In the next version, the system is allowed to cluster web documents whose formats are HTML or XML, as well as plain texts. More text formats are added as the subclasses which are derived from the class, Text, defined as abstract class. The XML is the standard document format where tags are defined based on the DTD (Document Type Definition) file, and the classes for processing them are available as Java Class Library.

The text clustering may be developed as a module which is attached to another system, as well as an independent program. Even if this program is only a prototype version, the text clustering system was implemented as an independent program, in this chapter. In 2006, Jo combined the text categorization and the text clustering into the automatic management tool [25]. The text clustering may be attached to the information retrieval system for displaying texts which are relevant to the query as clusters. We consider the two destinations of upgrading the prototype program: an independent commercial program and a module to another program.

11.5   Summary and Further Discussions

In this chapter, we presented the Java source code in implementing the text clustering system. We defined the classes and the interface which are involved in implementing the program. We explained the method implementations which are included in the classes. We demonstrated the process of clustering texts by the system and mentioned the directions for upgrading the system. In this section, we make some further discussions from what we studied in this chapter.

There is the possibility of adding other programs as modules to the system. Cluster naming for identifying clusters symbolically may be added to the system. The text categorization system which was developed in Chap. 7 may be modified into a module and added to the system for arranging texts which are added subsequently. The text summarization which will be covered in Chap. 13 is added to the program for clustering texts by their summaries, rather than their full texts for improving the clustering speed. The text segmentation may be added to the program for clustering subtexts like independent full texts.

We need to add more clustering algorithms as the approaches to the system in upgrading the system. The single pass algorithm, the divisive algorithm, and the growing algorithm which were covered in Sect. 6.1 may be added by defining the classes which are specific to the interface, "TextClusterAnalyzer." The k means algorithm and its variants are added as the most popular approaches. The Kohonen Networks and SOM are considered as neural based approaches. In implementing the text clustering system, we consider the combination schemes of multiple approaches, such as voting, gate mixture, and adaboost.

The class, "TextClusterAnalyzer," may be defined as an abstract class, instead of the interface. In this chapter, the class, "TextClusterAnalyzer," is defined as the interface, and the cluster algorithms will be defined as the classes which are specific to the interface. The class, "TextClusterAnalyzer," is defined as an abstract class, and the classes which indicate clustering algorithms are defined as the subclasses which inherit properties and methods from the abstract class. The construct implementation is the difference of the abstract class from the interface. The case where methods implemented in the abstract class, but not implemented in its subclasses may be possible.

The results from clustering texts are displayed as the textual forms in the message box, in the current version. We need to customize the graphic user interface for visualizing text clusters more graphically to enable the browsing. The text clustering is displayed initially, and when a user clicks a cluster, it displays a list of texts which belong to the cluster, together with their titles. If clicking a particular text, its information and full text are displayed. Therefore, we may consider the advanced interface in upgrading the current version.

Chapter 12
Text Clustering: Evaluation

This chapter is concerned with the schemes of evaluating text clustering systems or approaches. We introduce the basis for evaluating the clustering performances, in Sect. 12.1, and mention the three types of evaluation schemes in Sect. 12.2. In Sect. 12.3, we describe the clustering index which was proposed by Jo and the process of evaluating results from the fuzzy and hierarchical clustering as well as the simple clustering. We cover how to install the parameter tuning in the clustering algorithms in Sect. 12.4, and make the summary and further discussions on this chapter, in Sect. 12.5. In this chapter, we explain the schemes of evaluating text clustering systems and mention the clustering algorithms with the parameter tuning.

12.1 Introduction

Before mentioning the specific evaluation schemes, we consider the directions, the outlines, and the policies of evaluating clustering results. Because the target labels of examples are not available, it is not easy to evaluate clustering results. Even if using the labeled examples for evaluating them, there are many cases of matching clusters with the target categories. The direction of evaluating clustering results is to maximize the similarities of examples within each cluster and to minimize the ones of examples between different clusters. In this section, we present the outline of evaluating clustering results distinguished from doing classification results.

It is more difficult and complicated to evaluate the text clustering performance than the text categorization performance. The F1 measure or the accuracy depends on how to match the target labels with clusters in using the labeled texts in the test collection. The results from evaluating the text clustering depend on schemes of computing similarities among texts, when using unlabeled texts. Various schemes of

T. Jo, *Text Mining*, Studies in Big Data 45,
https://doi.org/10.1007/978-3-319-91815-0_12

evaluating results from clustering texts were previously proposed, but no standard scheme is not available. No desired number of clusters is usually available in evaluating the clustering results.

Even if no standard evaluation measure is available, the direction of implementing the text clustering systems is available. The similarities among items within each cluster which are called cohesions or intra-cluster similarities should be maximized. The similarities among clusters which are called discriminations or inter-cluster similarities should be minimized. In other words, we must avoid both the wrong partition which means that a desirable cluster is partitioned into several clusters and the wrong merge which means that really discriminated clustering is merged into a cluster. In previous literatures, various evaluation metrics of clustering systems have been proposed [7].

Let us mention the three views of evaluating clustering results. The external view is mentioned as the view where labeled examples are used for evaluating the clustering results. The internal view may be considered as the view where unlabeled examples are used for evaluating the results depending on similarities of individual items. The relative view exists as the view where alternative approaches are evaluated based on results from a particular one. The views of evaluating clustering results will be described in Sect. 12.2.2, in detail.

Let us explain briefly the process of evaluating results from clustering data items. It is assumed that labeled examples are prepared as a test collection. The examples are clustered into subgroups as many as the number of predefined categories. We compute the intra-cluster similarities as many as clusters and the inter-cluster similarities of all possible pairs of clusters and average over them. As a single metric integrating the both metrics, in 2006, Jo proposed the clustering index, and we describe it in detail, in Sect. 12.2.3, [25].

12.2   Cluster Validations

This section is concerned with the measures and the frames of evaluating clustering results. In Sect. 12.2.1, we mention the measures which are involved in the clustering evaluation, and present the desired clustering results. In Sect. 12.2.2, we describe the internal evaluation process where unlabeled examples are used for the clustering evaluation. In Sect. 12.2.3, we mention the relative one where clustering results are evaluated based on the pivoted ones. In Sect. 12.2.4, we explain the external evaluation process which is opposite to the internal one.

12.2.1   Intra-Cluster and Inter-Cluster Similarities

This section is concerned with the basic measures which are involved in evaluating clustering results. It is assumed that the data items are encoded into numerical vectors, and the similarity metric between them is defined. There are two kinds of similarities for evaluating clustering results: intra-cluster similarity and inter-cluster similarity. The direction of clustering data items is to maximize the intra-cluster similarity and minimize the inter-cluster similarity. In this section, we describe the scheme of computing the both similarities and the evaluation policy.

The first step of evaluating clustering results is to prepare a text collection which is the clustering target. When labels of texts are not available in the test collection, we use the evaluation measures which are defined based on the internal validation and the relative one. If the labels are available, the evaluation of clustering results follows the external validation. The labels are hidden during clustering items, but are presented in the evaluation step. The number of clusters is set by the number of target categories in the external validation, arbitrary in the internal validation, and by the number of clusters resulted from the pivot approach in the relative one.

Let us mention the process of computing the intra-cluster similarity for each cluster. It is assumed that a list of individual items within the cluster, C_i, is given as a set of numerical vectors, $\{\mathbf{x}_{i1}, \mathbf{x}_{i2}, \ldots, \mathbf{x}_{i|C_i|}\}$, and the similarity between the two vectors, $\mathbf{x}_{ik}$ and $\mathbf{x}_{im}$ is notated by $sim(\mathbf{x}_{ik}, \mathbf{x}_{im})$. The intra-cluster similarity to the cluster, C_i is computed by Eq. (12.1),

$$intrasim(C_i) = \frac{2}{|C_i|(|C_i| - 1)} \sum_{k=1}^{|C_i|} \sum_{m=k}^{|C_i|} sim(\mathbf{x}_{ik}, \mathbf{x}_{im}) \tag{12.1}$$

where $|C_i|$ is the cardinality of the cluster, C_i. The intra-cluster similarity to the entire clusters, $C = \{C_1, C_2, \ldots, C_{|C|}\}$ is computed by averaging the intra-cluster similarities of the clusters, as expressed in Eq. (12.2),

$$intrasim(C) = \frac{1}{|C|} \sum_{i=1}^{|C|} intrasim(C_i) \tag{12.2}$$

where $|C|$ is the number of the entire clusters. When a too large number of small sized clusters is given as results, they are overestimated by the intra-cluster similarity.

We need one more measure which is called inter-cluster similarity for evaluating clustering results. The results from clustering data items given as a set of clusters, $C = \{C_1, C_2, \ldots, C_{|C|}\}$, and notate the two clusters, C_i and C_j, as follows:

$$C_i = \{\mathbf{x}_{i1}, \mathbf{x}_{i2}, \ldots, \mathbf{x}_{i|C_i|}\}$$
$$C_j = \{\mathbf{x}_{j1}, \mathbf{x}_{j2}, \ldots, \mathbf{x}_{j|C_j|}\}$$

The inter-cluster similarity between the two clusters, C_i and C_j is computed by Eq. (12.3),

$$intersim(C_i, C_j) = \frac{1}{|C_i| \times |C_j|} \sum_{k=1}^{|C_i|} \sum_{m=1}^{|C_j|} sim(\mathbf{x}_{ik}, \mathbf{x}_{im}) \tag{12.3}$$

The inter-cluster similarity which reflects the entire results is computed by Eq. (12.4)

$$intersim(C) = \frac{2}{|C|(|C|-1)} \sum_{i=1}^{|C|} \sum_{j=i}^{|C|} intersim(C_i, C_j) \qquad (12.4)$$

The intra-cluster similarity and the inter-cluster similarity are computed, avoiding the two extreme cases: a single group of all data items and single skeletons as many as data items.

Let us consider the ideal clustering results. Each cluster is composed with items which are as similar as possible with each other, toward maximizing the intra-cluster similarity. The items in a cluster should be discriminated with the ones in the other clusters as strong as possible, toward minimizing the inter-cluster similarity. Texts of each target label should be same to ones in its corresponding cluster as the most desirable clustering results in using labeled examples for evaluating the clustering results. In the subsequent subsection, we explore the types of evaluating them, and describe the clustering index which is based on the above metrics in detail in Sect. 12.3.

12.2.2   Internal Validation

This section is concerned with the paradigm of evaluating clustering results, which is called internal validation. It is referred to the style of evaluating clustering results depending on the similarities among texts. Unlabeled texts are prepared and a similarity metric between texts is defined. The evaluation metric may be defined based on the intra-cluster similarity and the inter-cluster similarity without any external information. In this section, we explain and demonstrate the paradigm with a simple example.

Figure 12.1 illustrates the raw texts and their representations as what are prepared for the clustering evaluation. The assumption underlying in this kind of evaluation is that no external information such as target labels of texts is added. The raw texts are encoded into numerical vectors by the process which were described in Chaps. 2 and 3. The similarity metric between raw texts or their representations is defined as the preliminary task. The clustering results may be evaluated differently depending on how to define the similarity metric.

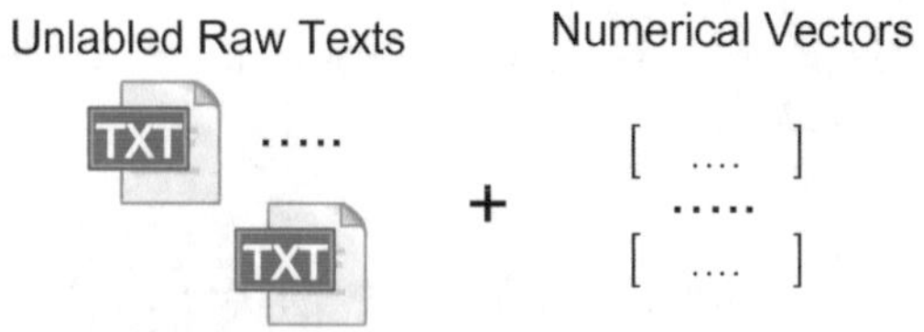

Fig. 12.1 Raw texts and text representations

Fig. 12.2 Example for
computing intra-cluster
similarity and inter-cluster
similarity

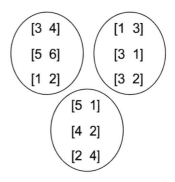

The process of evaluating the clustering results, following the paradigm, was described in [7]. The clusters are generated by a clustering algorithm. The intra-cluster similarities and the inter-cluster similarities are computed, corresponding to clusters and cluster pairs, respectively. By averaging them, the intra-cluster similarity and the inter-cluster similarity which represent the entire results are computed. The evaluation results are different, depending on whether we use the similarity between text representations or between raw texts.

The intra-cluster similarity and the inter-cluster similarity are computed in the example which is illustrated in Fig. 12.2. The intra-cluster similarities of the three clusters are computed as 0.9819, 0.8140, and 0.7963, respectively, and they are averaged as 0.8640 as the intra-cluster similarity to the entire results. The inter-cluster similarities of the three pairs are computed as 0.8910, 0.8561, and 0.8519, and they are averaged over 0.8693. If both measures are close to 1.0, the cluster cohesions are very good but the discriminations among clusters are poor. If two clusters have their high both measures, they should be merged with each other.

Let us mention some existing evaluation measures based on the internal validation and refer to [7] for its detail description. The simplest evaluation measure is Dunn's index which is based on the ratio of minimum distance between clusters to the largest cluster size. The silhouette index is the average over instances of clusters. The Hubert's correlation is computed from the similarity matrix where each column corresponds to a cluster, each row corresponds to an input vector, and each entry indicates the similarity between a cluster and an input vector. The evaluation results depend on the schemes of computing the similarity between clusters in the three evaluation metrics.

12.2.3 Relative Validation

In this section, we mention another paradigm for evaluating the clustering results, which is called relative validation. In advance, we decide most desirable clustering results. Cluster algorithms are evaluated by comparing their results with the most desirable one. Instead of absolute values, the clustering algorithms are evaluated by how much results from clustering data items are close to the desirable one. In this section, we describe the type of clustering evaluation in detail.

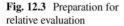

Fig. 12.3 Preparation for relative evaluation

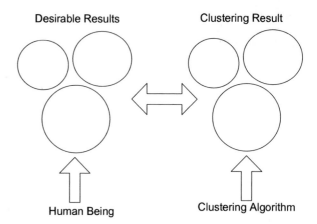

Figure 12.3 illustrates the preparation of data items for doing the relative validation. The desirable clusters of data items are constructed by subjective under the assumption that they are unlabeled. We gather results from clustering data items by clustering algorithms. By comparing the desirable results and generated ones with each other, the clustering algorithms are evaluated. In reality, no completely desirable clustering results do not exist and strong subjective bias always exist in constructing the desirable results.

Let us mention the frame of evaluating clustering results based on the relative validation, rather than a specific process. In the evaluation paradigm, it is assumed that the number of clusters is fixed in both the desired results and the generated ones. We construct the contingency table where each column corresponds to a cluster in the desired results and each row does to one in the generated results. We make the summation over values in the diagonal positions in the table, and the ratio of summation of diagonal elements to the summation of all entries becomes the evaluation metric. This type of evaluation scheme is characterized as the strong dependency on subjectivity of building the desired clustering results.

Figure 12.4 illustrates a simple example for calculating the evaluation measure which was mentioned above. The three clusters of the left part of Fig. 12.4 are given as the actual results from clustering data items and those of the right part of Fig. 12.4 are given as the desired ones which are defined in advance. The actual results and the desired ones are compared with each other; for example, the two items, and, belong to the both groups, and the two clusters in the both parts have one shared items. The total number of items is nine as shown in Fig. 12.4, and four items among them are shared by both sides. The evaluation metric is computed as 0.4444, by the relative validation.

In 2006, the relative validity was mentioned as the evaluation paradigm which is the comparative evaluation of clustering algorithms with the base one, by Brun et al. [7]. A new clustering evaluation measure which is based on the relative validity was proposed by Vendramin et al. in 2009 [94]. However, Halkidi et al. did not mention the relative validity among the three kinds of evaluation paradigms [21].

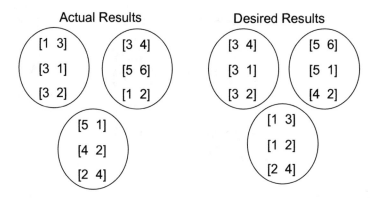

Fig. 12.4 Example of relative evaluation

The evaluation metrics which are based on the internal validity and the external validity are used more frequently than this type. In 2007, Jo and Lee proposed the evaluation measure which is based on the external validity, called clustering index [44].

12.2.4 External Validation

This section is concerned with the third evaluation paradigm which is called external validation. It is the style of evaluating clustering results by adding external information. The labeled examples are prepared as the test collection, their labels are hidden during clustering, and the similarity between items is based on their label consistencies during the evaluation. The external validation should be understood for studying the clustering index which is covered in Sect. 12.3. In this section, we describe the external validation as the frame of evaluating the clustering results.

Figure 12.5 illustrates that the labeled texts are prepared for evaluating the clustering results based on the external validation. The group of labeled texts is almost identical what is prepared for the relative validation which was shown in Fig. 12.3. The difference between the external validation and the relative validation is that in the relative validation, no label is initially given in the text collection, whereas the labels are given explicitly in the external validation. We use the standard text collections which were mentioned in Sect. 8.2 for evaluating the clustering results. However, there are two conditions for the relative validation: the collection which consists of unlabeled text and the desirable clusters which are made by subjectivity.

The labeled examples are used for evaluating the clustering results under this evaluation paradigm. The labels are hidden while data items are clustered and clusters are generated independently from their labels by a clustering algorithm. We compute the intra-cluster similarity and the inter-cluster similarity based on their

Fig. 12.5 Preparation for external evaluation

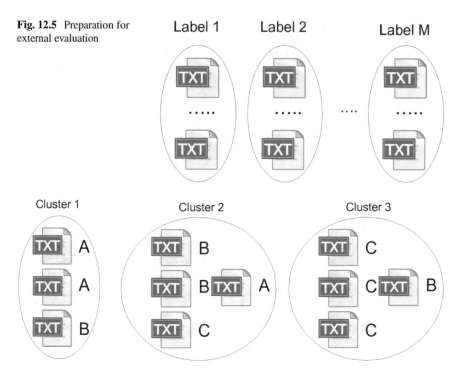

Fig. 12.6 Example of external evaluation

target labels; the similarity between data items is given as a binary value: zero as a similarity between differently labeled items and one as one between identically labeled ones. The final evaluation measure is computed by combining the two measures with each other. The difference from the internal validation is to use target label or other external information about data items for the evaluations.

A simple example is presented in Fig. 12.6, in order to demonstrate the external validation. Each text is labeled independently of its own cluster. The inter-cluster similarity and the intra-cluster similarity are computed by their labels from the clustering results, instead of their cosine similarities. The intra-cluster similarity of the left cluster is 0.3333, and the inter-cluster similarity between the left one and the middle one is 0.3333. The labels of texts are initially given as the external in formation in this type of evaluation paradigm.

Let us mention some evaluation metrics which are based on the external validation in [7]. As a metric, we may mention Hubert's correlation which computes the correlation between labels and clusters. As another metric, the Rand statistics is the position of vector pairs which agree in both labels and clusters to the total vector pairs. The Jaccard coefficient is the porition of vectors which belong to same clusters and labels to identically labeled ones, as one more evaluation metric. The Folks and Mallow index is the geometric version of the rand statistics.

12.3 Clustering Index

This section is concerned with the clustering index which was initially proposed in 2006 [25]. In Sect. 12.3.1, we describe the process of computing the clustering index. In Sect. 12.3.2, we explain the process of evaluating the crisp clustering results using the clustering index. In Sect. 12.3.3, we cover the scheme of evaluating the fuzzy clustering results. In Sect. 12.3.4, we consider the case of evaluating the hierarchical clustering results using the clustering index.

12.3.1 Computation Process

This section is concerned with the process of computing the clustering index which was initially proposed in 2007 [44]. The metric is based on the external validity which is mentioned in Sect. 12.2.4 and where labeled texts should be used. The intra-cluster similarity and the inter-cluster similarity are computed based on target labels of data items. The two measures are combined into a single metric which is called clustering index, like the case in F1 measure. In this section, we describe the process of computing the clustering index from clustering results.

The intra-cluster similarity is computed from the clustering results, and it is assumed that the labeled texts are used, following the external validity. The similarity between two texts is computed by their target labels, as expressed, in Eq. (12.5),

$$sim(\mathbf{x}_{ik}, \mathbf{x}_{ik}) = \begin{cases} 1 & \text{if their labels are same} \\ 0 & \text{otherwise} \end{cases} \qquad (12.5)$$

The intra-cluster similarity of the cluster, C_i, is computed by Eq. (12.1) which is mentioned in Sect. 12.1. The intra-cluster similarity of the entire clustering results, $C = \{C_1, C_2, \ldots, C_{|C|}\}$, is computed by Eq. (12.2), by averaging the intra-cluster similarities. The intra-cluster similarity is one of the two metrics for evaluating the clustering results.

Let us consider the alternative measure, the inter-cluster similarity, to the intra-cluster similarity. The data clustering has two requirements: maximization of intra-cluster similarity and minimization of inter-cluster similarity. The similarity between two data items is computed based on their target labels by Eq. (12.5), and inter-cluster similarity between two clusters is computed by Eq. (12.3). The inter-cluster similarity to the entire clustering results is computed by Eq. (12.4). The maximal discrimination among clusters means the minimized inter-cluster similarity to the clusters.

The two measures which are computed by the above process, the intra-cluster similarity and the inter-cluster similarity, are integrated into a single metric. The inter-cluster similarity is reversed as $1.0 -$ inter_cluster similarity. They are integrated into Eq. (12.6), called clustering index, following the style of integrating the precision and the recall into the F1 measure,

$$CI = \frac{2 \cdot \text{intra\_cluster similarity} \cdot (1.0 - \text{inter\_cluster similarity})}{\text{intra\_cluster similarity} + (1.0 - \text{inter\_cluster similarity})} \qquad (12.6)$$

The clustering index which is shown in Eq. (12.6) is proportional to the intra-cluster similarity, but anti-proportional to the inter-cluster similarity. The inverse of the inter-cluster similarity is called discrimination among clusters.

The clustering index was defined in Eq. (12.6) as the metric for evaluating clustering results. Because the clustering index is defined so, following the external validity, the preparation of labeled examples is required for evaluating any text clustering algorithm. The inter-cluster similarity is replaced by the discriminality, so Eq. (12.6) is modified into Eq. (12.7),

$$CI = \frac{2 \cdot \text{intra\_cluster similarity} \cdot \text{discriminality}}{\text{intra\_cluster similarity} + \text{discriminality}} \qquad (12.7)$$

The two measures, the intra-cluster similarity and the discriminality, correspond to the recall and the precision which are involved in the F1 measure. By computing the two metrics based on the cosine similarity, instead of target labels, we may use the clustering index for tuning the clustering results.

12.3.2 Evaluation of Crisp Clustering

This section is concerned with the process of evaluating the crisp clustering results by the clustering index. In Sect. 12.3.1, we studied the process of computing the clustering index. The simple clustering results which were illustrated in Fig. 12.7 were evaluated by the clustering index. In the subsequent sections, we use the clustering index for evaluating more complicated clustering results. In this section, we demonstrate the process of evaluating the crisp clustering results using the clustering index.

Fig. 12.7 Results from crisp clustering

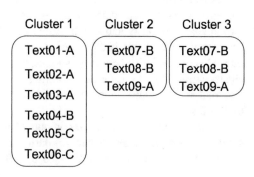

Figure 12.7 presents a simple example of clustering results. A, B, and C which are associated by texts are target labels which are initially given in the test collection. The intra-cluster similarities of the three clusters are computed as 0.2667, 0.3333, and 0.3333, respectively. The inter-cluster similarities of cluster pairs: cluster 1 and 2, cluster 1 and 3, and cluster 2 and 3 are computed as 0.2777, 0.2777, and 0.5556, respectively. The clustering index is computed as 0.4163, from the both values, the averaged intra-cluster similarity, 0.3111, and the averaged inter-cluster similarity, 0.3703, by Eq. (12.6).

Let us consider the case of binary clustering where a group of data items is divided into two clusters. For each cluster, we compute the intra-cluster similarity, and by averaging the two intra-cluster similarities, we compute one to the two clusters. The inter-cluster similarity between the two clusters is computed as the final one. The clustering index to the results from the binary clustering is computed by Eq. (12.6). At least, the fact that two clusters are available is the condition for computing the clustering index.

Let us consider the multiple clustering results which are given more than two clusters. The intra-cluster similarity is computed for each cluster by the same process in the case of the binary clustering, and the inter-cluster similarity over clustering results is computed by averaging the inter-cluster similarities of cluster pairs. From the m clusters $\frac{m(m-1)}{2}$ all possible pairs are generated and for each pair, and an inter-cluster similarity is computed. The average over the inter-cluster similarities of all possible pairs of clusters is one over clustering results and the clustering metric, clustering index, is computed by Eq. (12.7). The process of evaluating results from the multiple clustering results is identical to the case of binary clustering except generating all possible pairs of clusters for computing the inter-cluster similarities.

Let us mention the labeled text collections for evaluating text clustering systems. NewsPage.com which was mentioned in Sect. 8.2.1 was used for evaluating both text categorization systems and text clustering systems by Jo [27, 28]. Reuter21578 which was mentioned in Sect. 8.2.3 has been standard test collection for evaluating text categorization systems [85]. The above collections of labeled texts are applicable for evaluating text clustering systems by the clustering index which was mentioned in this section.

12.3.3 Evaluation of Fuzzy Clustering

This section is concerned with the process of evaluating results from fuzzy clustering, using the clustering index. In Sect. 12.3.2, we already mentioned the process of evaluating the crisp binary and multiple clustering. In Sect. 9.3.2, we mentioned the fuzzy clustering where each item is allowed to belong to more than one cluster. The fuzzy clustering results are evaluated by decomposing it into binary crisp clustering results and removing overlapping. In this section, we describe the process of computing the clustering index to the results from the fuzzy clustering.

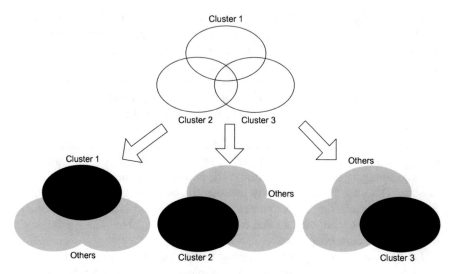

Fig. 12.8 Decomposing fuzzy clustering into binary clusterings

Figure 12.8 illustrates the process of decomposing the fuzzy clustering results
into binary clustering ones. A particular cluster is caught and its data items which
span over the current cluster and others are removed. Two exclusive clusters are
given as results from this decomposition after removing the overlapping ones. The
n binary clustering results are derived from the n overlapping clusters for evaluating
them, using the clustering index. Items which belong to the current and another
cluster are given into the current one, in the decomposition process.

Let us mention the process of computing the clustering index of binary clustering
which is decomposed from the fuzzy clustering. Because the given problem is
reduced to exclusive binary clustering problems, the clustering index is computed
by the process which was mentioned in Sect. 12.3.3. The inter-cluster similarity
between two clusters and the intra-cluster similarities of the two clusters are
computed and the intra-cluster similarity to the binary clustering is computed by
averaging them. The clustering index is computed by Eq. (12.7). In the fuzzy
clustering, the clustering indices as many as clusters are computed as many as
clusters.

We need to integrate the clustering indices of the decomposed binary clusters into
one to the entire fuzzy clustering. The way of integrating them is to average over
the clustering indices as many as clusters. The mean clustering index becomes as
the metric for evaluating the fuzzy clustering results. We use the data items which
are labeled with only one category for evaluating the crisp clustering system. If
using the data items which are labeled exclusively with one, the results may be
underestimated.

Let us mention the alternative metric of evaluating the fuzzy clustering results. It is assumed that the fuzzy clustering results are given as the item-cluster matrix which consists of membership values of data items to the clusters, as follows:

$$\begin{bmatrix} \mu_{C_1}(x_1) & \mu_{C_2}(x_1) & \dots & \mu_{C_{|C|}}(x_1) \\ \mu_{C_1}(x_2) & \mu_{C_2}(x_2) & \dots & \mu_{C_{|C|}}(x_2) \\ \vdots & \vdots & \ddots & \vdots \\ \mu_{C_1}(x_n) & \mu_{C_2}(x_n) & \dots & \mu_{C_{|C|}}(x_n) \end{bmatrix}$$

The desired item-cluster matrix is constructed by target labels of data items which is given in the test collection, as follows:

$$\begin{bmatrix} d_{11} & d_{12} & \dots & d_{1|C|} \\ d_{21} & d_{22} & \dots & d_{2|C|} \\ \vdots & \vdots & \ddots & \vdots \\ d_{n1} & d_{n2} & \dots & d_{n|C|} \end{bmatrix}$$

The differences between individual entries are computed by Eq. (12.8),

$$\sqrt{\sum_{i=1}^{n} \sum_{j=1}^{|C|} (d_{ij} - \mu_{C_j}(x_i))^2} \qquad (12.8)$$

The desired item-cluster matrix is hidden during the clustering process.

12.3.4 Evaluation of Hierarchical Clustering

This section is concerned with the scheme of evaluating the hierarchical clustering results, using the clustering index. In Sect. 12.3.3, we mentioned the scheme of evaluating the fuzzy clustering results. The issue in evaluating the hierarchical clustering results is that the results are underestimated by the lower intra-cluster similarities in the higher clusters and the higher inter-cluster similarities among nested ones. The two measures should be adjusted, depending on the cluster levels. In this section, we explain the process of computing the clustering index, cluster by cluster, and adjusting it depending on the cluster levels.

Figure 12.9 illustrates simple results from doing the hierarchical clustering. The labeled examples are used for evaluating the clustering algorithm following the external validity. The categories are category A under which A-1 and A-2 exist and category B. Category A corresponds to the cluster with its nested clusters, and category B corresponds to one without any nested one. It is not each to the hierarchical clustering results.

Fig. 12.9 Results from doing
hierarchical clustering

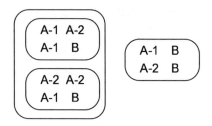

The clustering index is calculated from the example which is shown in Fig. 12.9 with the two separated views: the view of the two clusters in the higher level and the view of the three clusters as ones without nested ones. In the former view, the inter-cluster similarity and the intra-cluster similarity are 0.5 and 0.7736, respectively; the clustering index is 0.6074. In the latter view, the inter-cluster similarity and the intra-cluster similarity are 0.3165 and 0.1667, respectively, so the clustering index becomes 0.2683. The difference between the two clustering index values is outstanding; the reason is that the intra-cluster similarities of nested ones are underestimated. Even if the clustering algorithm is really applied to the hierarchical clustering task, it is more desirable to validate its performance to the flat clustering task.

Let us consider the process of evaluating the hierarchical fuzzy clustering by the clustering index. The results from this clustering type are decomposed into the flat clustering ones, level by level. For each flat fuzzy clustering which corresponds to its own level, it is decomposed into several binary clustering, by the process which was mentioned in Sect. 12.3.4. The clustering indices of the binary clustering results are computed and they are averaged as the clustering to the flat fuzzy clustering to each level. The average over clustering indices of the flat fuzzy clustering in all levels as the general clustering index to the entire results from the hierarchical fuzzy clustering.

If the results from the hierarchical clustering are decomposed into the flat clustering results by level, we need to consider various schemes of assigning weights to clustering indices of results from each flat clustering. In the above scheme, equal weights are identically assigned to clustering indices of levels; the clustering indices of levels are averaged as the clustering index of the entire results from the hierarchical clustering, as mentioned above. In order to prevent the results from being underestimated in the specific levels, higher weights are assigned to clustering index of general level, while lower weights are assigned to clustering indices of specific levels. Weights may be assigned to clustering index in the next level with only constant portion to weight to the current level. As an alternative way, to the next level, the weight is assigned with a negative exponential power to the weight to the current level.

12.4 Parameter Tuning

This section is concerned with the process of computing the clustering index, using unlabeled data items. In Sect. 12.4.1, we explain the process of computing the clustering index, based on unlabeled data items. In Sect. 12.4.2, we mention the scheme of modifying the simple clustering algorithm into the versions with the parameter tuning. In Sect. 12.4.3, we describe the k means algorithm where the parameter tuning is installed. In Sect. 12.4.4, we mention the process of applying the evolutionary computation to the clustering based on the clustering index.

12.4.1 Clustering Index for Unlabeled Documents

This section is concerned with the process of computing the clustering index using unlabeled data items. We described previously the process of computing the clustering index for evaluating data clustering results based on target labels in Sect. 12.3.1. It is assumed that data items are unlabeled and computing similarities between data items depend on their input vectors. We use the clustering index which is computed from unlabeled data items for not evaluating results, but proceed the data clustering. In this section, we explain the process of computing the clustering index from current clusters of unlabeled data items.

The intra-cluster similarities are computed from the clusters of unlabeled items. It is assumed that items are represented into numerical vectors, and the similarity between two numerical vectors is computed by the cosine similarity which is given in Eq. (6.1). The intra-cluster similarity is computed by Eq. (12.1). The average over intra-cluster similarities of clusters becomes one over entire results from data clustering. The process of computing the intra-cluster similarity is same to that which was mentioned in Sect. 12.3.1, except using Eq. (6.1), instead of Eq. (12.5).

The inter-cluster similarities are computed among clusters of unlabeled items. It is assumed that items are encoded into numerical vectors and the cosine similarity which is expressed in Eq. (6.1) is given as the similarity metric. All possible pairs of clusters are generated and the inter-cluster similarity is computed for each pair. The average over the inter-cluster similarities of pairs becomes the inter-cluster similarity over the entire results. We consider the similarity measure which is computed by Eqs. (6.2)–(6.4).

Let us mention the process of computing the clustering index from the clusters of unlabeled texts. The intra-cluster similarity over entire results is computed by process which is described in the second paragraph. The inter-cluster similarity is computed by the process which was described in the above paragraph. The clustering index is computed by Eq. (12.6) involving the intra-cluster similarity and the inter-cluster similarity. The process of computing the clustering index based on the unlabeled texts is same to that based on labeled ones, except using the similarities among input vectors.

The clustering index which is computed by the process is used for tuning the hyper parameters of clusters, rather than for executing the clustering results. The hyper parameters mean those which are decided externally and arbitrarily for executing the algorithm. The number of clusters becomes a hyper parameter of clustering algorithm: the k means algorithm and the Kohonen Networks. The hyper parameter of the single pass algorithm is the similarity threshold, rather than number of clusters. The hyper parameters may be automatically decided by computing the clustering index from the current results from clustering data items.

12.4.2   Simple Clustering Algorithm with Parameter Tuning

This section is concerned with the parameter tuning for the simple clustering algorithms. In Sect. 12.4.1, we studied the process of computing the clustering index by unlabeled items. We use the clustering index for evaluating current quality of clustering results, in order to decide continuing or terminating. The clustering results is optimized by tuning parameters using the clustering index, but it takes more time for proceeding the clustering as the payment. In this section, we describe some clustering algorithms which is installed with the parameter tuning based on the clustering index.

Let us mention the trials of parameter tuning in the AHC algorithm in Sect. 10.2.1. Because every cluster has one item, computing the intra-cluster similarity may be omitted, initially. The clustering index is computed by the process which is described in Sect. 12.4.1, just after comparing both of them with each other. If the results after merging clusters are better than those before doing them, the iteration continues, and otherwise, the results move back to the previous one, terminating the iteration.

The parameter tuning is considered for the divisive algorithm which is mentioned in Sect. 10.2.2. The process of computing the inter-cluster similarity is omitted in the initial status where one group of all items is given. The clustering index is computed after dividing the cluster into two clusters. The results before and after the division are compared with each other, and if the results become better afterward, the division will be continued. When some clusters have only one item in the late stage, the process of computing the intra-cluster similarity is omitted.

A single pass algorithm is mentioned as the fast clustering algorithm in Sect. 10.2.3, and the parameters are tuned by the clustering index. The similarity threshold is given as its hyper parameter, and the value is fixed while proceeding clustering. The similarity threshold is updated automatically during execution. The clustering indices of the both cases are computed: the case of creating one more cluster and the case of arranging an item into one of existing clusters; the similarity threshold is incremented in the latter case, and decremented in the former case. When installing the process of parameter tuning in this algorithm, its execution becomes very slow as the payment for the better clustering quality.

It is actually popular to use the clustering algorithms without the parameter tuning for real tasks. It is more reliable to cluster data items, tuning the parameter, based on the clustering index. The speed of clustering data items is degraded as the payment. If the clustering performance is improved not much, the parameter turning is not recommendable for clustering data items. The parameter turning effect is very variable depending on the application domain.

12.4.3 K Means Algorithm with Parameter Tuning

This section is concerned with the modified version of the k means algorithm which is installed with the parameter turning. Deciding the number of clusters and initializing the mean vectors are requirements for using the k means algorithm. What is decided in advance is automatically optimized by installing the parameter tuning in the current version. The parameter tuning which is based on the clustering index is applied for optimizing the initial vectors and representative ones. In this section, we describe the process of clustering data items by the modified version.

The quality of results from clustering data items depends on the initial mean vectors. The mean vectors are initialized by selecting data items as many as clusters in the initial version. We may evaluate the selected data items as the initial vectors by computing the inter-cluster similarity. If the inter-cluster similarity is higher than the threshold, we select another vectors as the initial mean vectors. Otherwise, we proceed the clustering.

By increasing the clustering index based on the hill climbing, instead of averaging over numerical vectors, we may update the representative vector in each cluster. The mean vectors are initialized by the above process, the data items are arranged by their similarities with the mean vectors, and the cluster index is computed to the current results. The mean vectors are updated by adding a random value between $-\epsilon$ and ϵ, the data items are arranged by those with updated ones, and the clustering index is computed to the results. The two clustering indices before and after updating are compared with each other, and if the clustering index is better, the updated one is taken. Otherwise, it moves back to the previous results, and it makes another trials.

Let us mention the process of computing the clustering index from the clusters of unlabeled texts. The intra-cluster similarity over entire results is computed by process which is described in the second paragraph. The inter-cluster similarity is computed by the process which was described in the above paragraph. The clustering index is computed by Eq. (12.6) involving the intra-cluster similarity and the inter-cluster similarity. The process of computing the clustering index based on the unlabeled texts is same to that based on labeled ones, except using the similarities among input vectors.

The clustering index which is computed by the process is used for tuning the hyper parameters of clusters, rather than for executing the clustering results. The hyper parameters mean those which are decided externally and arbitrarily for executing the algorithm. The number of clusters becomes a hyper parameter of clustering algorithm: the k means algorithm and the Kohonen Networks. The hyper parameter of the single pass algorithm is the similarity threshold, rather than number of clusters. The hyper parameters may be automatically decided by computing the clustering index from the current results from clustering data items.

12.4.4 Evolutionary Clustering Algorithm

This section is concerned with the process of applying the evolutionary computation to the text clustering. The evolutionary computation is the optimization scheme of constructing and evolving the solution populations based on the Darwin's theory, and the genetic algorithm where each solution is represented into a bit string is adopted among the evolutionary computations. In the genetic algorithm, the population of solution candidates is constructed at random, and the population evolves by the first evaluation, crossover, and mutation, into better solutions. It is assumed that the given problem is the binary clustering which is the process of clustering a group of data items into two subgroups; the clustering index is set the fitness value, a bit, zero or one, indicates a cluster identifier, and each bit string position means an item identifier, in applying the genetic algorithm to the data clustering. In this section, we describe the scheme of applying the evolutionary computation to the data clustering.

The two clusters of items are encoded into a bit string as a solution candidate. The position of each bit string indicates an item identifier and a bit value means a cluster identifier to which the item belongs to. The length of bit string becomes the number of data items; zero means that the item belongs to cluster 1 and one means that it belongs to cluster 2. The bit string is given as a genotype and the two clusters which correspond to a bit string are given as the phenotype. In the multiple or hierarchical clustering, we may another representation of clusters; we adopt the genetic programming, instead of the genetic algorithm, in the hierarchical clustering.

We need to evaluate the fitness of each solution candidate for evolving the current population. A bit string is decoded into two clusters; 0 and 1 in the bit string is interpreted into an item in cluster 1 and one in cluster 2, respectively. The intra-cluster similarities of cluster 1 and 2, and the inter-cluster similarity between two clusters are computed. By averaging the intra-cluster similarities, the intra-cluster similarity to the results and the clustering index are computed. The clustering index is used as the fitness value of the given solution candidate.

Let us explain the process of optimizing the binary clustering results by the genetic algorithm. The initial population which consists of bit string as the solution candidates is constructed at random. Two selection candidates are selected at random, other solution candidates which are called off-springs are generated from the selected ones by recombinant operators such as cross-over, some of them are mutated with a small probability. The current population is evolved into the next population which consists of better solution candidates; the solution candidates are evaluated by the fitness value and low fitness valued ones are removed. It reaches the population which consists of solution candidates with their high fitness values by iterating the above process.

The genetic algorithm was mentioned as an instance of the evolutionary computation, and let us mention other instances. The evolutionary strategy is the evolutionary computation instance where solution candidates are given as numerical vectors. The genetic programming is one where solution candidates are represented into trees. The evolutionary programming allows solution candidates to be represented into any structured data. The differential evolution, the cultural algorithm, and the co-evolution are additional evolutionary computation instances.

12.5 Summary and Further Discussions

We described the schemes of evaluating clustering results and modifications of existing clustering algorithms using them, in this chapter. We mentioned the three types of evaluating clustering results: the internal evaluation, the relative one, and the external one. We asserted that clustering index is the integration of the inter-cluster similarity and the intra-cluster similarity in the style of F1 measure, and describe the process of computing it from the clustering results. We mentioned how to use the clustering index for tuning parameters as well as evaluating results, and modified some clustering algorithms by installing it. In this section, we make some further discussions from what we study from this chapter.

Let us consider the process of evaluating results from cluster naming, separated from the data clustering. It is assumed that each cluster is identified with its symbolic name which reflects its contents. The cluster naming is intended for browsing, inherently, so easiness in browsing texts depending on cluster names becomes the direction for evaluating the cluster naming. The relevancy of cluster name to texts becomes another direction. The cluster naming evaluation is left as the next research topic in future.

Many metrics are defined for evaluating the clustering results in the literature. Performances of clustering algorithms cannot be confirmed by only one evaluation metric. We need to use several metrics for validating the approaches to data clustering. In displaying the empirical validations of clustering algorithms, evaluation metric values are presented as a set or they are integrated into one value by summing weighted measures. The clustering index which is mentioned in Sect. 12.3 becomes the integration of the inter-cluster similarity and the intra-cluster similarity.

It is difficult and complicated to evaluate the results from the multiple viewed clustering. In this clustering type, we should accommodate multiple versions of results from clustering texts. Each version is evaluated by the clustering index, and clustering indices of all versions are averaged into a single clustering index. The maximum or minimum clustering index may be selected alternatively as the evaluation metric. It is left to define schemes of evaluating the clustering results as a remaining task.

There are two ways of computing the clustering index. One way is to compute similarity by target labels of items and the other is to compute the similarity by the cosine similarity. The attributes of the input vector need to be discriminated among each other and the similarity may be computed, depending on the weighted attributes. The clustering index is computed by this process and used for tuning parameters. The way of parameter turning is also left as a remaining task.

Part IV
Advanced Topics

Part IV is concerned with the advanced topics on text mining. This part covers the text summarization and the text segmentation as the additional tasks of text mining. We mention the automatic predefinition of topics from a corpus, which is called taxonomy generation. In the last chapter, we describe the process of managing texts by combining the two text mining tasks with each other. Therefore, this part is intended to explore the additional two text mining tasks: the taxonomy generation and the automatic text management.

Chapter 13 is concerned with the text summarization which is the alternative task to the text categorization and the text clustering. This chapter begins with describing functionally the text summarization with its view into a binary classification. Afterward, we describe the schemes of applying the machine learning algorithms to the text summarization which is viewed into a classification task. We also present hybrid tasks where the text summarization is combined with other text mining tasks. Even if the text summarization is described as a classification task, it should be distinguished from the topic-based text categorization which was covered in Part II.

Chapter 14 is concerned with the text segmentation as another text mining task. This chapter begins with interpreting the task into a binary classification like the case of the text summarization. We explain the scheme of applying the machine learning algorithms to the text segmentation which is mapped into a classification task. Like the case in the text summarization, we also mention the hybrid tasks where the text segmentation is combined with other text mining tasks. Even if the text segmentation is mapped into a classification task, it is necessary to distinguish it from the topic-based text categorization.

Chapter 15 is concerned with taxonomy generation which is a composite text mining task. The taxonomy generation consists of the text clustering and the cluster naming, so cluster naming is explained in the functional view and the policies are defined for doing the task. Afterward, we describe in detail the schemes of naming clusters symbolically. We also mention ontology which is the expanded taxonomy organization and distinguish them from each other with respect to their goals. In this chapter, we cover the taxonomy generation for defining an organization frame.

Chapter 16 is concerned with the process of managing texts automatically by combining the text categorization and the text clustering with each other. The automatic text management has its two modes: the creation mode and the maintenance mode; each of them is explained in detail. We describe the scheme of implementing the automatic text management system, using the two text mining tasks. We also mention the system expansion, by adding additional tasks, such as the text summarization and text segmentation. This chapter focuses on the integration of the text mining tasks for the automatic text management.

Chapter 13
Text Summarization

This chapter is concerned with the text summarization task in terms of its function, methods, and implementation. We define the text summarization which is an instance of text mining task, in Sect. 13.1, and explore the types of text summarization, in Sect. 13.2. In Sect. 13.3, we describe the simple and state-of-the-art approaches to text summarization. We cover the hybrid tasks which are the combinations of text summarization with other tasks, in Sect. 13.4, and make the summarization and the further discussions on this chapter, in Sect. 13.5. In this section, we describe the text summarization as the additional task to text classification and clustering which are covered in the previous parts.

13.1 Definition of Text Summarization

Text summarization refers to the process of extracting a summary automatically from a given full text or full texts. The manual summarization means the process of rewriting or restating the full text into its brief version. Automatic summarization actually means to select automatically important portions which are paragraphs or sentences from a full text or full texts. In this chapter, we view the text summarization into a binary classification where each paragraph or sentence is classified into "summary" or "non-summary." In this section, we describe briefly the both kinds of text summarization in the general view.

Summarization is referred to the process of rewriting or restating a particular content into its brief version by understanding it. The summary which is generated by human reflects its entire content, accurately. It costs much and variable time depending on the understanding and intelligence level for generating a summary from a text. It is very tedious or almost impossible to summarize more than 100 texts, manually and individually. In spite of more accuracy in generating summary

© Springer International Publishing AG, part of Springer Nature 2019
T. Jo, *Text Mining*, Studies in Big Data 45,
https://doi.org/10.1007/978-3-319-91815-0_13

by human being, we need to introduce the automatic summarization, in order to solve the above problem.

It is not easy to implement an application program which summarizes texts automatically, like human being. The automatic text summarization should be defined differently from the manual text summarization which is mentioned above. It means the process of selecting essential parts of full text as some sentences or paragraphs. The automatic text summarization is interpreted into a classification task where each paragraph or sentence is classified into summary or non-summary, in this chapter. The paragraphs or sentences which are classified into summary are generated as the output of this task.

Similar or more advanced tasks are derived from the automatic text summarization. The text summarization may be expanded into the multiple text summarization which summarizes several texts into a single summary. The text summarization is modified into query-based one which summarizes a given text, biased toward the given query. By integrating summaries of texts in a cluster, its prototype text which represents the cluster may be generated. The summaries which are generated from texts may be used for improving the performance of the information retrieval and other text mining tasks.

The text summarization is interpreted into a binary classification task, in this study. A given text is segmented into paragraphs by the special character, carriage return, as the preprocessing, and the paragraphs are encoded into numerical vectors by the process which is described in Chaps. 2 and 3. In advance, we gather sample paragraphs which are manually labeled with summary or non-summary, and the adopted machine learning algorithm learns them for building its classification capacity. The novice text is partitioned into paragraphs and they are classified into one of the two labels. Therefore, the text summarization is mapped into the binary classification where each paragraph is classified into summary or non-summary.

13.2 Text Summarization Types

In this section, we explore the types of text summarization depending the dichotomy criteria. In Sect. 13.2.1, we examine the differences between the manual summarization and the automatic one. In Sect. 13.2.2, we mention the type of text summarization depending on the given input: a single text or multiple texts. In Sect. 13.2.3, we introduce the hierarchical text summarization which provides potentially the functions: zoom in and out. In Sect. 13.2.4, we consider the query-based text summarization which requires the query as well as the full text.

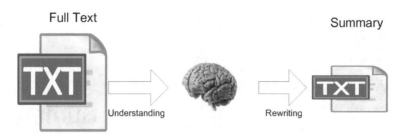

Fig. 13.1 Manual text summarization

13.2.1   Manual vs Automatic Text Summarization

This section is concerned with the type of text summarization executed by a human or a computer program. The text summarization is regarded as an important task in the area of literature science, as well as computer science. The manual text summarization is defined as the process of rewriting the content of full text into its brief version, and the automatic text summarization is defined as the process of selecting essential parts of full text as its brief version. Because the computer program is not able to summarize a text like a human, both kinds of text summarization are defined differently. In this section, we examine the two types of text summarization tasks with respect to their differences.

The manual text summarization is illustrated in Fig. 13.1. A full text is initially given as the input. The full text is scanned and understood by the human being. Its contents are rewritten into its brief version. The summary which is the results from the manual text summarization consists of several sentences which are not given in the original full text.

The automatic text summarization is illustrated in Fig. 13.2. The full text is given as the input like the case in the manual text summarization. The full text is partitioned into sentences or paragraphs by the punctuation mark or the carriage return. Some paragraphs or sentences among them are selected as the essential part. Selecting some paragraphs as the summary is preferred to doing sentences.

The comparisons of the two kinds of text summarizations are illustrated in Table 13.1. The manual summarization is defined as the process of understanding and rewriting the full text into its brief form, while the automatic one is defined as that of selecting some paragraphs or sentences as the essential part. As the preliminary task, in the manual summarization, understanding the full text is required, whereas in the automatic one, partitioning the full text into paragraphs or sentences is required. The automatic text summarization is necessary for summarizing individually a large number of texts. It is not able to expect better qualities of summaries from the automatic summarization.

Let us consider the mixture of two types of text summarization as well as the two independent ones. As shown in Table 13.1, there is trade-off between the two types of text summarization; this motivates for proposing the mixture. In a

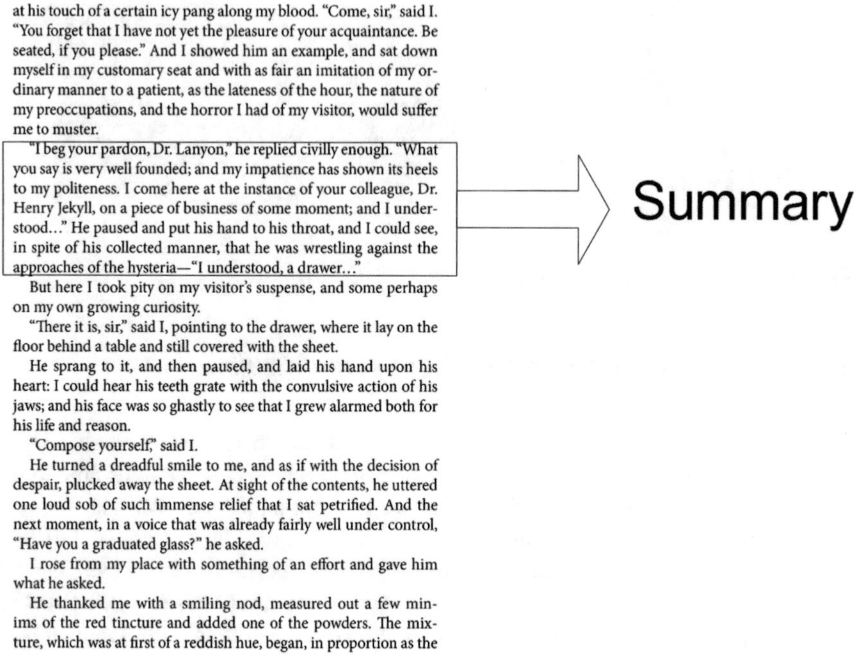

at his touch of a certain icy pang along my blood. "Come, sir," said I.
"You forget that I have not yet the pleasure of your acquaintance. Be
seated, if you please." And I showed him an example, and sat down
myself in my customary seat and with as fair an imitation of my or-
dinary manner to a patient, as the lateness of the hour, the nature of
my preoccupations, and the horror I had of my visitor, would suffer
me to muster.

"I beg your pardon, Dr. Lanyon," he replied civilly enough. "What
you say is very well founded; and my impatience has shown its heels
to my politeness. I come here at the instance of your colleague, Dr.
Henry Jekyll, on a piece of business of some moment; and I understood..." He paused and put his hand to his throat, and I could see,
in spite of his collected manner, that he was wrestling against the
approaches of the hysteria—"I understood, a drawer..."

But here I took pity on my visitor's suspense, and some perhaps
on my own growing curiosity.

"There it is, sir," said I, pointing to the drawer, where it lay on the
floor behind a table and still covered with the sheet.

He sprang to it, and then paused, and laid his hand upon his
heart: I could hear his teeth grate with the convulsive action of his
jaws; and his face was so ghastly to see that I grew alarmed both for
his life and reason.

"Compose yourself," said I.

He turned a dreadful smile to me, and as if with the decision of
despair, plucked away the sheet. At sight of the contents, he uttered
one loud sob of such immense relief that I sat petrified. And the
next moment, in a voice that was already fairly well under control,
"Have you a graduated glass?" he asked.

I rose from my place with something of an effort and gave him
what he asked.

He thanked me with a smiling nod, measured out a few min-
ims of the red tincture and added one of the powders. The mixture, which was at first of a reddish hue, began, in proportion as the

Fig. 13.2 Automatic text summarization

Table 13.1 Manual vs automatic summarization

	Manual	Automatic
Definition	Rewriting briefly	Paragraph or sentence selection
Preliminary task	Understanding	Partition
Mass summarization	Difficult	Possible
Writing quality	Good	Plain

particular mixture of them, the essential paragraphs are generated by the automatic
summarization as the draft, and it is edited into a complete summary, manually. In
another mixture, multiple versions of summary which are automatically generated
by several algorithms are presented for users, and they select one among the
candidates. Only the automatic summarization is actually useful than the hybrid
one, when a lot of text should be summarized, individually.

13.2.2   Single vs Multiple Text Summarization

This section is concerned with the types of text summarization in another view.
In the real world, several texts may be summarized as well as a single text. The

Fig. 13.3 Single text
summarization

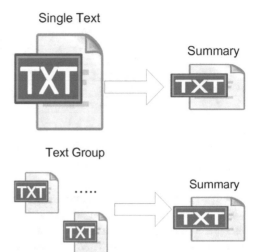

Fig. 13.4 Multiple text
summarization

summarization of multiple texts is called multiple text summarization, whereas that
of a single text is called single text summarization. The criteria for deciding one of
the two types is whether the input is a single text or a group of texts. In this section,
we explain the two types of text summarization, and compare them with each other,
and combine them into a hybrid one.

The single text summarization is illustrated in Fig. 13.3. Only single article is
initially given as the input. Essential paragraphs or sentences are selected as its
summary. The first or last paragraph usually summarizes the entire content in the
case of summarizing a single text. Because the single text summarization is simple,
it is feasible to use a heuristic approach to the task.

The multiple text summarization is referred to the process of summarizing more
than one text into a summary, as illustrated in Fig. 13.4. A group of articles which
have their different contents is given as the input, instead of a single article. The
articles are integrated into a single text by concatenating them and essential paragraphs or sentences are selected as the summary. The multiple text summarization
is further divided into the homogeneous multiple text summarization where texts in
the same or similar topics are given as the input, and the heterogeneous multiple
text summarization where texts in different topics are given. The multiple text
summarization is used for constructing a prototype text which represents the group,
artificially.

Table 13.2 presents the differences between the two kinds of text summarization.
A singe text is given as the summarization target in the single text summarization,
whereas a group of texts is given in the multiple text summarization. In the single
text summarization, the summary of a single text indicates its abstract, whereas
in the multiple text summarization, the summary indicates the script of the texts.
The cohesion of texts in the group becomes the important issue in the multiple text
summarization. There are two ways of summarizing multiple texts; one way is to

Table 13.2 Single vs multiple summarization

	Single	Multiple
Input	Single text	Text group
Output	Abstract	Group script
Cohesion	Not issue	Important issue
Steps	Partition + selection	Partition + selection + integration

extract summaries from individual texts and integrate them with each other and the other is to integrate the individual texts into a large single text and summarize it at a time.

Let us consider the combination of the two types as a hybrid one. A group of texts is initially given as the input. The individual texts are summarized; each text is associated with its own summary. Among the summaries, some are selected as the summary of the text group. Therefore, this is the hybrid case where the multiple text summarization is performed by means of single text summarizations.

13.2.3   Flat vs Hierarchical Text Summarization

This section is concerned with the flat text summarization and the hierarchical one. The former is one where any intermediate summary is not allowed, whereas the latter is one where it is allowed. For example, in summarizing a research paper, we consider the two kinds of summary: its abstract and its extended summary. Based on the contents, zooming in and out is possible for viewing a text in the hierarchical text summarization. In this section, we compare the two kinds of text summarization with each other and describe the hierarchical text summarization in detail.

The flat text summarization is illustrated in Fig. 13.5. The top of Fig. 13.5 shows the single text summarization, and the bottom shows the multiple text summarization. There are only two kinds of versions of text or texts in the flat summarization: full text version and summarized version. No intermediate version between them exists in this type of text summarization. There are only two levels of zooming in and out of a text or texts into the summary or the full text.

Figure 13.6 illustrates the three types of hierarchical text summarization. A single text is segmented into subtexts based on their contents and each subtext is summarized into its own summaries, and they are integrated into the brief form as shown in the top of Fig. 13.6. In the middle of Fig. 13.6 is shown a text group clustered into subgroups, each subgroup is summarized into their own summary, and they are integrated into the general summary. The bottom of Fig. 13.6 shows the gradual summarization of a single text. The top and the bottom of Fig. 13.6 show the hierarchical and single text summarizations, and the middle shows the hierarchical and multiple text summarization.

Fig. 13.5 Flat text
summarization

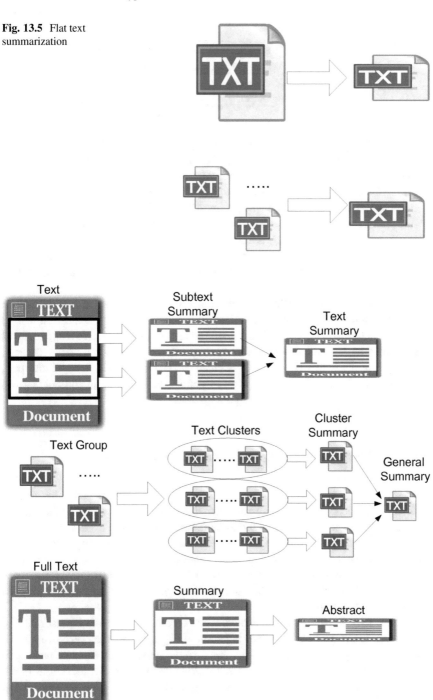

Fig. 13.6 Hierarchical text summarization

Table 13.3 Flat vs
hierarchical summarization

	Flat	Hierarchical
Suitable input	Single text	Text group or long text
Multiple summary	Less suitable	More suitable
Zooming	Limited	Possible
Output	Summary	Summary tree

Table 13.3 illustrates the comparisons of the two types of text summarization which was mentioned in this section. The flat text summarization is applicable to the text with its medium length, whereas the hierarchical text summarization applicable to a long text or a text group. In the flat summarization, only a single version of summary is expected whereas multiple versions of summary are expected in the hierarchical summarization. In the flat summarization, only two zooming levels, summary and full text, are available whereas much more zooming levels are available in the hierarchical summarization. The goal of the flat summarization is to generate only a summary, whereas the goal of the hierarchical summarization is to zoom in and out a text or some texts.

It is possible to interpret the text summarization into a regression task rather than a classification one. The text summarization is viewed into a classification task in both the flat one and the hierarchical one. The text summarization is mapped into a regression where a continuous normalized value between zero and one is assigned to each paragraph. The value which is close to zero indicates a nonessential part, whereas one which is close to one is an essential one. Zooming in and out is given to a text by increasing and decreasing the threshold.

13.2.4   Abstraction vs Query-Based Summarization

This section is concerned with the two types of text summarization by another dichotomy criteria. One is the abstraction which summarizes a text in the general view and the other is the query- based summarization which does it, focusing on the given query. The abstraction is viewed as the process of selecting paragraphs as the essential parts which reflect the entire content. The query-based summarization which selects paragraphs relevant to the query is regarded as the task which is close to the information retrieval. In this section, we explain and compare the two types of text summarization.

Figure 13.7 illustrates the abstraction as a type of text summarization. The abstraction is the process of generating some paragraphs or some sentences as the abstract of the given full text; it is identical to the general definition of automatic text summarization. A single text or multiple texts are given as the initial input, they are partitioned into paragraphs, and essential paragraphs are selected as the abstract. The abstraction is characterized by the flat, automatic, and written text summarization. The assumption underlying in this type is that the same part is always generated from the same full text or full texts.

Fig. 13.7 Abstraction

Full Text

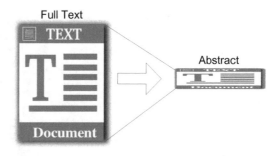

Figure 13.8 illustrates the query-based text summarization. Depending on the given query, a different portion of the given full text is generated as its summary. The query and the full text are given as the input, the relevancies of paragraph to the query are computed within a full text, and highly relevant ones are selected as the summary. A fixed number of paragraphs to any query or a variable number of them may be generated, depending on the policy. The query-based summarization is viewed as an instance of information retrieval within a given text.

Table 13.4 shows the comparison of the two types of text summarization. In the abstraction, only text is given as the input, whereas in the query-based summarization, a text and a query are given as the input. The fixed essential part is generated as the output in the abstraction, whereas the most relevant part to the query is generated in the query-based text summarization. The abstraction is intended for previewing a full text, whereas the query-based text summarization is intended for retrieving interesting parts. In the abstraction, its performance is evaluated by matching the desired summary and the actual one, whereas in the query-based text summarization, its performance is evaluated by the precision and the recall.

Let us consider the hybrid text summarization into which the two types of text summarization are combined with each other. It is not guaranteed that the queries are provided by all users and express their information need, exactly; user may expect the summary from a full text without providing their queries. The system needs to provide both the abstract in the case of no query and the query-based summary in the case of a query. If no query is provided, it is more desirable to provide the abstract instantly rather than waiting for a query. In order to avoid computing matching value between a query and portions during the summarization process, each query is associated with its relevant portions of full texts and its abstract is prepared, in advance.

13.3 Approaches to Text Summarization

This section is concerned with the representative approaches to the text summarization. In Sect. 13.3.1, we mention some heuristic approaches. In Sect. 13.3.2, we interpret the text summarization into a classification task. In Sect. 13.3.3, we describe the scheme of gathering sample paragraphs. In Sect. 13.3.4, we explain the scheme of applying the machine learning algorithms to the task.

Fig. 13.8 Query-based text
summarization

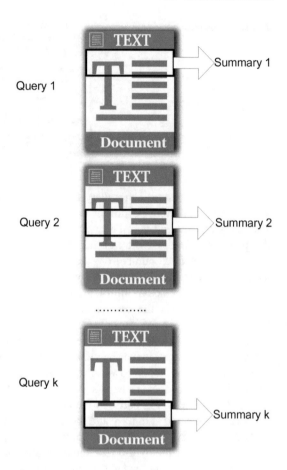

Table 13.4 Abstract vs query-based summarization

	Abstraction	Query-based summarization
Input	Text	Text + query
Summary	Essential part	Query focused part
Goal	Preview	Relevancy improvement
Evaluation	Summary matching	Recall + precision

13.3.1   Heuristic Approaches

This section is concerned with the simple and heuristic approaches to the text
summarization. The approaches in this kind are used without mapping the task into
a classification one. A text is partitioned into paragraphs and each of them is decided
by keywords or phrases. However, the approaches are weak to text manipulations
which are behaviors of changing text contents, intentionally. In this section, we
mention some heuristic approaches before discussing state-of-the-art ones.

Let us mention the simplest scheme of summarizing a text, automatically. A text is partitioned into paragraphs by the carriage return. The first or the last paragraph may be extracted as the summary, absolutely. The scheme is simple but very fragile to artificial text manipulations. Sometimes, there is possibility that a medium paragraph may be an essential part.

Let us mention another heuristic scheme of summarizing a text. In advance, we define the key phrases which indicates a summary, such as "in conclusion," "in summarization," and "finally." The paragraph or the sentence which includes one of the above phrases is selected as the summary. However, it is not guaranteed that all texts have one of the above phrases; this scheme is not applicable to the texts which have no key phrase. If a text has no phrases, we need to consider another scheme.

Let us mention one more scheme of summarizing a text based on its keywords. It is assumed that the text is initially associated with a list of keywords whether they are generated automatically or manually. In this scheme, the paragraphs where keywords are concentrated are extracted as the summary. It is more probable to generate intermediate paragraphs as the summary rather than the first or the last, in this scheme. If a text is not associated with its keywords, this scheme is not applicable.

Let us mention the heuristic schemes of text summarization which are applicable to the query-based one. It was covered in Sect. 13.2.4. In this scheme, query is given as the input, paragraphs or sentences are selected by the query concentrations, and the selected ones are generated as the summary. In the advanced version, the similarity between a paragraph and a query may be computed. If there is no relevant paragraph, nothing is extracted.

13.3.2 Mapping into Classification Task

Figure 13.9 illustrates the process of mapping the text summarization into the binary classification task. A text is partitioned into paragraphs by the carriage return. Each paragraph is classified into summary or non-summary. As the summary, paragraphs which are classified into summary are extracted. In this section, we explain the classification which is mapped from the text summarization and its comparison with the topic-based text categorization.

The text summarization is interpreted into a classification task to which the machine learning algorithms are applicable, as shown in Fig. 13.9. We gather sample paragraphs which are labeled with one of the two categories: summary or non-summary. The sample paragraphs are encoded into numerical vectors and the machine learning algorithms learn them for building its classification capacity. A novice text is given as a list of novice paragraphs, and each of them is classified into one of the two categories. The binary paragraph classification becomes the core task of the text summarization.

Let us explain the process of extracting the summary from a full text, using the above binary classification. A full text is given as the input and is partitioned

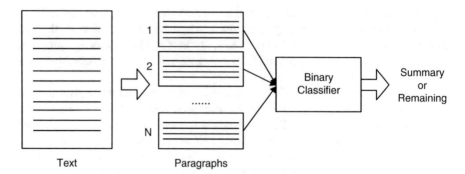

Text   Paragraphs

Fig. 13.9 Process of mapping text summarization into binary classification

into paragraphs by the carriage return. The paragraphs are encoded into numerical vectors, and classified into summary or non-summary. The paragraphs which are classified with summary are extracted as the output. Results from classifying the paragraphs may be different, depending on the selected features and the adopted machine learning algorithm.

Even if both the tasks belong to the classification tasks, we need to distinguish the text summarization from the topic-based text classification. In the text summarization, an individual paragraph is classified, whereas in the topic-based text classification, an individual article which consists of more than one paragraph is classified. The classification task which is mapped from the text summarization is an instance of binary classification where each paragraph is classified into summary or non-summary, whereas the topic-based text classification is usually an instance of multiple classification where more than two topics are predefined as categories. The classification which is derived from the text summarization is always a flat classification, whereas the topic- based text classification is sometimes given as a hierarchical classification task. In the text summarization, we need to consider the domain for gathering sample examples, whereas we do not need to consider the domain for doing so in the text categorization.

As shown in Fig. 13.10, it is possible to map the text summarization into a regression task as well as a classification one. Each paragraph is estimated with its essence score which is the importance degree in the given text. In this case, sample paragraphs are labeled with their essence scores, instead of one of the two categories, summary and non-summary. The process of assigning essence scores to sample ones is very dependent on subjectivity. The text summarization which is mapped into a regression rather than a classification may be more useful for zooming a text in or out.

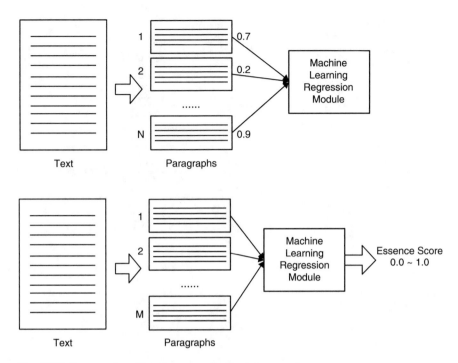

Fig. 13.10 Process of mapping text summarization into regression

13.3.3   Sampling Schemes

This section is concerned with the scheme of gathering sample labeled paragraphs for performing the text summarization. In Sect. 13.3.2, we interpreted the task into the binary classification to which the machine learning algorithms are applicable. The text summarization was mentioned as a domain specific task where each paragraph is classified into summary or non-summary, depending on the domain. Sample labeled paragraphs are gathered within a domain, in order to do so. In this section, we describe the scheme of sampling paragraphs for learning the machine learning algorithms.

Figure 13.11 illustrates the process of gathering sample paragraphs domain by domain, called domain-specific sampling. It is assumed that the text collection is partitioned into domains based on their contents, initially. In each domain, each text is partitioned into paragraphs and each paragraph is labeled with summary or non-summary, by scanning them, manually. Keeping the balanced distribution over two classes, we select labeled paragraphs and encode them into numerical vectors by the process which was mentioned in Chap. 3. In each domain, labeling manually individual paragraphs is a very tedious job for implementing the text summarization system.

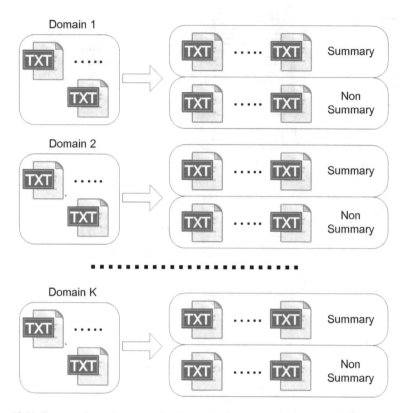

Fig. 13.11 Process of sample paragraphs domain by domain

In order to gather sample paragraphs domain by domain, we need to accompany the text categorization system in the scheme, called topic- based sampling as shown in Fig. 13.12. In advance, the text categorization system is constructed by doing the preliminary tasks and learning the sample texts, in advance. Texts in the corpus are classified into one of the predefined topics. Topic by topic, we gather manually the sample paragraphs which are labeled with summary or non-summary. The text summarization is carried out by the two classifications: classifying entire text into one of the predefined topics and classifying paragraphs into summary or non-summary.

The text summarization by clustering-based sampling is illustrated in Fig. 13.13. The entire texts in the corpus are clustered into subgroups of content-based similar ones by a cluster analyzer. In each cluster, individual texts are portioned into paragraphs and they are labeled manually into summary or non-summary. In the process of text summarization, a text is arranged into a cluster by its similarity with the cluster prototypes. This scheme is characterized as the text summarization which is supported by the text clustering.

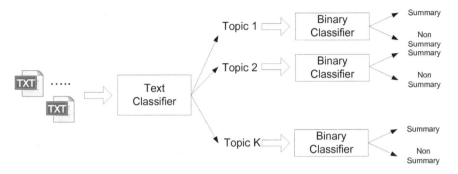

Fig. 13.12 Topic-based sampling

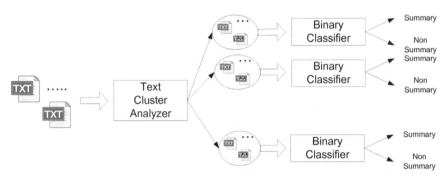

Fig. 13.13 Cluster-based sampling

The degree how much the domain is specific is called granularity and we mention the trade-off between the broad domain and specific domain. In the broad domain as the large granularity, it is easy to gather sample paragraph, but the reliability of classifying them is not good. In the specific domain where the granularity is small, the reliability is better, but it becomes difficult to gather sample paragraphs. Currently, it is impossible to measure the domain granularity which is the scope of gathering sample paragraphs as a quantitative value. However, the granularity is optimized by results from classifying validation paragraphs which are separated from the sample one.

13.3.4 Application of Machine Learning Algorithms

This section is concerned with the scheme of applying the machine learning algorithms to the text summarization task. Previously, we interpreted the text summarization into a classification task and mentioned the scheme of gathering sample paragraphs. The text collection is divided into sub-collection by topics or domains, and a machine learning-based classifier is allocated to each topic

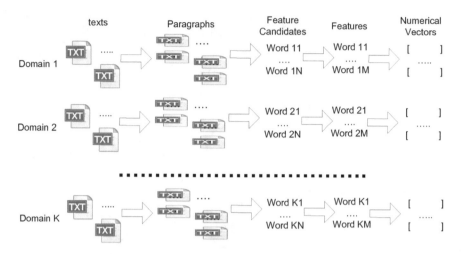

Fig. 13.14 Encoding paragraphs into numerical vectors

or domain. Novice texts are arranged into its relevant domain or topic and its paragraphs are classified into summary or non-summary. In this section, we explain how to apply machine learning algorithms to the text summarization task.

Figure 13.14 illustrates the process of encoding paragraphs into numerical vectors. The text collection is divided into sub-collections which correspond to domains and each sub-collection is mapped into paragraph sub-collections. The feature candidates are extracted by indexing the paragraphs and some among them are selected as features. Each paragraph is represented into a numerical vector by assigning values to features. The process of encoding paragraphs into numerical vectors is same to that of encoding texts so.

Figure 13.15 illustrates the process of training the machine learning algorithms with the sample paragraphs. A machine learning algorithm is allocated as a classifier to each domain. The numerical vectors which are generated by the process which is shown in Fig. 13.14 and associated with paragraphs are labeled manually, by summary or non-summary. A set of training paragraph in each domain is divided into two groups: the summary group and the non-summary group. The machine learning algorithm which corresponds to its own domain builds its own classification capacity by learning the sample paragraphs.

Figure 13.16 illustrates the process of classifying the paragraphs into summary or non-summary. The text is tagged with its own domain and it is partitioned into the paragraphs. Each paragraph is classified by the classifier which correspond to the domain into one of the two categories. The paragraphs which are classified with summary are extracted as the essential part. In the process in Fig. 13.16, it is assumed that the domain is known.

A long single text which spans over more than one domain may be summarized. It may be segmented into subtexts by the text segmentation process which is covered in

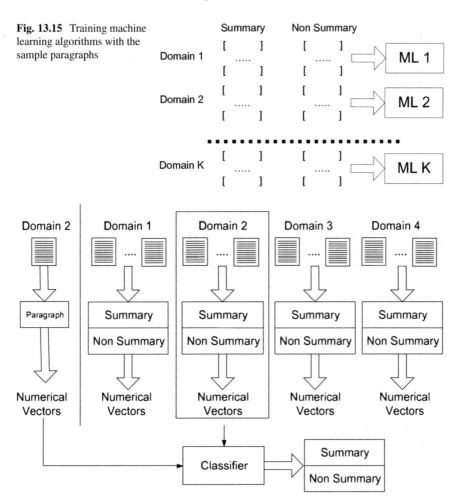

Fig. 13.15 Training machine learning algorithms with the sample paragraphs

Fig. 13.16 Classifying paragraph into summary or non-summary

Chap. 14. Each subtext tends to span over one domain; each subtext is tagged with its own domain by the text categorization or the text clustering. The individual subtexts are summarized by extracting the essential paragraphs one by one. However, in this case, the summaries are ones to the subtexts, but not ones to the entire full texts.

13.4   Combination with Other Text Mining Tasks

This section is concerned with the combination of the text summarization with other tasks and consists of four sections. In Sect. 13.4.1, we mention the summary-based classification which is the combination of the text summarization and the

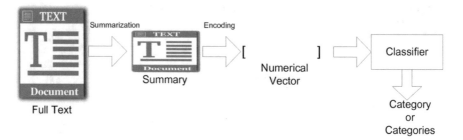

Fig. 13.17 Summary-based classification

text categorization with each other. In Sect. 13.4.2, we explain the task which is called summary- based clustering which is the hybrid task of the text summarization and the text clustering. In Sect. 13.4.3, we cover the topic-based summarization as another type of combined task. In Sect. 13.4.4, we describe the text expansion which is the opposite task to the text summarization.

13.4.1   Summary-Based Classification

This section is concerned with the hybrid task of the text categorization and the text summarization which is called summary- based classification, as presented in Fig. 13.17. The summary is extracted as a paragraph or some paragraphs from a full text and encoded into a numerical vector, instead of the full text. The summary is classified into one or some of the predefined topics. Because the parts which deal with other topics are removed, we may expect the better performance of text categorization by the summary than by the full text. In this section, we describe the hybrid task which combines the text summarization and the text categorization, with each other.

Let us consider the preliminary tasks for executing the text classification based on the summary. The categories are predefined as a list or a tree, and corresponding sample labeled texts are gathered for doing the text categorization. Texts are clustered by a clustering algorithm into subgroups, independently of the predefined categories. For each cluster, each text is partitioned into paragraphs and sample paragraphs which are labeled with summary or non-summary are gathered. Under the assumption of not decomposing the text classification into binary classifications, we prepare a single classifier for executing the text classification and classifiers as many as clusters for executing the text summarization; each cluster indicates the domain within which each paragraph is decided to be the summary, or not.

Let us mention the process of classifying a text based on its summary, after preparing the classifiers. Each text is arranged into its nearest cluster and it is partitioned into paragraphs. They are encoded into numerical vectors, and each of them is classified into summary or non-summary. The paragraphs which are classified into summary are extracted as the text summary, and it is encoded into a numerical vector, and it is classified into one or some among the predefined categories. Only summary is used for assigning a topic to the text, instead of the full text.

The typical example of summary-based classification is the spam mail filtering which depends on email titles. The spam mail filtering was mentioned in Sect. 5.4.1, as the process of removing spam mails among the arriving ones. The email consists of its title and contents and tends to be decided whether it is a spam or a ham by its title. The email whose title consist of special characters and punctuation marks tend to be classified into spam. The spam mail filtering by titles is more risky of misclassification than that by full texts.

Let us consider some issues in classifying texts by their summaries. If nonessential paragraphs are selected as the summary, there is more possibility of classifying a text into its wrong topic. A topic should be assigned to text before doing the text summarization in case of applying the machine learning algorithms. The features should be set for encoding paragraphs into numerical vectors for text summarization, separately from the feature set for classifying the summary. If the text classification is decomposed into binary classifications, we need classifiers as many as topics for text categorization and ones as many as domains for text summarization.

13.4.2 Summary-Based Clustering

This section is concerned with the summary-based clustering which is illustrated in Fig. 13.18. The summary-based clustering is viewed as the hybrid task of the text clustering and the text summarization. The summary is extracted from each text in the group as a surrogate, and it is encoded into numerical vector. A group of texts is clustered into subgroups based on their summaries. In this section, we describe this hybrid task where texts are clustered by their summaries in detail.

Let us consider some benefits from doing the text clustering by summaries, instead of full texts. It takes less time for encoding summaries into numerical vectors than for doing full texts, so the higher clustering speed is expected. Because the summary is regarded as the essential text part, we expect the clustering quality to be improved by avoiding noises from nonessential parts. A corpus is characterized with both results: text summaries and text clusters. The summaries in each cluster become its potential scripts as a guide for browsing data items.

Let us explain the process of clustering texts by their summaries. A group of texts is given as the input and each text is mapped into its summary. The summaries are encoded into numerical vectors and they are clustered into subgroups by a clustering algorithm. It is assumed that the heuristic scheme which was mentioned in Sect. 13.3.1 is used. The issue in the summary-based clustering is that numerical vectors which are encoded from summaries are sparser than those done from the full texts.

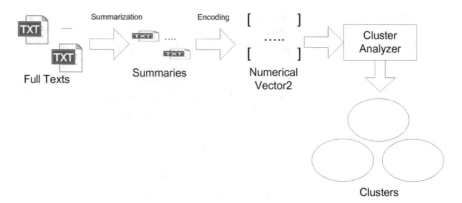

Fig. 13.18 Summary-based clustering

Let us mention the title-based clustering which is a variant of the summary-based clustering. It is assumed that entries of texts and their titles are given as associated forms. The titles are encoded into numerical vectors and they are clustered into subgroups by a clustering algorithm. However, because each title is given as a very short text, we need to expand the word list with their relevant ones in encoding titles into numerical vectors. The title-based clustering is used for making the preview of clustering texts, before executing the main clustering which is the process of clustering texts by their full texts.

If a short text or a title consists of very few words, we need to expand the text by adding its associated text. For each text, its similarity with other texts in the external corpus and most similar ones are taken as its associated ones. Titles and summaries of the similar texts are added to the original text. Texts each of which is added by the associated ones are encoded numerical vectors, and they are clustered into subgroups. By paying the degraded clustering speed, it is expected that the numerical vectors become less sparse.

13.4.3   Topic-Based Summarization

This section is concerned with another hybrid task of the text categorization and the text summarization, shown in Fig. 13.19. The hybrid task which is mentioned in Sect. 13.4.2 is the text categorization in which the text summarization is used as the mean. However, the task which is covered in this section is the text summarization type in which the text categorization is used as the mean. The two hybrid tasks of the text categorization and the text summarization are different from each other. In this section, we describe the hybrid task which is the special type of text summarization.

Let us point out the differences from the hybrid task of the text categorization and the text summarization which was covered in Sect. 13.4.2. Both the tasks in

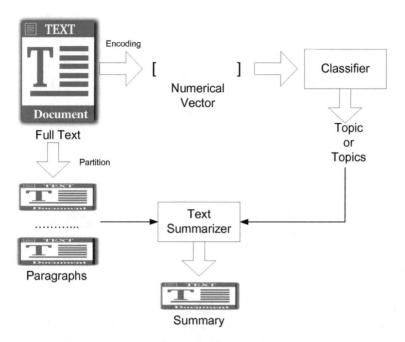

Fig. 13.19   Topic-Based Summarization

Sect. 13.4.2 and this section are the hybrid ones of the both tasks. The hybrid task in Sect. 13.4.2 is intended for the text categorization whereas one in this section is intended for the text summarization. The task in Sect. 13.4.2 is the text categorization which is reinforced by the text summarization, whereas one in this section is the text summarization which considers involved topics. In both the tasks, a single text is given as the input, but the outputs are different from each other.

The process of summarizing a text with the mixture of the text categorization is illustrated in Fig. 13.19. When a single text is given as the input, it is encoded into a numerical vector for the text classifier and it is partitioned into paragraphs for the text summarization. The text is classified into one or some of the predefined topics and the classified labels are transferred to the summarization module. The summary of the given text is decided based on the classified labels as well as paragraph contents by the summarization module. The process of summarizing a text which is presented in Fig. 13.12 is a particular instance of the topic-based summarization which is shown in Fig. 13.19.

A topic is given as a single word or two words, so we need to add more words which are associated with the topic. Word association rules are extracted from a corpus by the process which is described in Chap. 4. If a word which hits the association rule list is given as topic, its associated words are added to the summarization model which is presented in Fig. 13.19. The words which indicate a topic directly and their associated ones guide us for selecting paragraphs as summary. If topic and its associated words are included in the paragraph, it is selected as the summary.

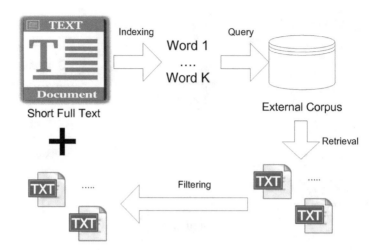

Fig. 13.20 Text expansion

The text classification results may influence on summarizing texts strongly in this scheme. If a given text is misclassified, there is possibility of generating wrong summaries by the wrong category. Wrong associated words are retrieved by the misclassification. The wrong topic and its associated words guide wrongly for deciding whether the current paragraph is a summary, or not. Therefore, this scheme is proposed under the assumption that the current text classification system is strongly reliable.

13.4.4 Text Expansion

This section is concerned with the text expansion which is the opposite task to the text summarization, as shown in Fig. 13.20. The text expansion is defined as the process of expanding a text with its relevant texts, and it is usually applicable to a short text. A text is indexed into a word list and among the words, some are selected as a query. The texts which are relevant to the query are retrieved as the associated texts, and some among them are added to the given texts. In this section, we describe the text expansion which is the opposite task to the text summarization.

The associated texts mean the ones which are strongly relevant to the given text with respect to its contents. Even if the text expansion and the text summarization are opposite to each other, they are intended to optimize the text length. The text summarization becomes the tool of cutting down the text length, and the text expansion becomes that of increasing the text length. The assumption underlying in applying the text expansion is that a text consists of only title or only one paragraph. The text summarization is not applicable to such texts.

Let us explain the process of expanding a text with its associate ones as its additional part. It is assumed that a text is given as its title, its single paragraph, and it short text. The original text is indexed into a list of words, and the words are used as the query for retrieving relevant texts. The texts which are relevant to the query are retrieved and most relevant ones are selected among them as associated ones. Their full texts or some parts are added to the original ones.

There are two ways of accessing texts: search and browsing. Search is to retrieve relevant texts by the query, and browsing is to access texts in the top–down direction, from the general cluster to its specific ones. Texts are accessed automatically through pass from the general cluster and to the specific cluster based on similarity of the original text with clusters. Most similar text is selected within the most specific cluster as the associated one. It requires the hierarchical organization of texts in the corpus for accessing texts by browsing.

Let us consider some schemes of getting the associated texts for doing the text expansion. The similarities of the original text with texts are computed and most similar ones are retrieved as associated texts. The entire corpus is clustered into a large number of small groups and texts in each groups are associated with each other. Words are represented into text sets and text association rules are extracted from them. We may use commercial search engines for retrieving associated texts, giving an original text as a query.

13.5 Summary and Further Discussions

In this chapter, we mentioned the types and the schemes of the text summarization, and hybrid tasks. We presented the four dichotomies of dividing the text summarization into the two opposite types. We mentioned some heuristic schemes of the text summarization and explained the scheme of applying the machine learning algorithms by interpreting the task into a classification task. We mentioned the hybrid tasks in which the text summarization is combined with other tasks. In this section, we make some further discussions from what we study in this chapter.

The text summarization may be interpreted into the regression rather than the classification. In this chapter, we viewed the text summarization into a binary classification where each paragraph is classified into summary or non-summary. The regression which is mapped from the text summarization is the task which estimates a continuous value of each paragraph as its essential degree. In this case, we construct the sample paragraphs which are labeled with a continuous value, instead of one of the two categories, in order to apply a machine learning algorithm. Because the final output of text summarization is actually some paragraphs as the text summary, it is recommended to view it as a classification.

The multiple viewed text summarization may be considered as one more type of this task. Different summaries in the manual text summarization are generated differently, depending on subjective; the text summaries are characterized as subjective answers. The multiple viewed text summarization is one which accommodates different versions of summary which are generated from the text by a particular approach with its different parameters or multiple different approaches. Different levels are assigned to each paragraph among ones which are generated in this type of text summarization. However, a single version of summary is expected in summarizing a relatively short text.

We need to customize the summary based on user's interests, rather than providing an abstract. We mentioned the query-based summarization where the summary is generated differently from the identical text, depending on the query. Because the query is not the exact expression of users' information need, the summary is generated based on users' profiles as well as the query. The different summary versions are generated, depending on users' interests from the same text as the results. It is applicable to only long texts; the same summary version is generated in a short text, whatever query is given.

A single text, multiple texts, and a corpus may be visualized into a tree or a network. A text may be viewed as a hierarchical structure where the root node is given as a text and terminal nodes are given as words. Paragraphs and sentences are given as intermediate nodes in the hierarchical structure. The network where nodes are given as words and edges are given semantic relationships visualizes a text or a text group. Its visualization shows the hierarchical relations among words.

Chapter 14
Text Segmentation

This chapter is concerned with text segmentation which is another text mining task. We define the text segmentation conceptually in Sect. 14.1 and explore the types of text segmentation in Sect. 14.2. In Sect. 14.3, we describe the simple and state-of-the-art approaches to text segmentation, in detail. We cover the tasks which are derived from the text segmentation, in Sect. 14.4, and make the summarization and the further discussions on this chapter, in Sect. 14.5. In this chapter, we describe the text segmentation with respect to its functions, methods, and derivations.

14.1 Definition of Text Segmentation

Text segmentation refers to the process of segmenting a text based on its content or topics into several independent texts. This task is usually applied to a long text which deals with more than a topic. A text is segmented into paragraphs in the case of a spoken text, and it is segmented into topic-based subtexts in the case of a written text. In this chapter, the latter is focused on, and it is interpreted into a classification task, where each paragraph pair is classified into boundary or continuance. In this section, we describe briefly the text segmentation in its general view.

Let us consider the case of segmenting a spoken text into its paragraphs. The speech which is given as a sound signal is transformed into a text which consists of characters, by speech recognition techniques. Boundaries of paragraphs and sentences are not available in the text which is generated from the speech by a speech recognizer. We need to segment the text into paragraphs or sentences as the additional step, in implementing the speech recognition system. This kind of text segmentation which partitions a text that is mapped from the speech into paragraphs is out of scope of this chapter.

Let us consider another case of segmenting a written text into its subtexts based on its topics and contents. It is assumed that a full text may be partitioned into

© Springer International Publishing AG, part of Springer Nature 2019
T. Jo, *Text Mining*, Studies in Big Data 45,
https://doi.org/10.1007/978-3-319-91815-0_14

paragraphs by the carriage return. We need to analyze two adjacent paragraphs, in order know how much the paragraphs are different from each other. We make the decision on the boundary or the continuance between the two paragraphs, depending on their content-based difference. The text segmentation may be interpreted into a binary classification where each adjacent paragraph pair is classified into one of the two cases.

We mention the two kinds of text segmentation and compare them with each other. The text which is generated by the speech recognition is not segmented into paragraphs, initially, whereas the written text is segmented into paragraphs initially. Segmenting a text into paragraphs is the primary role in the spoken text, whereas segmenting a text into topic or content-based subtexts is the role to the written text. Noises in the speech text come from wrong speech recognitions, whereas noises in the written text are caused by omitting periods or carriage returns between paragraphs. This comparison is concluded that approaches to both kinds of text segmentation should be developed differently.

We may use the results from segmenting a text based on its contents for reinforcing the information retrieval systems. Because a long text contains many words which are matching with the given query, potentially, it tends to be retrieved more frequently, for users. It is inconvenient to access long texts which are given as relevant ones, it is tedious to find relevant parts of each long text. If a long text is segmented into independent subtexts, only relevant subtexts are retrieved depending on the given query. Therefore, the text segmentation may provide the convenience for users of information retrieval systems by providing only their interesting parts.

14.2   Text Segmentation Type

In this section, we explore the types of text segmentation, depending on the dichotomy criteria. In Sect. 14.2.1, we examine the differences between the spoken text segmentation and the written one which were mentioned briefly in Sect. 14.1. In Sect. 14.2.2, we mention the ordered text segmentation and the unordered one, depending on which the paragraphs are ordered, or not. In Sect. 14.2.3, we describe the exclusive text segmentation and the overlapping one, depending on whether each paragraph is allowed to belong to two adjacent subtexts, or not. In Sect. 14.2.4, we introduce the hierarchical text segmentation which builds a hierarchical structure of a given text, and compare it with the flat text segmentation.

14.2.1   Spoken vs Written Text Segmentation

This section is concerned with the types of text segmentations depending on spoken and written texts. The text segmentation is generally referred to the process of partitioning a text into several independent ones. It contains both partitioning a text

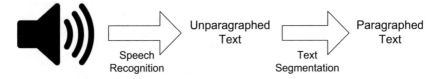

Fig. 14.1 Spoken text segmentation

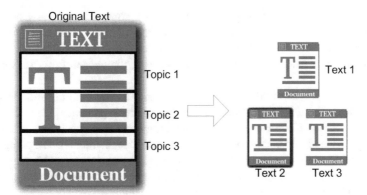

Fig. 14.2 Written text segmentation

into paragraphs or sentences and doing a text into subtexts based on its contents. In this chapter, we focus on the latter in context of text mining tasks. Before doing that, in this section, we compare the two types of text segmentations with each other.

The spoken text segmentation is illustrated in Fig. 14.1. The spoken text is given as the input which is a sound signal. The sound signal is transformed into a text which consists of characters by the speech recognition. The text which is generated by the speech recognition is segmented into paragraphs. Process of segmenting a written text into subtexts based on its contents will be covered subsequently.

Figure 14.2 illustrated the written text segmentation which segments a full text into content-based subtexts. A single full text is given in its written form. A full text is partitioned into paragraphs, and boundary or continuance is decided between adjacent paragraphs. A text is divided into subtexts by the positions which is decided as boundary. This kind of text segmentation is applicable to a long text which covers multiple topics.

The comparisons of the two types of text segmentation which are mentioned in this section are illustrated in Table 14.1. No paragraph mark exists in the text which is generated by the speech recognition, whereas in any written text, paragraph marks are initially given. We need to divide the text from the spoken

Table 14.1 Spoken vs
written text segmentation

	Speech	Written
Paragraph mark	Not available	Available
Results	Paragraph list	Subtext list
Noise	Much	Less
Initial input	Sound signal	Text

one into paragraphs, before generating topic or content-based subtexts. In the
speech recognition process, very much noise is anticipated, depending on voices
and pronunciations. In segmenting the spoken text, the sound signals are given as
the initial input, whereas in doing the written one, the written full text is initially
given.

The text segmentation may be applied one more time in dealing with the spoken
texts. A text which is not partitioned into paragraphs is generated from the spoken
one which is initially given as a sound signal through the speech recognition. The
text is partitioned into paragraphs by the spoken text segmentation. The paragraphed
one is divided into subtexts based on their topics by one more text segmentation.
The two kinds of text segmentation may be applied to the optical recognition of
texts which are given as images as well as the spoken texts.

14.2.2   Ordered vs Unordered Text Segmentation

This section is concerned with the two types of text segmentation in another
view. There exists the possibility of partitioning a text or texts into paragraphs
and clustering them; this case is called the unordered text segmentation. Placing
topic-based boundaries between adjacent paragraphs belongs to the ordered text
segmentation. Paragraph clusters become results from the unordered text segmentation, and ordered subtexts become those from the ordered one. So in this section,
we explain the two types of text segmentation, compare them with each other, and
combine them into one.

Figure 14.3 illustrates an example of the ordered text segmentation. The text
which consists of eight paragraphs is given as the initial input in this example. The
boundaries are placed, between paragraph 2 and 3, between paragraph 5 and 6, and
between paragraph 7 and 8, as shown in Fig. 14.3. The four subtexts are generated,
keeping the order, as follows:

- Subtext 1: Paragraph 1 and Paragraph 2
- Subtext 2: Paragraph 3, Paragraph 4, and Paragraph 5
- Subtext 3: Paragraph 6 and Paragraph 7
- Subtext 4: Paragraph 8

The paragraph order is maintained in the subtexts as shown in Fig. 14.3.

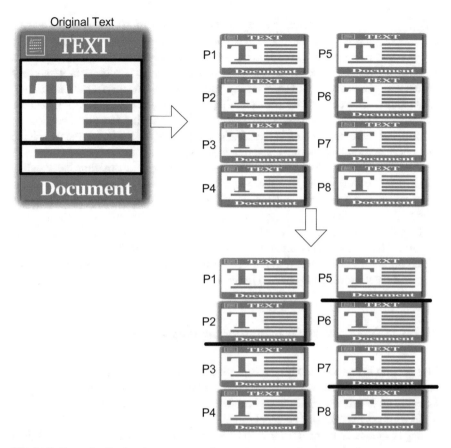

Fig. 14.3 Example of ordered text segmentation

Table 14.2 Ordered vs unordered text segmentation

	Ordered	Unordered
Output	Temporal subtexts	Subtext clusters
Paragraph pair	Adjacent	Not
Virtual text	X	O
Complexity	Linear	Quadratic

The unordered text segmentation is referred to the process of segmenting a text into subgroups which are paragraph clusters regardless of the order. A text is mapped into a group of paragraphs. The paragraphs are organized into subgroups based on their content similarities. Each cluster of paragraphs is given as subtext, ignoring the order of paragraphs. The unordered text segmentation is viewed as the paragraph clustering to a long text.

Table 14.2 illustrates the difference between the two types of text segmentation. In the ordered text segmentation, the subtexts which are ordered temporally are generated, whereas in the unordered one, clusters of content-based similar paragraphs are generated. In the ordered text segmentation, each subtext consists of adjacent paragraphs, whereas in the unordered one, each subtext consists of temporally independent paragraphs. In the former, each subtext becomes an actual text which consists of ordered paragraphs, whereas in the latter, each subtext is given as a virtual text which consists of independent paragraphs. To the number of paragraphs, it takes the linear complexity for executing the ordered text segmentation because two adjacent paragraphs are classified sequentially by sliding two-sized window on the full text, whereas it takes the quadratic complexity for executing the unordered text segmentation because the clustering task takes such complexity.

The task of text ordering is considered as the process of ordering independent texts logically into a text. An unordered list of independent texts is given as the input and a single text which consists of them in the logical order is built as the output. All possible pairs of given texts are generated, it is decided whether the texts in the pair should be ordered or reversed, and they are ordered based on the labels of pairs. The text which is generated from the text ordering becomes a virtual text which does not exist actually. The text ordering is viewed as the reverse task to the text segmentation which is covered in this chapter.

14.2.3   Exclusive vs Overlapping Segmentation

This section is concerned with the exclusive text segmentation and the overlapping one. The former is one where no overlapping between subtexts is allowed, and the latter is one where any overlapping is allowed. In the overlapping text segmentation, a paragraph spans over two adjacent subtexts in the ordered text segmentation, and it belongs to more than one cluster in the unordered one. The exclusive segmentation is used in the technical domain where each paragraph focuses on topic, clearly, while the overlapping one is used in informal domain where each paragraph may span over more than one topic. In this section, we describe the two types of text segmentation and compare them with each other.

Figure 14.4 illustrates the process of segmenting a text exclusively into subtexts. The full text is initially partitioned into eight paragraphs. As the results from the text segmentation, the system generates the four subtexts: paragraph 1 and 2, paragraph 3, 4, and 5, paragraph 6 and 7, and paragraph 8. No paragraph spans over two subtexts, in observing the subtexts which are shown in Fig. 14.4. The text segmentation which is presented in Fig. 14.4 belongs to the ordered text segmentation.

Figure 14.5 demonstrates the overlapping text segmentation by a simple example. The full text is partitioned into eight paragraphs like the case in Fig. 14.4. The process generates the four subtexts as follows:

- Subtext 1: Paragraph 1, Paragraph 2, and Paragraph 3
- Subtext 2: Paragraph 3, Paragraph 4, and Paragraph 5
- Subtext 3: Paragraph 5, Paragraph 6, and Paragraph 7
- Subtext 4: Paragraph 7 and Paragraph 8

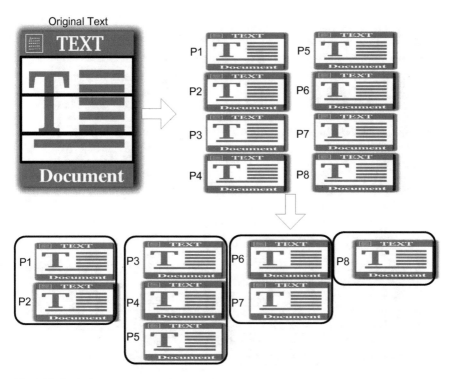

Fig. 14.4 Exclusive text segmentation

The overlapping paragraphs are available as follows: paragraph 3 between subtext 1 and 2, paragraph 5 between subtext 2 and 3, and paragraph 7 between subtext 3 and 4. The condition of the overlapping segmentation is that at least one paragraph spans over two adjacent subtexts.

Table 14.3 illustrates the comparisons of the two types of text segmentation. The status between two adjacent paragraphs is continuance or boundary in the exclusive text segmentation, whereas the status may be given as a continuous value between zero and one in the overlapping text segmentation. In the exclusive text segmentation, a paragraph belongs to only one subtext, whether it is the ordered text segmentation or the unordered one. In the overlapping one, a paragraph may belong to two adjacent subtexts in the ordered text segmentation, and to more than two subtexts in the unordered one. The exclusive text segmentation is simpler than the overlapping one because of its less complexity.

Let us consider the overlapping text segmentation where a paragraph is allowed to belong to two subtexts. The principle of writing article is that a paragraph deals with a topic consistently. In the technical articles in the domains such as science, medicine, and engineering, a paragraph deals with its own topic frequently so it spans over more than two topics rarely. Because an essay is an article which is written very freely with considering the organization very little, many paragraphs span

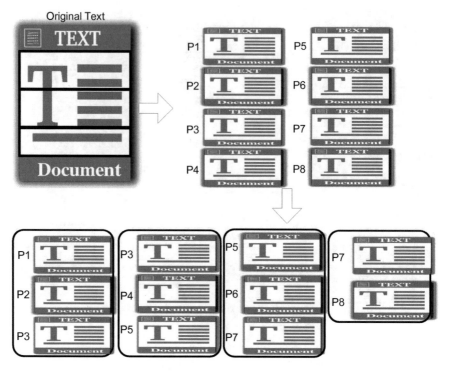

Fig. 14.5 Overlapping text segmentation

Table 14.3 Exclusive vs overlapping text segmentation

	Exclusive	Overlapping
Inter-paragraph status	Binary	Continuous
Ordered segmentation	Only one subtext	Two adjacent subtext
Unordered segmentation		More than two subtexts
Complexity	Linear	Quadratic

over more than one topic. Therefore, the exclusive text segmentation is practical for dealing with technical articles, whereas the overlapping text segmentation is so for doing with essays.

14.2.4   Flat vs Hierarchical Text Segmentation

This section is concerned with the two types of text segmentation by one more dichotomy criteria. The text segmentation is divided into the flat segmentation and the hierarchical one, depending on whether nested segmentation is allowed or not. Only one time, a text is segmented into subtexts in the flat segmentation, while a subtext is segmented into its nested ones in the hierarchical one. The list of subtexts

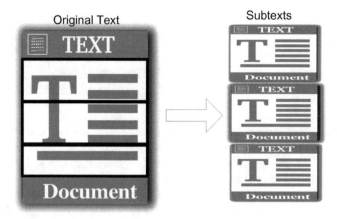

Fig. 14.6 Flat text segmentation

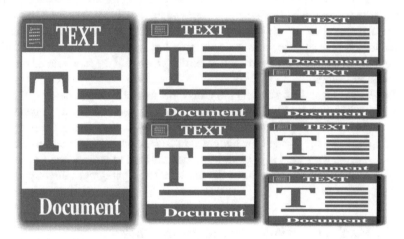

Fig. 14.7 Hierarchical text segmentation

is given as result in the flat segmentation, while a tree of subtexts is given as result in the hierarchical one. In this section, we explain and compare the two types of text segmentation.

The flat text segmentation is illustrated in Fig. 14.6. It is one where any nested subtext in a subtext is not allowed. In this type, a list of subtexts is given as output. The exclusive and flat clustering of paragraphs correspond to the flat and unordered text segmentation. No nested subtext in any subtext is the condition of the flat segmentation.

The hierarchical text segmentation is illustrated in Fig. 14.7. The full text is segmented into two subtexts in the first level, and each subtext is segmented into two nested subtexts in the second level, as shown in Fig. 14.7. The hierarchical segmentation can be implemented by setting boundary condition strictly in the

Table 14.4 Flat vs
hierarchical text segmentation

	Flat	Hierarchical
Results	Subtext list	Subtext tree
Boundary level	Single	Multiple
Subtext zooming	Limited	Possible
Nested subtext	Not available	Available

general level, and loosely in the specific level. The classifiers may be allocated level by level, by viewing the text segmentation into classification as mentioned in Sect. 14.3.2. A tree structure of subtexts rather than a list of ones is generated as the output in this type of text segmentation.

Table 14.4 illustrates the comparisons of the two types of text segmentation. In the flat text segmentation, a list of subtexts is generated as the output, whereas in the hierarchical text segmentation, a tree of subtexts is generated. In the flat text segmentation, the boundary and the continuance are given as the paragraph pair status, whereas in the hierarchical text segmentation, multiple degrees of boundary between paragraphs are given as the status. Zooming in and out in the flat text segmentation is limited because multiple boundary degrees are not allowed. In the flat text segmentation, no nested subtext exists, whereas in the hierarchical one, any nested subtext does.

The hierarchical text segmentation may be used for implementing the text browsing within a long text. As mentioned above, a tree of subtexts is generated. The root of subtext tree stands for the entire text and the leaf stands for individual paragraphs. A text is divided for finding wanted portions within a text in the top–down direction. This kind of text segmentation belongs to topic-based or semantic one which is distinguished from partitioning a text into paragraphs directly.

14.3   Machine Learning-Based Approaches

This section is concerned with some approaches to the text segmentation. In Sect. 14.3.1, we mention some heuristic approaches to the task. In Sect. 14.3.2, we interpret the text segmentation into the classification of paragraph pairs. In Sect. 14.3.3, we describe the process of encoding paragraph pairs into numerical vectors. In Sect. 14.3.4, we explain the scheme of applying the machine learning algorithms to the text segmentation.

14.3.1   Heuristic Approaches

This section is concerned with the simple and heuristic approaches to the text segmentation. The approaches in this kind are used without mapping the text segmentation task into a classification one. A text is partitioned into paragraphs and the boundary or the continuance between adjacent paragraphs is decided by the difference of words between them. Because the approaches in this kind depend strongly on lexical analysis of words, the semantic relations among words are not considered. In this section, we mention some heuristic schemes of text segmentation before covering the state-of-the-art approaches.

Let us mention the simplest approach to the text segmentation. An entire text is partitioned into paragraphs, and each one is mapped into a set of characters. The two sets of characters which represent paragraphs are compared with each other and the boundary between them is decided by the difference between the two character sets. In a variant of this method, each paragraph is mapped into an ordered list of characters and two paragraphs are compared by character frequencies. However, there is possibility of covering identical topic from different expressions.

Let us mention another heuristic scheme of segmenting a text. Each paragraph is indexed into a set of words. The two sets of words which represent the adjacent paragraphs are compared with each other; when their difference is more than the threshold, the boundary is put between them. This scheme belongs to the lexical comparison, together with the previous scheme. In both schemes, a case of the same meaning of different words is not considered.

Let us mention one more scheme of segmenting a text, based on the numerical vectors which are representing its paragraphs. The features which are attributes of numerical vectors are defined and the paragraphs in the text are encoded into numerical vectors. The similarity or distance between two numerical vectors which represent adjacent paragraphs is computed. When the similarity is less than threshold or the distance is greater than one, the boundary is put between the paragraphs. Important issues of using this method are how to define the features and how to assign values to them.

Let us mention the scheme of segmenting a text by classifying paragraphs based on their topics. As the requirement for using this scheme, the text categorization system should implemented and installed. Paragraphs are classified into one of the predefined topics, and the boundary is put between two adjacent paragraphs which are classified differently from each other. We need the preliminary tasks such as predefining categories and allocating sample labeled paragraphs. The domain of texts each of which we try to segment and the scope of predefined categories should match with each other.

14.3.2 Mapping into Classification

This section is concerned with the process of mapping the text segmentation into the binary classification, as shown in Fig. 14.8. A text is partitioned into paragraphs and adjacent paragraph pairs are generated by sliding window on the ordered paragraph list. Each adjacent paragraph pair is classified into boundary or continuance. The topic-based boundary is set between paragraphs in the pair which is labeled with boundary. In this section, we explain the classification which is mapped from the text segmentation and its comparison with other tasks.

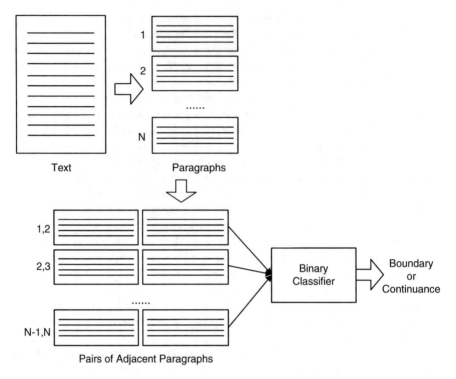

Fig. 14.8 Process of mapping text segmentation into binary classification

We illustrate the binary classification of two adjacent paragraphs which is mapped from the text segmentation in Fig. 14.8. As sample examples, we gather the adjacent paragraph pairs which are labeled with one of the two categories: boundary and continuance. The paragraph pairs are encoded into numerical vectors, and the classification capacity is built by learning the sample examples. A novice text is given as a list of adjacent paragraph pairs, each pair is encoded into a numerical vector and it is classified into boundary or continuance. The binary classification of paragraph pair becomes the core part in implementing the text segmentation system.

Let us explain the process of segmenting a text into subtexts based on their topics, using the above binary classification. A text is given as the input, text is partitioned into paragraphs, and a list of two adjacent paragraphs is generated by sliding the two-sized window. The adjacent paragraph pairs are encoded into numerical vectors, and each of them is classified into boundary or continuance. The boundary is put between the adjacent paragraph pairs which are classified with boundary, and text is segmented by the boundary into subtexts. The adjacent paragraph pair which is classified into continuance becomes a subtext which deals with an identical topic.

Table 14.5 illustrates the comparisons among the three tasks: the topic-based classification, the text summarization, and the text segmentation. In the text categorization, more than two topics are predefined, whereas in the text summarization

Table 14.5 Text classification vs summarization vs segmentation

	Classification	Summarization	Segmentation
Label	Topic	Summary or not	Boundary or continuance
Task	Multiple classification	Binary classification	
Domain	Independent	Dependent	
Entity	Article	Paragraph	Paragraph pair

and the text segmentation, only two categories are predefined. The topic- based text classification is usually given as a multiple classification, whereas the text summarization and the text segmentation are given as a binary classification. In the text categorization, an entity is labeled with topics independently of domain, whereas in the text summarization and the text segmentation, an entity is labeled with one of the two categories, depending on a domain. In the text categorization, an article which consists of more than one paragraph is a classification target, whereas in the text summarization and the text segmentation, a single paragraph and an adjacent paragraph pair is a classification target, respectively.

It is possible to map the text segmentation into a regression task as well as a classification task. The regression which is mapped from the text segmentation is the process estimating a boundary score to each adjacent paragraph pair. In order to gather sample paragraph pairs, assigning manually the boundary scores to sample ones is very dependent on subjectivity. By mapping the task into a regression, the granularity of segmenting a text into subjects should be controlled. In the regression, it is more difficult to gather sample paragraph pairs which are labeled with boundary scores than in the classification.

14.3.3   Encoding Adjacent Paragraph Pairs

Figure 14.9 illustrates the overall process of encoding paragraph pairs into numerical vectors. The adjacent paragraph pairs are generated by sliding a window on the paragraph list. In the collection, the texts are indexed into a list of words as feature candidates, and only some are selected among them as features. Adjacent paragraph pairs are encoded into numerical vectors whose attributes are the selected features. In this section, we review the feature extraction and selection and explain the two schemes of encoding them.

Let us review the process of extracting and selecting features in encoding paragraphs. The text collection is given as the source and the texts in the collection are indexed into a list of words. Only some words are selected by the criteria which is mentioned in Chap. 2, as features. We may consider other factors such as grammatical and posting attributes as well as words, themselves, as features. However, in this section, the scope of features is restricted to only words.

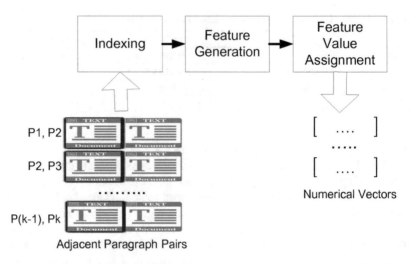

Fig. 14.9 Process of encoding paragraph pairs

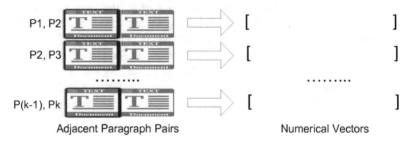

Fig. 14.10 Single encoding scheme

Figure 14.10 illustrates the first scheme of encoding adjacent paragraphs into numerical vectors. The adjacent paragraph pair is given as a text in this scheme, and it is assumed that features were already selected. A text is encoded into a numerical vector by the process which was mentioned in Chap. 3. If elements in the numerical vector are distributed with balance, this case is classified with more probability into boundary. The machine learning algorithms are applied as the binary classifier of paragraph pairs to the text segmentation.

The alternative scheme of encoding paragraph pairs is presented in Fig. 14.11. In this scheme, it is assumed that an adjacent paragraph pair is given as two texts. The two paragraphs in each pair are encoded into two independent numerical vectors as shown in Fig. 14.11. In this encoding scheme, we may consider some heuristic approaches to the text categorization based on Euclidean distance or cosine similarity between two vectors. We need to modify machine learning algorithms to this task to accommodate dual numerical vectors.

We need to introduce some features from other domains as well as the current domain. The features from the current domain are called internal features, whereas

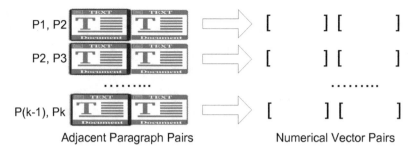

P1, P2

P2, P3

P(k-1), Pk

Adjacent Paragraph Pairs   Numerical Vector Pairs

Fig. 14.11 Pair encoding scheme

ones from other domains are called external features. External features may become important factors for deciding boundary or the continuance between paragraphs; nonzero values of external features imply the higher possibility of deciding the boundary, and zero values do that of deciding the continuance. Note that more topics or domains may be nested in the current domain; two types of boundaries may be put between paragraphs: the transition between specific topics and the transition from the topic in the current domain to one in another domain. Sometimes, a topic or topics which are out of the current domain are mentioned in a text.

14.3.4   Application of Machine Learning

This section is concerned with the scheme of applying the machine learning algorithms to the text segmentation. In the previous sections, we mentioned the process of mapping the text segmentation to a classification task and encoding paragraph pairs into numerical vectors. A text collection is divided into sub-collections by domains and machine learning-based classifier is allocated to each domain. The classifier is learned from the representations of sample paragraph pairs, and novice paragraph pairs are represented into numerical vectors and classified into boundary and continuance. In this section, we describe in detail the process of applying the machine learning algorithm to the classification task which is mapped from the text segmentation.

Figure 14.12 illustrates the process of encoding paragraph pairs into numerical vectors. The text collection is divided into sub-collections which correspond to their own domains, and adjacent paragraph pairs are generated by sliding a window on the paragraph list. The feature candidates are extracted from the collection of paragraph pairs and some among them are selected as features. Each paragraph pair is encoded into a numerical vector by assigning values to features and it is labeled into boundary or continuance, manually. The process of encoding paragraph pairs into numerical vectors is the same to that of encoding texts or paragraphs so.

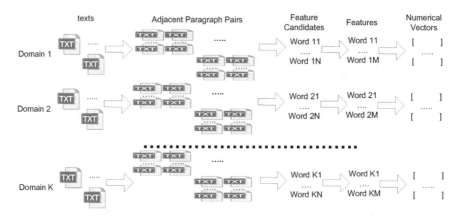

Fig. 14.12 Encoding paragraph pairs into numerical vectors

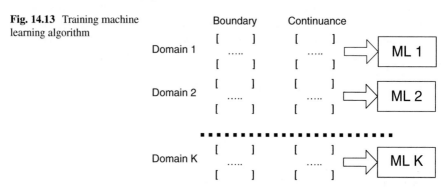

Fig. 14.13 Training machine learning algorithm

Figure 14.13 illustrates the process of training the machine learning algorithm by training examples which are generated by the process which is shown in Fig. 14.12. A machine learning algorithm is allocated to its own domain, and labeled numerical vectors which represent paragraph pairs are prepared as training examples. A set of training examples is divided into the two groups: the boundary group and the continuance group, as shown in Fig. 14.13. The classification capacity is constructed by learning the prepared training examples to domain by domain. Tagging a domain to texts or classifying them into one of domains is necessary for doing the text segmentation.

Figure 14.14 illustrates the process of classifying the paragraph pairs into boundary or continuance. The text which is given as input and tagged with its own domain is partitioned into paragraphs and a classifier which corresponds to the domain is nominated. Adjacent paragraph pairs are extracted by sliding the two-sized window on the paragraphs, and encoded into numerical vector. Individual paragraph pairs are classified into one of the two categories and boundaries are marked between paragraphs in the pair which is classified into boundary. Subtexts are generated by the marked boundaries from the full text and they are treated as independent texts.

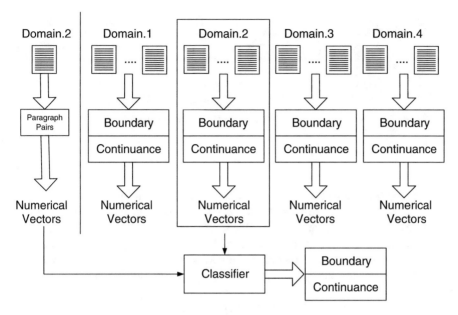

Fig. 14.14 Classifying paragraph pairs into boundary or continuance

14.4 Derived Tasks

This section is concerned with tasks which are derived from the taxonomy generation and consists of the four subsections. In Sect. 14.4.1, we mention the process of analyzing the full text, paragraph by paragraph, based on its involved topics. In Sect. 14.4.2, we explain the information retrieval where subtexts relevant to the query are retrieved, instead of full texts. In Sect. 14.4.3, we consider the subtext synthesization which assembles subtexts into a full text, as the opposite to the text segmentation. In Sect. 14.4.4, we mention virtual texts which is the output from the subtext synthesization.

14.4.1 Temporal Topic Analysis

This section is concerned with the process of spotting ordered topics, as shown in Fig. 14.15. A text is partitioned into ordered paragraphs, and one of the predefined topics is assigned to each paragraph. A list of topics which are assigned to paragraphs becomes the ordered topic list to the given full text. We adopt the HMM (Hidden Markov Model) for considering topics of previous paragraphs, as well as properties of current one. In this section, we describe the HMM as the learning algorithm, briefly and explain its application to the topic analysis.

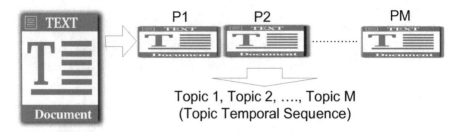

Fig. 14.15 Spotting ordered topics

The HMM (Hidden Markov Model) is the probabilistic model for defining a state sequence using the observation sequence, based on the conditional probabilities which are given previous states. We define the two sets: the state set, $S = S_1, S_2, \ldots, S_{|S|}$, and the observation set, $O = O_1, O_2, \ldots, O_{|O|}$. The state transition probabilities are defined as the matrix where rows correspond to the list of current states and the columns correspond to the list of next states,

$$
\mathbf{A} = \begin{bmatrix}
P(S_1|S_1) & P(S_2|S_1) & \ldots & P(S_{|S|}|S_1) \\
P(S_1|S_2) & P(S_2|S_2) & \ldots & P(S_{|S|}|S_2) \\
\vdots & \vdots & \ddots & \vdots \\
P(S_1|S_{|S|}) & P(S_2|S_{|S|}) & \ldots & P(S_{|S|}|S_{|S|})
\end{bmatrix}
$$

The probabilities that observations are given are defined as the matrix where the rows correspond to the list of states and the columns correspond to the list of observations,

$$
\mathbf{B} = \begin{bmatrix}
P(O_1|S_1) & P(O_2|S_1) & \ldots & P(O_{|O|}|S_1) \\
P(O_1|S_2) & P(O_2|S_2) & \ldots & P(O_{|O|}|S_2) \\
\vdots & \vdots & \ddots & \vdots \\
P(O_1|S_{|S|}) & P(O_2|S_{|S|}) & \ldots & P(O_{|O|}|S_{|S|})
\end{bmatrix}
$$

The initial probabilities are defined as a vector: $\Pi = \begin{bmatrix} P(S_1) & P(S_2) & \ldots & P(S_{|S|}) \end{bmatrix}$.

Let us consider the tasks to which the HMM is applied. The probability of the observation sequence, $O_i O_{i+1} \cdots O_{i+j}$, is evaluated based on the above definitions. The observation sequence is given as the input and the state sequence with its maximum likelihood to the observation sequence is found. The set of observation sequences is given as training examples and the matrices, $\mathbf{A}$ and $\mathbf{B}$, and Π are estimated by learning them. The process of generating the topic sequence from the text corresponds to the second task.

If applying the HMM to the ordered topic spotting, the topic sequence is given as the state sequence, according to the paragraph order, and the word sequence is given as the observation sequence. Representative words are extracted from each paragraph, and the word sequence becomes the observation sequence. By analyzing the corpus statistically, we define the parameters: **A**, **B**, and *Π*. We find the state sequence as the ordered topic sequence, by computing the maximum likelihood of given observed sequence to all possible state ones. The second and the third task of HMM may be also involved in the ordered topic spotting.

We need to compare the ordered topic spotting and the unordered one with each other. The list of topics which are assigned to a text is interpreted into a topic list which is covered in the text in the unordered one, while it is interpreted as the temporal sequence of topics in the ordered one. The supervised learning algorithms are used as the approaches in the unordered one, while the HMM is used as the approach in the ordered one. The unordered topic spotting belongs to the fuzzy classification, while the ordered one belongs to the temporal sequence analysis. The unordered one is used for organizing texts based on their topics, while the ordered one is used for analyzing contents within a text.

14.4.2 Subtext Retrieval

This section is concerned with the subtext retrieval as shown in Fig. 14.16. The full texts which are relevant to a given query are retrieved in the traditional information retrieval. Even if the subtext is referred to a text part such as a paragraph, a sentence, and a word, it means here that the text part which deals with the consistent topic as results from text segmentation. The process of retrieving relevant subtexts instead of full texts is called subtext retrieval. In this section, we describe the task of subtext retrieval.

The information retrieval is referred to the process of retrieving texts which are relevant to the query. The query is given by a user which expresses his or her information needs and we need the query processing by refining and expanding the query to its more specific ones. We compute relevancy between a query and a text and texts are ranked by their relevancy. The texts which are relevant to a query are retrieved and presented for users. Long texts tend to be overestimated of their relevancy because of many words.

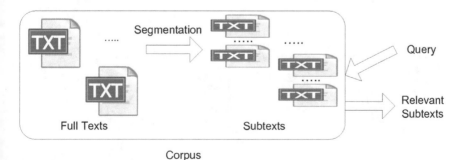

Fig. 14.16 Process of retrieving subtexts

Let us explain the information retrieval which is specialized for retrieving subtexts. Each text in the corpus which is the source is segmented into paragraphs which are subtext instances. The paragraphs are indexed and encoded into structured forms. The paragraphs which are relevant to the query are retrieved as independent texts. It is expected to prevent from overestimating the relevance of long texts by doing so.

Instead of paragraphs, a text is partitioned into topic-based subtexts. The text collection is divided into sub-collections by domains or topics. In each sub-collection, paragraph pairs are gathered and classified into continuance and boundary by the corresponding classifier. Between each paragraph pair which is classified with boundary, a mark is put as the boundary; a text is divided by the boundary into subtexts. These subtexts are treated as independent ones, in executing the information retrieval.

We need the subtext classification and the subtext clustering, in order to reinforce the subtext retrieval. The subtexts are extracted based on topics from the full text by the text segmentation. The subtexts are treated as the independent data items, and the information retrieval is applied to them. There is possibility of executing the clustering and the classification to the subtexts as well as the full texts. Subtexts which are partitioned from a long text and short full texts become the information retrieval source.

14.4.3   Subtext Synthesization

This section is concerned with the subtext synthesization which is shown in Fig. 14.17. It is the process of making the full text artificially by assembling subtexts. The full text fragments are actually concentrated on a particular topic. The texts are assembled as shown in the right part of Fig. 14.17, within each cluster into a single text after clustering them. In this section, we describe the process of synthesizing subtexts into a full text.

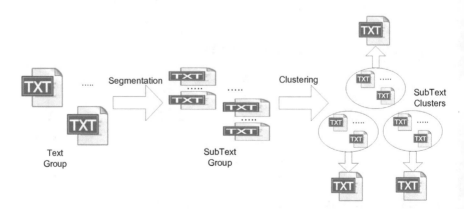

Fig. 14.17 Subtext synthesization

In this section, it is assumed that a subtext is given as a paragraph. A text group is mapped into a paragraph group by partitioning each text into paragraphs. Paragraphs are encoded into numerical vectors like independent texts and they are clustered into subgroups of similar ones by a clustering algorithm. Because many paragraph permutations may exist in each cluster, a paragraph cluster never become a single full text. We should find the optimal paragraph permutation for synthesizing a text.

It is assumed that a text domain is a technical area and each full text consists of the three parts: introduction, body, and conclusion. We need to classify paragraphs into one of the three parts, before synthesizing them into a full text. In the sampling process, the first and the last paragraph in each full text are labeled with introduction and conclusion, respectively, and intermediate ones are labeled with body. The labeled paragraphs are encoded into numerical vectors and the supervised learning is trained by the sample paragraphs. If assuming that paragraphs from known full texts and ones whose sources are not known exist together, the former becomes sample paragraphs and the latter becomes novice ones.

The subtexts are classified into one of the above three categories and they are assembled into a full text. In the given cluster, the subtexts which are classified into introduction are taken as the first full text part. Subtexts which are classified into body are attached to the first part. The subtext synthesization is finalized by attaching subtexts which are classified into conclusion. There are many cases of permutations of subtexts which are classified into body, in the full text medium.

Let us consider some issues in assembling subtexts into an article. We must decide whether subtexts which are classified into body are put after one which is classified into introduction, after one which is done into body, or before one which is done into conclusion. We must decide the number and the order of subtexts which are classified into body for building the logical full text. We need to decide the relevancies of subtexts which are classified into conclusion before attaching it. The distribution over the three categories of subtexts is never balanced in any cluster.

14.4.4 Virtual Text

This section is concerned with virtual texts which are the ones synthesized automatically by subtexts. It is assumed that there are two kinds of texts: actual texts which are manually written by human being and virtual texts which are assembled by subtexts from different sources. Users tend to have interests spanning over several texts, rather than concentrating on a particular text, so they need to synthesize his or her interesting parts into a single text. We may consider various schemes of making the virtual texts: replacing a particular paragraph by another from external sources, and synthesizing subtexts from difference sources into a text. In this section, we compare the two kinds of texts with each other and describe the scheme of making virtual texts.

Let us compare the actual text and the virtual texts with each other. The actual texts are ones which are written or edited manually and entirely by a human being, and the virtual texts are ones which are automatically by a computer. The virtual text consists of subtexts from different actual texts. If a sentence or a paragraph may be replaced by one from a different text, the text becomes a virtual text which satisfies the minimum requirement. The process of making a virtual text by assembling only interesting subtexts for users is called text customization.

Let us mention some schemes of generating virtual texts. A paragraph or a sentence may be replaced by one from another text. A group of texts is partitioned into one of paragraphs and they are clustered into subgroups, as mentioned in Sect. 14.4.3. A short text is expanded by adding its associated texts as mentioned in Sect. 13.4.4. A long text is divided into subtexts based on their topics as independent texts.

The virtual texts which are customized to users are built after retrieving relevant texts. User never scans entirely individual texts which are relevant to the query. The scanned parts of relevant texts are assembled into one text as more desirable text. Subtexts are gathered from relevant text which were clicked by a user, they are partitioned into paragraphs, and they are assembled into a single text. The important issue is to determine which is the introduction, the body, or the conclusion, in building a virtual text.

Let us consider the automatic text editing for making and modifying virtual texts. Editing a text is a manual task, requiring scanning and understanding texts. Deletion of irrelevant sentences and addition of associated sentences from other texts are operations for editing texts, automatically. Virtual texts are usually not logical, so they need to be edited, manually or automatically. Much more research is needed for editing automatically the virtual texts to be more logical.

14.5   Summary and Further Discussions

In this chapter, we described the text segmentation with respect to the types, the approaches, and the derived tasks. We presented the dichotomies for dividing the text segmentation into the two opposite types. We mentioned some heuristic schemes of text segmentation and the applications of machine learning algorithms to it by viewing it into a classification task. We covered the tasks which are derived from the text segmentation and related with it. In this section, we make some further discussions from what we studied in this chapter.

In this chapter, we covered the text segmentation and mentioned the text expansion which is opposite to it in Sect. 14.4.4. Textual data is always given with very variable lengths in a collection; in the collection of news articles, only title is given as a very short text and a series of news is given as long text. We may use the tasks for normalizing the text length; the text segmentation is applicable to a long text and the text expansion is applicable to a short text. The long text may be divided into subtexts which are treated as independent texts by the text segmentation and the short text is expanded by adding its associated text by the text expansion. We need to define the criteria for deciding one of the two tasks to each text.

The text segmentation system may be implemented by the text classification system. A text is partitioned into paragraphs and they are encoded into numerical vectors. The paragraphs are classified into one or some of the predefined categories. Two adjacent paragraphs which are labeled with the same category or nonempty intersection of two sets of labeled categories are treated as continuous; otherwise, they are treated as boundary. In the crisp classification, a text may be segmented into small-sized subtexts and in the fuzzy classification, a text may be segmented into large- sized ones.

We may consider the text segmentation based on the word clustering results. The words which are collected from a corpus are clustered into subgroups by their meaning. Two paragraphs are indexed into their own lists of words and the similarity of the word list with word clusters is computed. Each paragraph is associated with the word cluster whose similarity is maximum, and if two lists of words are associated with different word clusters, the boundary is put between the two paragraphs. The semantic word organization is used for segmenting a text based on the paragraph meaning, in this case.

We need to trim virtual texts for improving its readability. The virtual text which is made by replacing a paragraph with another is relatively more readable. The paragraphs in the virtual text are made by synthesizing them from their different sources that should be trimmed. Trimming such as removing unnatural sentences and adding some phrases will improve the readability. The criteria or the metric should be defined for evaluating the quality of virtual texts before proposing the schemes of doing that.

Chapter 15
Taxonomy Generation

This chapter is concerned with the taxonomy generation which is corresponding to the automatic category predefinition. We define the taxonomy generation in its general view, in Sect. 15.1, and mention some of its relevant tasks in Sect. 15.2. In Sect. 15.3, we describe the schemes of taxonomy generation in detail. We cover the process of maintaining taxonomies in Sect. 15.4, and make the summarization and further discussions on this chapter in Sect. 15.5. In this chapter, we describe the taxonomy generation which especially supports the text categorization.

15.1 Definition of Taxonomy Generation

Taxonomy generation refers to the process of generating topics or concepts and their relations from a given corpus. The task of taxonomy generation consists basically of listing topics or categories from the corpus and linking each of them to its relevant texts. A hierarchical structure or network of topics or categories is built as an advanced form of taxonomy generation. The taxonomy generation is used for defining the classification frame automatically and it may be used by itself as a knowledge base. In this section, the taxonomy generation is described in its functional view before discussing its process in detail.

Let us consider reasons of generating taxonomies from a given corpus. In order to perform text categorization tasks, we must define and update categories as a list or tree. For the semantic processing, an ontology as a kind of knowledge representation is constructed manually or semiautomatically. A single text, several texts, or a corpus many be visualized graphically based on their contents. By building the semantic structures among words, we may analyze contents of texts or corpus, in order to organize them.

Let us consider some schemes of generating taxonomies from a given corpus. An entire corpus is indexed into a list of words and some among them are selected as

© Springer International Publishing AG, part of Springer Nature 2019
T. Jo, *Text Mining*, Studies in Big Data 45,
https://doi.org/10.1007/978-3-319-91815-0_15

taxonomies. The texts in the corpus are clustered into subgroups, and each subgroup is named with its most relevant words. Association rules are extracted from viewing a text as a word set and construct taxonomies based on them. Taxonomies are organized as a graph where each node indicates a word, and each edge indicates a semantic relation between words.

We may express taxonomies as various graphical configurations from a simple list to a complicated graph. A list of concepts and categories is the simplest form of taxonomy organization for predefining categories automatically as the preliminary task of text categorization. A hierarchical structure of concepts or categories from the abstract level to the specific level is suitable for predefining them for the hierarchical text categorization. A network of concepts or categories together with their semantic relations is a typical form of taxonomy organization. The information about relation, properties and methods of each concept, and axioms which are constraints may be added to the above network of taxonomies.

Even if other goals are available, it is assumed that the taxonomy generation is mainly intended to define the classification frames automatically. Only a list of unnamed clusters which result from clustering texts is not sufficient for automating the preliminary tasks for the text categorization. Prior knowledge about the given domain is required for doing the preliminary tasks manually. The list of important concepts which is generated from corpus by the taxonomy generation is the classification frame which should be defined for doing the text categorization. The relations among them as well as the important concepts are also generated as the output of the taxonomy generation.

15.2   Relevant Tasks to Taxonomy Generation

This section is concerned with the text mining tasks which are relevant to the taxonomy generation, in order to provide the background for understanding it. In Sect. 15.2.1, we describe the keyword extraction and its relevancy to the text generation. In Sect. 15.2.2, we mention the task where a single word or a bigram is classified into one or some of predefined topics, instead of text. In Sect. 15.2.3, we explain the word clustering which is associated with the word classification, where words are clustered semantically into clusters. In Sect. 15.2.4, we cover the topic routing which is the reverse task to the text categorization where a topic is given as an input and a list of texts belonging to the topic is generated as the output.

15.2.1   Keyword Extraction

Keyword extraction is referred to the process of extracting important words which represent the entire contents of the given full text, as illustrated in Fig. 15.1. The full text is given as the input, and a list of words which are included in the text

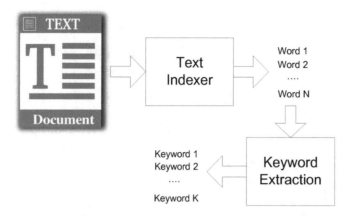

Fig. 15.1 Keyword extraction

is generated as the output. The full text is indexed into a list of words by the text indexer, and some among the words are selected as the keywords. The keyword extraction is viewed as the decision of whether each word which is indexed from text is a keyword or not. In this section, we describe the keyword extraction with its connection with the taxonomy generation.

Let us explain the keyword extraction in its functional view. A single full text is externally given as the input. The text is viewed as a list of words; a full text consists of a paragraph, a paragraph consists of sentences, and a sentence consists of words. Some words which reflect the entire full text content are generated in the keyword extraction as the results. Multiple texts or a corpus may be given as the input to the keyword extraction in some cases.

Let us describe the process of extracting keywords from a full text. A full text is given as the input, and it is indexed into a list of words. Each word which is extracted by the indexing process is decided into an important or an unimportant one. The words which are judged as important ones are generated as keywords. The decision which is mentioned above belongs to a binary classification task.

The mapping from the keyword extraction into a classification task is illustrated in Fig. 15.2. A text which is given as the initial input is indexed into a list of words, and the process of doing so was already mentioned in Chap. 2. Each word which is indexed from the text is classified into a keyword or a non-keyword, by the classifier. The words which are classified into keywords are selected and generated externally as the output. We need to consider the domain for collecting the sample labeled words.

Figure 15.3 illustrates the relevancy of the keyword extraction to the taxonomy generation. Individual texts in the collection are indexed into their own lists of words. In each text, among its indexed words, some words are selected as its keywords. The list of keywords which are gathered from the texts become taxonomies which are given as a flat list. Additional filtering steps may be needed for selecting further keywords.

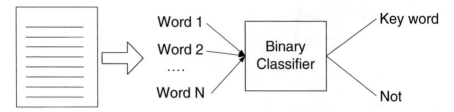

Fig. 15.2 Mapping keyword extraction into classification task

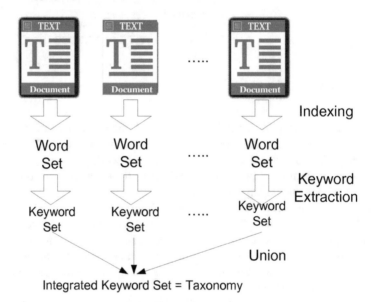

Fig. 15.3 Relevancy of keyword extraction to taxonomy generation

15.2.2   Word Categorization

Word categorization is referred to the process of assigning one or some among the predefined categories to each word. The text categorization means the process of classifying a text which is an article, whereas the word categorization is that of classifying a word. The POS (Part of Speech) tagging which is a typical instance of word categorization is to classify a word into one of the grammatical categories, such as noun, verb, adjective, and so on. The word categorization which is covered in this section is the task which is distinguished from the POS tagging and is to classify each word based on its topic or meaning. In this section, we explain the word categorization and mention its connection to the taxonomy generation.

Let us explain the process of encoding words into numerical vectors. A corpus is needed for doing it and text identifiers become features. Texts are selected from the corpus by their lengths as the features, and its frequencies in the selected texts are assigned to them as the feature values. The corpus may be represented into a

Table 15.1 Ordered vs unordered text segmentation

	Word categorization	Keyword extraction
Categories	Topics	Keyword or not
Type	Multiple classification	Binary classification
Entity	Word	Word
Classification criteria	Semantic	Importance

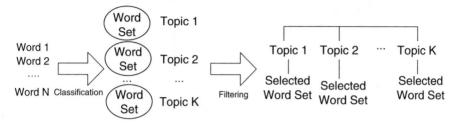

Fig. 15.4 Relevancy of word categorization to taxonomy generation

word–text matrix where each column stands for a text and each row for a word. The rows of the word–text matrix become numerical vectors which represent words, while the columns become ones which represent texts.

Let us explain the process of classifying words by a machine learning algorithm. The sample labeled words are encoded into numerical vectors by the process which was mentioned above. The classification capacity is constructed by learning the samples. The test words are encoded into numerical vectors, and they are classified into one or some of the predefined categories. Therefore, the word categorization is the same as the text categorization except that words are given as classification targets, instead of articles.

Even if both tasks belong to the classification, the differences between the topic-based word categorization and the keyword extraction are illustrated in Table 15.1. In the word categorization, the categories are predefined as a list or a tree, whereas in the keyword extraction, the two categories, keyword and non-keyword, are predefined. The word categorization usually belongs to the multiple categorization, whereas the keyword extraction always belongs to the binary classification. In both the tasks, words are given as classification targets. In the word categorization, words are classified based on their meanings, whereas in the keyword extraction, they are classified based on their importance degrees to the given text.

Let us mention the relevancy of the word classification to the taxonomy generation which is described in this chapter. The predefinition of categories and the allocation of sample words are manual preliminary tasks for the word categorization. As illustrated in the left part of Fig. 15.4, the words which are classified into one or some of the predefined categories are gathered. Through filtering, we select important words among the classified ones; texts are retrieved based on the selected ones. We may add one more task, word clustering, for automating the manual preliminary tasks.

	Cluster 1	Cluster2	Cluster3	Cluster4
Word 1	O	X	X	X
Word 2	X	O	X	X
Word 3	X	X	X	O
Word 4	X	O	X	X
Word 5	X	O	X	X
Word 6	O	X	X	X
Word 7	X	X	O	X
Word 8	X	O	X	X

Fig. 15.5 Word clustering

15.2.3   Word Clustering

Word clustering is referred to the process of segmenting a group of words into subgroups of similar ones, as illustrated in Fig. 15.5. In the broad view, we consider the lexical clustering where words are clustered by their spellings and the grammatical one where they are done by their grammatical functions. However, in this section, we restrict the scope to only semantic clustering where words are clustered by their meaning or topics. A corpus is necessary for encoding words into numerical vectors. In this section, we describe the word clustering in detail, compare it with the word categorization, and present its relevancy to the taxonomy generation.

The word clustering is compared with the word categorization before discussing it. The word categorization requires predefining categories and gathering labeled sample words as the preliminary tasks. To the word categorization, the supervised learning algorithms are applied, whereas to the word clustering, the unsupervised learning algorithms are applied. The word clustering proceeds depending on the semantic similarities among words; the lexical and grammatical ones are out of scope of this section. In both the tasks, words should be encoded into numerical vectors whose attributes are text identifiers.

Let us explain briefly the process of clustering words by their meanings. The words are encoded into numerical vectors whose attributes are text identifiers, referring to a corpus. The numerical vectors are clustered by a clustering algorithm into subgroups. The clusters of content-based similar words become the reference for doing the taxonomy generation. Clustering words by their spellings is regarded as a trivial task and is not counted in this section.

Figure 15.6 presents the synergy by combining the text clustering and the word clustering with each other. The left part of Fig. 15.6 shows the word clustering and the right part shows the text clustering. The important references which are necessary for doing the text clustering are derived from the word clustering. The results from clustering words are adjusted by ones from clustering texts. The

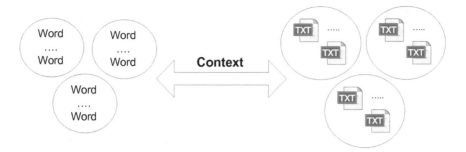

Fig. 15.6 Combination of word and text clustering

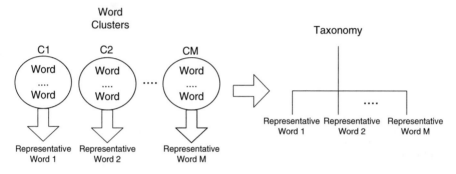

Fig. 15.7 Application of word clustering to taxonomy generation

decision whether the initial clustering is performed on words, texts, or mixtures becomes the issue of connecting the two tasks with each other.

Figure 15.7 illustrates the process of applying the word clustering to the taxonomy generation. The words are encoded into numerical vectors, referring to the corpus, and they are clustered by their semantic similarities. The representative words are extracted from clusters by the medoid schemes which are mentioned in Chap. 10. The representative words are given as a list of predefined categories and they are linked with their relevant texts. Extracting and selecting words from the corpus become the preliminary tasks.

15.2.4 Topic Routing

Topic routing is defined as the process of retrieving relevant texts to the given topic, as shown in Fig. 15.8. It is viewed as the inverse of text categorization where the input is a topic and the output is a list of texts. The task is interpreted as a special type of information retrieval where a topic is given as a query. The core operation in the topic routing is to compute the relevancy between a text and the topic. In this section, we explain the topic routing in its functional view, interpret it into a classification task, and connect it with the taxonomy generation.

Fig. 15.8 Topic routing

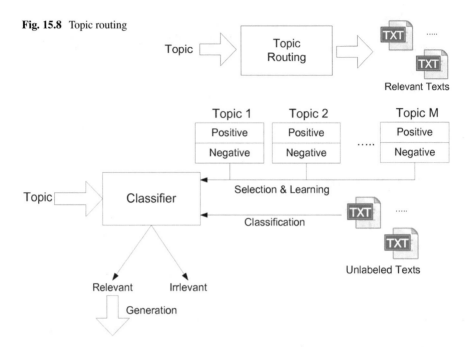

Fig. 15.9 Mapping topic routing into classification task

Let us review briefly the topic spotting which was mentioned in Sect. 5.4.4. The topic spotting is from viewing the fuzzy text categorization into the scene where some topics are spotted to each text. A full text is given as the input and its relevant topics are assigned to it through the fuzzy classification. In the literature, it was decomposed into binary classifications. The topic routing is derived by reversing the input and the output of the text spotting.

The topic routing is the process of finding relevant texts to the given topic. The topics are predefined as a list or a tree, and each of them is given as the input. Matching value between the topic and each text is computed, and highly matched texts are selected by ranking them. A list of texts which are relevant to the given topic is generated as the output. The topic routing is viewed as the task which is similar as the information retrieval, in that a topic becomes a query.

The topic routing is mapped into a binary classification task as shown in Fig. 15.9. The topics are predefined in advance, and classifiers and training examples are allocated to their corresponding topics. The classifier which corresponds to its own topic learns the training examples which are labeled with the positive class and the negative class. The topic is given as the input, texts in the corpus are classified by its corresponding classifier, and texts which are classified into the positive class are retrieved. To an alien topic which is out of the predefined ones, one more classifier is created and training examples should be prepared.

Figure 15.10 illustrates the process of implementing the taxonomy generation using the topic routing. A corpus is indexed into a list of words, and only essential

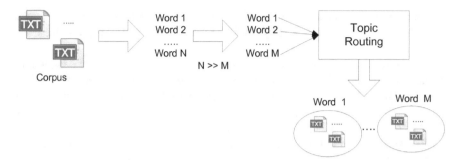

Fig. 15.10 Application of topic routing to taxonomy generation

Fig. 15.11 Process of generating taxonomies based on text indexing

words are selected among them. The selected words become the topics which are given as input in the topic routing, and the relevant texts are retrieved through the topic routing. The topic is associated with a list of texts. The criterion of selecting words as topics is very important for implementing the taxonomy generation.

15.3   Taxonomy Generation Schemes

This section is concerned with the schemes of generating taxonomies from a corpus. In Sect. 15.3.1, we explain the simplest scheme which is called the index-based scheme. In Sect. 15.3.2, we mention the popular scheme which is called the clustering-based one. In Sect. 15.3.3, we cover the taxonomy generation through the data association. In Sect. 15.3.4, we describe the process of generating taxonomies by analyzing semantic networks of words and texts.

15.3.1   Index-Based Scheme

Figure 15.11 illustrates the process of generating the taxonomies based on the text indexing. A corpus is given as the input for generating taxonomies. It is indexed into a list of words, some among them are selected as taxonomies, and relevant texts are arranged to each taxonomy. The selected words are categories or topics of a given text group. In this section, we explain the process of generating taxonomies from the corpus by this scheme.

The entire corpus is indexed into a list of words as the first step of the taxonomy generation. The texts in the corpus are concatenated into an integrated text, and it is tokenized into a list of tokens. Each token is converted into its root form by applying the stemming rules. The stop words are removed. The process of indexing a text is described in detail in Chap. 2.

We need to select some words among them as taxonomies. The selection criterion is the total frequencies of words in the given corpus, assuming that all of stop words were already removed. The average TF-IDF weight is the advanced criteria for selecting words. The posting information and the grammatical information are considered may be considered as the criteria for selecting so. The process of selecting words as taxonomies may be viewed as the classification task to which machine learning algorithms are applicable.

After selecting important words as taxonomies, we need to consider their semantic connections. Taxonomies are given as ones which are semantically connected by their networks, rather than independent ones. The connections among them are determined by their collocations. Two words are connected by rate of the number of texts which include both words multiplied by two to the summation of the number of texts which include either of them, which is called collocation rate. For deciding the connections among words, we may consider other factors: adjacency of two words within a text and number of co-occurrences of them in the given text.

It is assumed that semantic relations among words are available. The semantic operations were defined and their mathematical theorems were set by Jo in 2015, under the assumption [40]. The semantic similarity and the lexical one between two words are not always correlated. The semantic similarity between two words is computed based on their collocations. We need to define and characterize mathematically more advanced semantic operations on words.

15.3.2   Clustering-Based Scheme

This section is concerned with the process of generating taxonomies by clustering texts, as shown in Fig. 15.12. A corpus is given as the input, and texts in the corpus are clustered into subgroups. Each cluster is named by the process which was described in Sect. 9.4.1. The named clusters which are generated through the text clustering and the cluster naming become the output of taxonomy generation. In this section, we describe the scheme for generating taxonomies based on the text clustering, in detail.

A corpus which is given as a text collection is clustered into unnamed subgroups. Texts in the corpus are encoded into numerical vectors by the process which was mentioned in Chap. 3. Clusters of content-based similar texts are constructed by a clustering algorithm, such as the AHC algorithm and the k means algorithm.

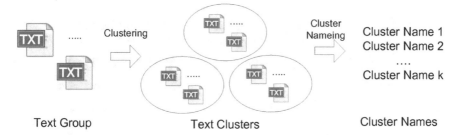

Fig. 15.12 Clustering-based taxonomy generation scheme

By policies, we decide the crisp or fuzzy clustering and the flat or hierarchical clustering. The process of clustering texts was explained in detail in Part III.

The cluster naming is the subsequent task to the text clustering for generating taxonomies. The cluster naming was mentioned in Sect. 9.4.1, in terms of its principle and process. For each cluster, TF-IDF weights of words are computed and words with their higher weights are extracted as the cluster names. Identifying clusters by numbers is out of the definition of cluster naming. The browsing should be enabled from the results from naming clusters.

Analyzing relations among clusters is an additional task after naming clusters. Clusters which are results from data clustering are not always independent ones; some clusters independent of others and others are related to each other semantically. We may analyze the links by computing the intercluster similarities which are mentioned in Chap. 12. As the results from analyzing the links among clusters, a graph where its vertices are cluster names and its edges are semantic links is derived. Taxonomies are generated in this scheme through the three steps: text clustering, cluster naming, and the link analysis.

Even if this scheme is expected to generate results with their good qualities, it takes much computation cost. It takes the quadratic complexity to the number of data items for clustering data items, and it also does the identical complexity for analyzing the links among clusters. It takes less complexity in clustering data items by the single pass algorithm which was described in Sect. 10.2.3, but it generates results with the worse quality as the payment. As the solution, we propose that texts are clustered by their summaries instead of their full texts by adding the task, text summarization. In order to do that, we need to summarize text individually in the given corpus.

15.3.3 Association-Based Scheme

This section is concerned with the taxonomy generation scheme based on the word association, as illustrated in Fig. 15.13. The corpus is given as the input and individual texts in it are indexed into a set of words. Association rules are extracted

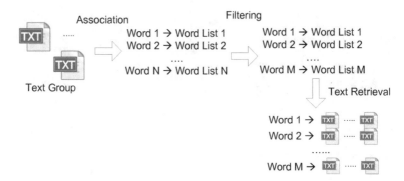

Fig. 15.13 Word association

from word sets by the process which was described in Chap. 4. Some association rules are filtered and the words in the conditional parts are given as the list of taxonomies. In this section, we review the process of extracting the association rules and explain the taxonomy generation scheme.

We mentioned the process of extracting association rules of words in Chap. 4. Individual texts in the corpus are indexed into word sets. Their subsets are generated and some of them are filtered by their supports. For each subset, all association rules are generated and some of them are filtered by their confidences. The association of a single word in the conditional part with a list of words in the causal part is given as an association rule.

As shown in Fig. 15.13, we need the additional filtering of association rules for generating taxonomies. Here, the association rules are those whose causal parts are subsets of the causal part with its maximum size in another association rule; when the causal part of the association rule is a subset of that in another association rule, the two association rules are treated as redundant ones. The TF-IDF of the word in the conditional part becomes a criterion for filtering association rules. The causal parts are pruned by cutting down some words for increasing their support in each association rule. The results from filtering association rules is input which is direct to the task of the taxonomy generation.

The conditional words of association rules which are pruned and filtered become the topics of corpus. We need to link texts to each association rule. The causal part and a text are encoded into numerical vectors, and the similarity between a word list in the causal part and a text is computed by the cosine similarity. If the similarity is more than the threshold, the text is linked with the causal part of the association rule. The final results of generating taxonomies from a corpus by this scheme are the effect of accomplishing the preliminary tasks for text categorization automatically.

It is possible to cluster association rules into subgroups of similar ones. The intersection between causal parts of association rules is used for defining the similarity metric between them. Using the AHC algorithm, the association rules are clustered based on the similarity metric. An alternative way is to encode association

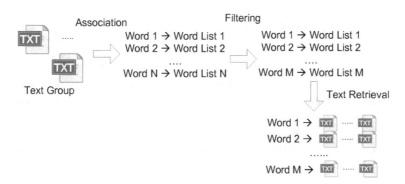

Fig. 15.14 Link analysis-based scheme

rules into numerical vectors and select representative one in each cluster, using the k means algorithm, or the k medoid algorithm. The taxonomies are generated as a hierarchical structure from the clustered association rules.

15.3.4 Link Analysis-Based Scheme

This section is concerned with the alternative scheme of taxonomy generation to what is covered in the previous sections, as shown in Fig. 15.14. In the previous sections, we already described the index-based scheme, clustering-based scheme, and association-based scheme. We define the connections among texts in the corpus as a network and select texts which play a role of hubs in the network. As shown in Fig. 15.14, taxonomies are generated by indexing selected texts which are called hub texts. In this section, we explain in detail the process of generating taxonomies by the scheme which is presented in Fig. 15.14.

Creating links among texts is to start building the text networks. Texts are encoded into numerical vectors, and cosine similarities among them are computed. If the similarity between the two texts is greater than or equal to the threshold, the link between texts is created. The threshold is given as an external parameter, and if the threshold is close to zero, dense links among texts are built. In the too dense networks, almost all of texts are selected as hubs, while in the very sparse networks, very few texts are selected.

The text network is constructed by creating links among texts. It is represented as a graph where vertices are texts and edges are links between them. The links which are created by the process which is mentioned above are bidirectional among texts. Each link is associated with a cosine similarity value as a weighted edge. The texts which are not connected with any other text are removed as the isolated one.

Taxonomies are generated from the text network which is constructed by the above process. The degree of each node which indicates a text is counted as the number of connected ones. Texts which correspond to the node with the degree which is more than or equal to the threshold are selected as hub texts. The hub texts are indexed into a list of words, and among them, some with higher weights are selected as taxonomies. Relevant texts are linked to each selected word.

Let us consider the undirected links among texts in the text network. The link which is created between texts implies the two texts which are similar with each other based on their contents. We may consider the directed link between texts where one is the source and the other is destination. Words which occur in both texts are called shared words, and the text which includes the shared words with less portion becomes the source text. Texts with a higher number of outgoing degrees are selected as hub texts in the directed text network.

15.4   Taxonomy Governance

This section is concerned with the operations which are involved in governing the existing taxonomy and consists of four subsections. In Sect. 15.4.1, we explain the schemes of maintaining the existing taxonomy. In Sect. 15.4.2, we mention the scheme for growing a single taxonomy into its advanced one by adding more texts. In Sect. 15.4.3, we consider the process of integrating multiple taxonomy organizations into a single taxonomy organization. In Sect. 15.4.4, we mention ontology that is the form expanded from the taxonomy organization.

15.4.1   Taxonomy Maintenance

This section is concerned with the maintenance of existing taxonomy to the additions of more texts and the deletions of some texts. In Sect. 15.3, we studied the schemes of creating a new taxonomy organization from a corpus. We need to maintain the created one to changing corpus by the operations: addition, deletion, and update. The operations which are involved in maintaining the taxonomy organization are division of a particular topic into several specific ones, merge of multiple topics into one topic, and addition of new topics. In this section, we explain the operations which are involved in managing the taxonomy organizations to the dynamic corpus.

Figure 15.15 illustrates the taxonomy division which is an operation for governing the taxonomy. Among M taxonomies, taxonomy i grows as a bigger one, as texts are accumulated. The taxonomy i is divided into taxonomy i and $i + 1$ by a clustering algorithm. The total number of taxonomies is increased by one through division. The intra-cluster similarity is considered as the alternative criteria to the number of texts for deciding the taxonomy division.

The taxonomy merge as another operation is illustrated in Fig. 15.16. In the figure, it is assumed that the two taxonomies, i and $i + 1$, are most similar with each other. The two taxonomies are merged into a taxonomy, and the number

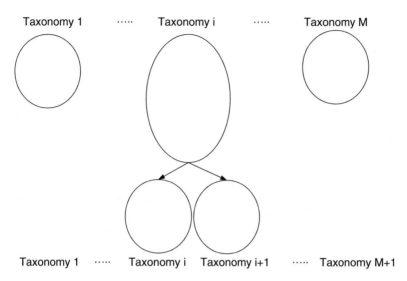

Fig. 15.15 Taxonomy division

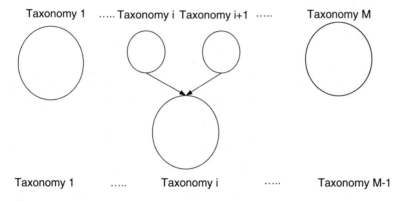

Fig. 15.16 Taxonomy merge

of taxonomies is decremented by one. A pair of taxonomies is selected by their similarity and the total number of texts in both the taxonomies. The intercluster similarity may be considered as the criterion for deciding the taxonomy merge.

Let us consider the taxonomy addition and deletion as well as taxonomy merge and division in governing taxonomies. There are cases that very sparse taxonomies which are iterated from others happen and texts need to be added to a new topic rather than to one or some of existing ones. If adding texts about a new topic or new topics, we need to create new taxonomies. If taxonomies have very few texts and are isolated from others, they need to be deleted. Here, the four operations, the taxonomy division, the taxonomy merge, the taxonomy creation, and the taxonomy deletion, become the main operations which are involved in maintaining taxonomies.

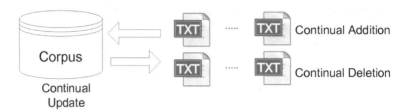

Fig. 15.17 Continual addition and deletion of texts

Let us consider the taxonomy script which is the brief version of taxonomy contents. It is very tedious to access some texts for knowing taxonomy contents. It is necessary to make a script for each taxonomy by summarizing texts in it. In Chap. 13, we mentioned the techniques of summarizing texts automatically. We use the script about taxonomy for previewing its contents.

15.4.2   Taxonomy Growth

This section is concerned with the taxonomy growth which is the gradual evolution of taxonomies. In this section, it is assumed that texts are added continually to the corpus which consists of taxonomies. In the given organization, taxonomies grow continually in the sizes by doing so. We need to maintain them by upgrading organization. In this section, we describe the taxonomy growth by adding texts, continually, and schemes of updating taxonomies.

In Fig. 15.17, we illustrate the continual addition of new texts and continual deletion of some texts. Taxonomies are constructed from the corpus by the schemes mentioned in Sect. 15.3. When adding more texts than deleting ones, taxonomies in the organization grow with respect to its size. The quality of organizing texts with the current taxonomies may be degraded by doing so. Therefore, we need to maintain the text organization by updating taxonomies.

Figure 15.18 illustrates the process of creating a new taxonomy to alien texts. There is a possibility that texts about new topics are given rather than ones about existing topics. The entire group of texts divided into the familiar and alien groups by the similarities with existing taxonomies. The texts in the alien group are indexed into a list of words and the new taxonomies are created from the list. It is important to define the criteria for doing the division.

Figure 15.19 illustrates the process of expanding the hierarchical taxonomy by one level. Taxonomy 1 and 2 are given as the big sized ones on the left side in Fig. 15.20. Taxonomy 1 is divided into taxonomy 1–1 and 1–2, and taxonomy 2 is divided into taxonomy 2–1 and 2–2. Taxonomy 1, 2, and 3 are given in the first level, and taxonomy 1–1, 1–2, 2–1, and 2–2 are given in the second level on the right side in Fig. 15.19. The divisive clustering algorithm which was mentioned in Sect. 10.2.2, is used for dividing a cluster into two clusters, at least.

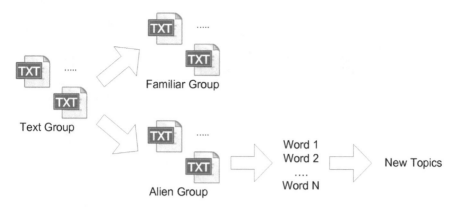

Fig. 15.18 Creating new taxonomy to alien texts

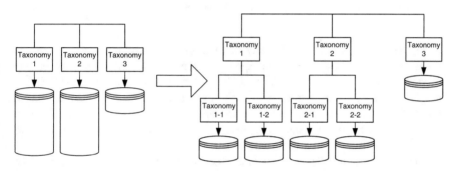

Fig. 15.19 Expanding hierarchical taxonomy by one level

We need to consider the taxonomy downsize by deleting texts continually. The taxonomy downsizing is the process of cutting down the level one by one by deleting texts continually. It is regarded as the opposite process to what is shown in Fig. 15.19. Taxonomy 1–1 and 1–2 are merged into taxonomy 1, and taxonomy 2–1 and 2–2 are merged taxonomy 2, for example. Because more texts are usually added than the ones which are deleted, the taxonomy growing is more frequent than the taxonomy downsizing.

15.4.3 Taxonomy Integration

This section is concerned with the taxonomy integration which is the process of merging existing taxonomy organizations into a single organization. Several taxonomies are constructed by different schemes which were mentioned in Sect. 15.3. We need to integrate them into a single organization of taxonomies. In this case, we consider the similarity between taxonomies as the important fact for doing so. In this section, we describe schemes for integrating taxonomy organizations which are made by different algorithms from difference sources.

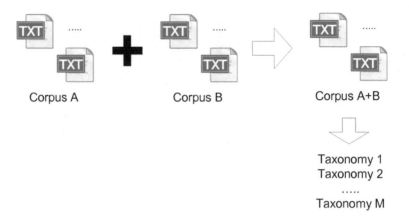

Fig. 15.20 Simple way of merging taxonomy organizations

Figure 15.20 shows the easy way of merging taxonomy organizations by the corpus integration. We gather the corpora which are associated with the taxonomy organizations which we try to merge. The gathered corpora are integrated into a single corpus by merging them with each other. Taxonomies are constructed from the merged one using one of the schemes which were mentioned in Sect. 15.3. In this merging scheme, it should be considered that the important structures in two existing taxonomy organizations may be lost.

Before integrating multiple taxonomy organizations, we need to consider the similarity between two taxonomies. A taxonomy is given as a single symbolic name and its linked text list; the taxonomy is viewed as a named cluster of texts. The similarity between two taxonomies is one between two clusters; the process of computing the cluster similarity was mentioned in Sect. 10.2.1. In the process of integrating taxonomy organizations, two taxonomies from different organizations are merged into one taxonomy. The taxonomy merge within the same organization which is presented in Fig. 15.16 is called intra-taxonomy merge, whereas the taxonomy merge between different organizations is called inter-taxonomy merge.

The process of integrating the taxonomy organizations is illustrated by a simple example in Fig. 15.21. The two organizations are merged; corpus A consists of taxonomies: business, information, society, and Internet, and corpus B consists of taxonomies: IoT, company, sport, and automobile. There are similar taxonomies between the two corpora; one has business and company, the other has Internet and IoT. The taxonomies, business and company, are merged into business, and the taxonomies, Internet and IoT, are merged into Internet, on integrating the two taxonomy organizations. The four taxonomies, information, society, sport, and automobile, are treated as the independent ones, in the integrated organizations.

The flat taxonomy organizations are integrated by simply merging similar taxonomies into one as shown in Fig. 15.21, but some issues should be considered in integrating hierarchical taxonomy organizations. We need to decide in which level the two similar taxonomies in the different levels of two organizations are merged.

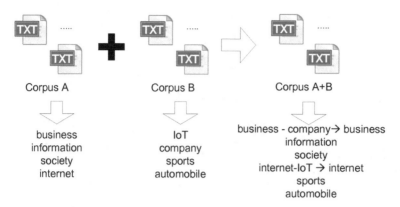

Fig. 15.21 Integrating taxonomy organizations

The issue is how nested taxonomies should be treated in merging their parents into one. Another issue is how the nested taxonomies whose parents are very different are merged. We need the manual edit of taxonomy organization into which two organizations are merged.

15.4.4 Ontology

Ontology is referred to the specification of what exists in the real word as a tree or a graph. In the graph representing an ontology, each node indicates a concept or an entity which exists in the world and each edge indicates the relation of two concepts or entities. More components which are not available in the taxonomy organization such as the script, the properties, and the methods are added, based on the object-oriented programming concept. Because the ontology is much more complicated than the taxonomy organization, it tends to be constructed manually or semiautomatically. In this section, we describe briefly the ontology with simple examples and their definition language.

Figure 15.22 presents a simple example of ontology. The department of computer science is given as a root node, the three concepts, graduate course and undergraduate course, are provided from it, and people are given as members. The two concepts, machine learning and neural networks, are given as instances of graduate courses, and the two concepts, data structures and Java programming, are given as instances of undergraduate courses. The concept, people, is divided into the three subgroups: students, faculty, and staff. The three levels of faculty are defined as full professor, associate professor, and assistant professor, at the bottom of Fig. 15.22.

The OWL (Web Ontology Language) is the standard language for defining an ontology. The ontology which is illustrated in Fig. 15.22 is expressed by the OWL, as shown in Fig. 15.23. The OWL consists of tags which indicate objects. The node in the ontology in Fig. 15.22 is given as a tag which consist of owl, thing, RDF, and name. We need the indent for writing nested ones in objects.

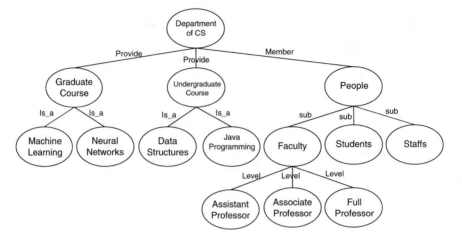

Fig. 15.22 An example of ontology

```
<owl: thing rdf=ID="department of CS">
    <owl:thing rdf: Provide = "Graduate Course">
        <owl:thing rdf: is_a = "Machine Learning"/>
        <owl:thing rdf: is_a = "Neural Networks"/>
    </owl:thing>
    <owl:thing rdf: Provide = "Undergraduate Course">
        <owl:thing rdf: is_a = "Data Structures"/>
        <owl:thing rdf: is_a = "Java Programming"/>
    </owl:thing>
    <owl:thing rdf: Member = "People">
        <owl:thing rdf: sub = "Faculty">
            <owl:thing rdf: Level = "Assistant Professor"/>
            <owl:thing rdf: Level = "Associate Professor"/>
            <owl:thing rdf: Level = "Full Professor"/>
        </owl:thing>
        <owl:thing rdf: sub = "Student"/>
        <owl:thing rdf: sub = "Staff"/>
    </owl:thing>
</owl:thing>
```

Fig. 15.23 Expressing ontology in OWL

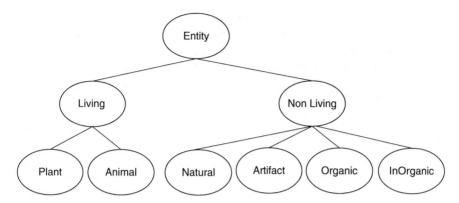

Fig. 15.24 Nodes in first level of WordNet

Figure 15.24 illustrates the nodes at the root level and the first level of the ontology, WordNet. The existing ontologies are built manually or semiautomatically, including WordNet, by Noy and Hafner [74]. The WordNet is the ontology type where entities are organized semantically; the entity which is what exists in the world is divided into living and nonliving. The living is divided into plants and animals, and the nonliving is divided into natural, artifact, inorganic, and organic. The WordNet is used for the semantic information retrieval where matching between the query and the text is computed based on not the spelling but the meaning.

Let us compare the ontology with the taxonomy organization with respect to their differences. The taxonomy organization is intended for building a frame of performing the text categorization, whereas the ontology is intended for building a knowledge base. The concepts called taxonomies are given as a list, a tree, or a graph in the taxonomy organization, whereas the concepts which are connected with others are given as forests or graphs in the ontology. In the taxonomy organization, the relation between taxonomies indicates a semantic similarity or a specification, whereas in the ontology, the relation indicates one of various types. The properties and the methods are added to each concept, assuming each concept is regarded as a class, in the ontology.

15.5   Summary and Further Discussions

In this chapter, we described the taxonomy generation with respect to the involved tasks, the generation schemes, and the maintenance schemes. We mentioned some tasks which are needed for executing the taxonomy generation such as the keyword extraction, the word classification, the word clustering, and topic routing. We explained the four schemes of creating taxonomy organization, newly. We considered schemes of maintaining the taxonomy organization to the cases where texts are added or deleted continually. In this section, we make some further discussions from what we studied in this chapter.

Let us consider the process of identifying clusters with their symbolic names. Text clusters are named by weights or frequencies of words in the texts in their corresponding cluster. In the word cluster, each cluster consists of words, so the representative word is selected among them, in order to name the cluster. In this case, the k medoid algorithm which was mentioned in Sect. 10.3.4 is applied to the word clustering. The word clusters are named with their representative word which was selected by the algorithm.

We mentioned several schemes of generating taxonomies in Sect. 15.3. We may consider the combinations of multiple schemes, rather than using a single scheme for generating taxonomies. Taxonomies are generated by each scheme, independently, and taxonomy organizations are integrated into a single version. Integrating intermediate outputs during the taxonomy generation is the alterative way of combining the multiple schemes. If using the combination of multiple schemes for the task, we expect more reliable results but degraded speed as the payment.

Text prediction is the process of estimating a full text, a summary, a title, or a topic in figure to the given temporal text sequence. The assumption underlying in the task is that texts have been added continually, following a pattern. Texts are arranged as a time series sequence, and they are encoded into numerical vectors. The numerical vector is estimated in the next step, and decoded into a virtual text. It is constructed by assembling existing paragraphs, based on the numerical vector.

The taxonomy organization consists of taxonomies and linked text lists in the simple version, but we need the additional components, to reinforce it. The semantic relations among taxonomies are defined in the organization, rather than a linear taxonomy list. The categories are defined on the relations among taxonomies and the relations should be discriminated by their strength. The summary of linked texts is given as the taxonomy script. Each taxonomy may be defined as a class which has the properties and methods, and taxonomy instances are created as objects.

Chapter 16
Dynamic Document Organization

This is concerned with the process of executing the dynamic document organization system which was initially proposed by Jo in 2006. We explore the dynamic document organization system with respect to its execution, in Sect. 16.1, and explain the online clustering which is the basis for implementing the system, in Sect. 16.2. In Sect. 16.3, we describe the dynamic organization which is the compound task of text categorization and clustering, in detail. We point out the issues in implementing the DDO (Dynamic Document Organization) system, in Sect. 16.4, and make the summarization and further discussions on this chapter, in Sect. 16.5. In this chapter, we describe the DDO system as the compound system.

16.1   Definition of Dynamic Document Organization

The system of DDO is a software agent or program for managing texts automatically based on their contents. This system is able to be implemented by only online clustering. However, in 2006, Jo validated that the system should be implemented by compounding the text clustering and the text categorization with each other for its better performance. The DDO system executes with the two operation modes: the creation mode and the maintenance mode, which will be explained in detail, later. Before describing its process in detail, in this section, we explore the functions of the DDO system.

Organizing and managing texts are viewed as a compound task of text clustering and categorization. The DDO system which is covered in this chapter is one which is distinguished from the text mining system which executes the text clustering and the text categorization, separately and independently. Depending on the given modules, one of the two tasks is automatically selected. This system starts with the maintenance model where texts are piled as a single group and maintains texts by automatic transition between the two operation modes. Human beings never intervene into the system for managing and organizing texts.

© Springer International Publishing AG, part of Springer Nature 2019   341
T. Jo, *Text Mining*, Studies in Big Data 45,
https://doi.org/10.1007/978-3-319-91815-0_16

Let us mention the role of text clustering in executing the automatic management of texts. The role is to provide a new organization or an updated one. Texts are continually added into a single group in the initial stage, and the texts are clustered into several subgroups. The quality of text organization is monitored and the organization is updated into one with its better quality. There are two types of updating the text organization: the soft reorganization which is merge and division of existing clusters and the hard organization which is the reorganization of entire texts, ignoring the previous one.

Let us explain the role of text categorization in executing the DDO system. In the initial stage, added texts are piled, continually. Afterward, the system learns the texts which are organized by the text clustering into clusters as sample texts, and constructs its classification capacity. Texts which are added subsequently are arranged after the organization; it is the main role of text categorization. The mode where texts are arranged into one of the clusters by the text categorization is called the maintenance mode.

Before discussing the proposed system, we need to distinguish between the mixture and the compound of two tasks. The fact that the two tasks are given independently as two menus is regarded as a mixture of them, whereas if the two tasks are coupled into one task, this case is regarded as the compound of them. In the mixture, human beings intervene for selecting one of the two tasks, whereas no intervention is available in the compound. In the mixture, both the tasks are evaluated independently of each other, whereas in the compound, both are evaluated as a single task. In the mixture, either of the two tasks is executed depending on user's selection, semiautomatically, whereas the human decision is not required in the compound.

16.2   Online Clustering

This section is concerned with the online clustering which is a way of managing text automatically. In Sect. 16.2.1, we explain the online clustering in its conceptual and functional view. In Sect. 16.2.2, we describe the version of k means algorithm which is modified to the online clustering. In Sect. 16.2.3, we mention the KNN which is the typical supervised learning algorithm but is modified into an online clustering version. In Sect. 16.2.4, we introduce the online fuzzy clustering which estimates memberships of examples to their clusters, rather than arranging them.

16.2.1   Online Clustering in Functional View

The online clustering is referred to the specific type of clustering where items are given as a stream, and the clustering is continued almost infinitely. The data clustering, which was covered entirely in Part III, is the offline clustering, assuming

Fig. 16.1 Offline clustering

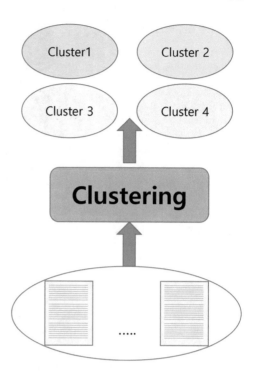

that all items to cluster are given at a time. In the real world, items which we try to cluster are never given at a time, and data items enter as an almost continual stream. We need to develop online clustering techniques by modifying existing clustering algorithms and creating a new one, for clustering data items which are given as a continual stream. In this section, we compare the offline and online clustering, and point out their issues.

The offline clustering is illustrated in Fig. 16.1. The clustering which was entirely covered in Part III belongs to this clustering type. The offline clustering is referred to the clustering where all data items are given at a time. The assumption underlying in the offline clustering is that no data item is available subsequently after clustering existing ones. The traditional clustering algorithms which were described in Chap. 10 are used for doing the offline clustering, so they need to be modified into their online clustering versions for processing big data.

The online clustering is illustrated as a diagram in Fig. 16.2. The assumption underlying in the online clustering is that data items are given as a continuous stream, and it is impossible to wait for all data items to be clustered. Before data items arrive, the number of clusters should be decided and each cluster should be characterized, initially. Whenever it arrives, it is arranged into one or some of clusters, and their characters are updated. The difference from the offline clustering is that clusters are updated, interactively and incrementally.

Let us consider some issues in expanding the offline clustering into the online one. Because data arrives as an almost unlimited stream especially in processing big

Fig. 16.2 Online clustering

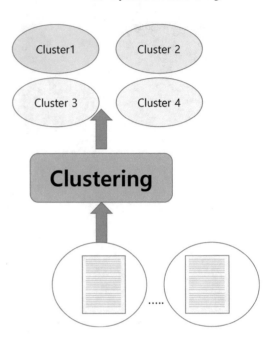

data, it is impossible to investigate the entire collection of data items for executing the data clustering. The results from the online clustering depend strongly on the order in which data items arrive. We need to adjust results from clustering a stream of data items for improving them; too frequent adjustment causes the heavy load of the system, and too infrequent one causes the poor quality of clustering results. We need to monitor a stream of data in queues for deciding when the results should be adjusted.

Because the existing clustering algorithms were developed under the assumption of the offline clustering, we need to modify them into their online versions. Each cluster should be updated interactively, whenever an item arrives, instead of updating it in batch after exploring all items. We may prepare the predicted data items which are made artificially depending on the knowledge about the application domain, called virtual examples, for initializing the cluster characteristics. The decremental clustering may be added for the situation of deleting some data from the collection. The online clustering may be expanded into one which deals with the data items whose values are variable.

16.2.2   Online K Means Algorithm

Let us mention the modified version of k means algorithm which is suitable for the online clustering. In Sect. 16.2.1, we already explained the online clustering in its functional view. The k means algorithm consists of computing mean vectors of clusters and arranging items into their clusters, depending on their similarities with cluster mean vectors. The modification from the offline version into the online one is to replace the batch updates of the cluster mean vectors by the interactive ones. So, in this section, we describe the online version of the k means algorithm and mention some of its variants.

We already explained the offline version of k means algorithm, in Chap. 10. All of data items which we try to cluster are given initially, and the cluster mean vectors are initialized at random. The data items are arranged into clusters by their similarities with the cluster mean vectors, and they are updated by averaging arranged data items. The arrangement and the update are repeated until the cluster mean vectors are converged. Refer to Chap. 10 for obtaining a more detailed description of the k means algorithm.

The k means algorithm is modified into the online version, by updating interactively mean vectors to each arrived data item. It is assumed that data items arrive as a stream. The cluster mean vectors are initialized at random or based on the prior knowledge about the application domain. Each arrived data item is arranged into one or some of the clusters, depending on its similarities with the mean vectors, and the mean vectors are updated. Updating the mean vectors interactively for each arrived data item is called at-a-time streaming processing and doing them after arranging several data items as a block is called micro-batched stream processing [68]. The issues of online clustering are the worse clustering performance than the offline clustering and the strong dependency on the order in which data items arrive.

It is possible to modify the variants as well as the traditional version. The fuzzy k means algorithm may be modified by updating the mean vectors whenever membership values of clusters of the arrived item are computed. The k medoid algorithm may be modified into its online version by nominating the medoids, whenever an individual item arrives. In order to optimize the mean vectors, we may add virtual examples which are derived from existing examples which arrived previously. An offline clustering may be attached, in order to organize the previous data items again.

Because the k means algorithm is the most simplified version of EM algorithm, we may modify the EM algorithm into its online version. For each cluster, a Gaussian distribution is defined with its initial mean vector and covariance matrix at random or on primary knowledge. Whenever an item arrives, the two steps, E-step and M-step, are executed; the E-step is to estimate probabilities that the item belongs to the clusters and the M-step is to update the parameters of Gaussian distributions which represent the clusters. We consider other distributions for modeling the clusters such as triangle distribution, trapezoid distribution, or uniform distribution, as well as the Gaussian distribution, in designing the EM algorithm.

16.2.3 Online Unsupervised KNN Algorithm

This section is concerned with the unsupervised version of KNN. The KNN algorithm was already mentioned in Chap. 6 as an approach to text categorization.

The supervised learning algorithm, KNN, is modified into its online clustering algorithm. It is possible to modify supervised learning algorithms into unsupervised ones, and vice versa. In this section, we will describe the KNN as an online clustering algorithm.

Let us review briefly the original version which was covered in Chap. 6 for providing the background for understanding the modified version. In advance, the training examples are never touched before any novice example is given. For each novice example, its similarities with the training examples are computed. Most similar training examples are selected as its nearest neighbors and its label is decided by voting their labels. In Chap. 6, we mentioned its variants as well as its original version.

In modifying the KNN into the online version, it is assumed that the labeled training examples are not available initially. The number of clusters is determined initially and virtual training examples are generated in each cluster, at random or by prior knowledge about the given application domain. When each training example arrives, its similarities with virtual examples are computed and its cluster is decided by voting clusters which the selected virtual examples belong to. In the supervised KNN, some of actual training examples are selected as the nearest neighbors, whereas in this version, some of virtual ones are selected. If the KNN is applied to the offline clustering, some items are selected at random as representative ones of clusters.

The variants of KNN may be modified into their unsupervised versions as well as its original version. The radius nearest neighbors may be modified by generating virtual examples in each cluster, at random or depending on the prior knowledge about the given application domain. The variant which assigns different weights to the selected nearest neighbors may be modified into unsupervised version by doing so, as mentioned above. The strong prior knowledge about the application domain becomes the requirement for discriminating attributes by their importance. It is considered to derive further more virtual examples by cross-over, mutation, and midpoint which are mentioned in the area of evolutionary computations [16].

The constraint clustering reminds us of the unsupervised KNN algorithm. The constraint online clustering means the clustering where some labeled examples are given initially, for clustering subsequent data items. The labeled examples are assigned to their corresponding clusters for using the unsupervised KNN algorithm, instead of virtual examples. If the number of labeled examples is actually insufficient, more virtual examples tend to be added to the clusters. The labeled examples in the constraint clustering become the prior knowledge for proceeding with the clustering.

16.2.4 Online Fuzzy Clustering

We mention the online fuzzy clustering in this section. The fuzzy clustering is one where each item is allowed to be arranged into more than one cluster. Actually, rather than arranging data items, membership values of each item in the clusters are computed. In the fuzzy clustering, when each item arrives, the membership values of clusters are computed and it is arranged into more than one cluster based on their membership values. In this section, we describe the fuzzy clustering in detail.

Let us review the fuzzy clustering which was mentioned in Chap. 9, before discussing the fuzzy online clustering. The fuzzy clustering which was called overlapping clustering is the clustering where each item is allowed to belong to more than one cluster. Even if the overlapping clustering and the fuzzy clustering are mentioned as the same one, they are different from each other with respect to their focuses. The overlapping clustering focuses on the fact that some items stand on more than two clusters as overlapping ones, whereas the fuzzy clustering focuses on the fact that membership values of clusters are computed as continuous values. In the case of visualizing clustering by different clusters, overlapping areas tend to be colored by mixtures of colors of clusters.

Let us expand the fuzzy clustering into the online version. It is assumed that data items arrive as a stream instead of being given at a time. For each item which arrives, its similarities with the cluster prototypes are computed, and it is arranged into clusters whose similarities are above the given threshold. As an alternative way, its similarities with them are computed and transformed into its membership values through normalization. Whenever a new item arrives subsequently, the clusters, which are overlapped with others, or the matrix, where each column corresponds to a cluster, each row corresponds to a data item, and an entry stands for its membership value of the cluster, are updated.

Let us mention the online fuzzy k means algorithm as a fuzzy online clustering tool. The number of clusters is decided and their mean vectors are initialized. The membership values of each item are computed by its cosine similarity with a mean vector which is expressed in Eq. (16.1),

$$
\mu_c(\mathbf{x}) = \frac{\mu \cdot \mathbf{x}}{||\mu|| \cdot ||\mathbf{x}||} \tag{16.1}
$$

where μ is the mean vector, and $\mu_c(\mathbf{x})$ is the membership value of the input vector, $\mathbf{x}$, to the cluster, C. Each mean vector is updated based on the membership value by Eq. (16.2),

$$
\mu(t + 1) = \frac{\mu(t) + \mu_c(\mathbf{x})\mathbf{x}}{1 + \mu_c(\mathbf{x})} \tag{16.2}
$$

Each item has its membership vector which consists of membership values to the clusters.

Let us mention the fuzzy constraint clustering as the alternative type to the fuzzy online clustering. The constraint clustering was already mentioned in Sect. 9.2.4 as one where some examples are clustered in advanced as labeled ones. Some examples are labeled with more than one category or have their own membership values of clustering. The cluster prototypes are initialized, referring to the fuzzy labeled examples. This type of clustering is identical to the constraint clustering, except that some examples are labeled with more than one class or have continuous membership values.

16.3   Dynamic Organization

This section is concerned with the dynamic organization of texts which was proposed by Jo in this PhD dissertation in 2006 [25]. In Sect. 16.3.1, we demonstrate the execution process of the system. In Sect. 16.3.2, we explain the maintenance mode of the system. In Sect. 16.3.3, we describe the creation mode of executing the system. In Sect. 16.3.4, we mention text mining tasks which are involved in implementing the system.

16.3.1   Execution Process

This section is concerned with the process of executing the dynamic document organization system which is implemented by Jo in 2006 [25]. The system has no text initially, and when it is installed in the field, it starts with the initial maintenance mode where texts are piled into a single group. Texts start to be clustered after piling them, as the initial transition into the creation mode. Subsequently added texts are arranged into clusters by classifying them in the maintenance mode. In this section, we explain briefly the two modes, the creation mode and the maintenance mode, of the dynamic document organization system.

Figure 16.3 illustrates the initial maintenance mode of the dynamic document organization system. When the system is installed initially, there is no document in it. Texts arrive into the system continually as the process of piling them. After piling so, we need to define the criteria for transitioning the current mode into the creation mode, automatically. The number of texts or intra-cluster similarity is the criterion for carrying out the transition.

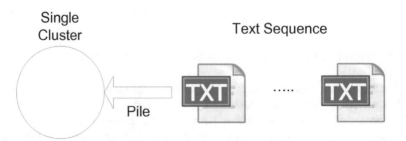

Fig. 16.3 Initial maintenance in DDO system

Classifying subsequently added texts into their own cluster identifiers is the alternative scheme of arranging them, in the maintenance mode. A supervised machine learning algorithm is adopted separately from the unsupervised one in the creation mode, and it learns the organized texts which are sample texts. The subsequently added texts are encoded into numerical vectors, and they are classified into one or some of the cluster identifiers. They are arranged into clusters whose identifiers are assigned to themselves. It is more desirable to arrange them by the supervised learning algorithm through the empirical validation which was performed by Jo in 2006 [25].

In order to decide whether the system transits into the creation mode, we need to evaluate the quality of organizing texts. Because target labels are not available in maintaining the system, the clustering index needs to be modified into the version which is suitable for the situation. The intercluster similarity and the intra-cluster similarity are computed for organizing texts, depending on the cosine similarities among texts. The clustering index is computed by Eq. (12.6), using the two measures. The cosine similarity depends on text representations, rather than full texts, themselves.

As the maintenance mode continues, the organization quality may be degraded, so we need the transition from the maintenance mode to the creation mode. The number of texts which are added subsequently becomes the direct transition criterion. The distribution over text clusters may become another criterion; the transition is caused by a very unbalanced distribution over clusters. The clustering index which is based on the intercluster similarity and the intra-cluster similarity may be the transition criterion. Once the tradition is decided by the criteria which are mentioned above, we need to decide one of the soft transition which adjusts clusters by the merge and division and the hard transition which clusters texts entirely, again.

16.3.2 Maintenance Mode

This section focuses on the maintenance mode of the DDO system. The maintenance mode is the status of DDO system for maintaining the organization to texts which are added and deleted, continually. The texts which are added subsequently are arranged into their own clusters, during the maintenance mode. We need to evaluate the current organization of texts, in order to decide whether it moves into the creation mode, or not. So, in this section, we explain the execution, the evaluation, and issues which are involved in the maintenance mode.

Texts which are added subsequently are arranged into their own clusters, depending on their distances or similarities. It is assumed that more texts arrive as a stream at the system. Their similarities as and distances from the prototype vectors of clusters are computed. They are arranged into clusters whose similarities are maximum and whose distances are minimum. It is assumed that the clusters which are built in the creation mode are hyperspheres.

Table 16.1 Non
decomposition vs
decomposition

	Non decomposition	Decomposition
Classifiers	Single	#Classes
Task	Multiple classification	Binary classifications
Adv	No overhead	More reliability
Dis	Less reliability	Overhead

Classifying subsequently added texts into their own cluster identifiers is the alternative scheme of arranging them, in the maintenance mode. A supervised machine learning algorithm is adopted separately from the unsupervised one in the creation mode, and it learns the organized texts which are sample texts. The subsequently added texts are encoded into numerical vectors, and they are classified into one or some of the cluster identifiers. They are arranged into clusters whose identifiers are assigned to themselves. It is more desirable to arrange them by the supervised learning algorithm through the empirical validation which was performed by Jo in 2006 [25].

In order to decide whether the system transits into the creation mode, we need to evaluate the quality of organizing texts. Because target labels are not available in maintaining the system, the clustering index needs to be modified into the version which is suitable for the situation. The intercluster similarity and the intra-cluster similarity are computed for organizing texts, depending on the cosine similarities among texts. The clustering index is computed by Eq. (12.6), using the two measures. The cosine similarity depends on text representations, rather than full texts, themselves.

Table 16.1 illustrates between the decomposition into binary classifications and non-decomposition in the maintenance mode. Without decomposing the classification into binary classifications, only one single classifier is used for arranging subsequent texts, whereas with doing so, the classifiers as many as clusters in the text organization are used. Without doing so, each text is usually arranged into only one cluster, whereas with doing so, it is usually arranged into more than one cluster. Without the decomposition there is no overhead as the advantage, whereas with it, it is more reliable in arranging texts in the maintenance mode as its advantage. However, note that decomposing the classification into binary classifications causes the system to become heavy.

16.3.3   Creation Mode

This section is concerned with the creation mode in operating the DDO system. The creation mode is the status where texts are organized or reorganized by clustering them. The system is initially given in the maintenance mode where texts are piled, and its status is transitioned into the creation mode by a number of piled texts. The maintenance mode after clustering texts in the creation mode becomes the status where subsequent texts are arranged into one of clusters through the classification task. In this section, we explain the creation mode and enumerate approaches which are adopted for executing the creation mode.

The system status is transitioned into the creation mode from the initial maintenance mode or subsequent one. The organized texts are flatted into a single group and they are clustered into subgroups of content-based similar texts, entirely. The creation mode may be implemented into various versions depending on which clustering algorithm is adopted. The cluster naming was implemented as the subsequent task but its performance was not evaluated [25]. We may consider the multiple text summarization for providing a script to each cluster and the virtual text generation as the additional tasks for upgrading the DDO system.

In 2006, Jo adopted various clustering algorithms in implementing the DDO system [25]. The single pass algorithm, which was mentioned in Sect. 10.2.3, was used as the fast approach. The AHC algorithm, which was covered in Sect. 10.2.1, was adopted as another approach to clustering text in the DDO system. The means algorithm, which was described in Sect. 6.3.1, was also used for executing the creation mode. The Kohonen Networks were adopted for performing the creation as the unsupervised neural networks.

The named clusters of content-based similar texts are generated as results from the creation mode. A list of cluster names becomes the results from predefining categories. Texts which are organized into clusters are sample texts for learning classifiers which are involved in the maintenance mode. The creation mode provides the automation of preliminary tasks for the text categorization as its effect. The quality of samples that are produced from the creation mode is less than that of the samples that are manually prepared.

Let us consider the transition from the second maintenance mode to the second creation mode. There are two types of text organization: the hard organization where entire texts are clustered at a time and the soft organization where some clusters are updated and adjusted. It takes too much time for performing the hard organization in the second creation mode. The soft organization is to merge some smaller clusters which are similar to each other and to divide a large cluster into several clusters. Selecting either of two types or both of them depends on the policy which we decided in advance.

16.3.4 Additional Tasks

This section is concerned with additional tasks that are necessary for reinforcing the DDO system. The text categorization, the text clustering, and the cluster naming are involved in implementing the DDO system by Jo in 2006 [25]. We need the text taxonomy generation, the text summarization, and the text segmentation which were covered in the previous chapters as the additional tasks to the system. The taxonomy generation is needed to transition from the maintenance mode to the creation mode, and the text summarization is needed for processing texts more efficiently in the system. In this section, we describe the additional tasks briefly and explain their roles of the DDO system.

Let us consider the taxonomy generation which was described in Chap. 15 as an additional task for implementing the DDO system. The taxonomy generation means the process of generating topics and their links to texts from a corpus. In Sect. 15.3, we mentioned the four schemes of doing it. In the current version, we adopted the clustering-based scheme for generating the taxonomies. Another scheme of generating may be adopted to the current version.

Let us consider the text summarization as another additional task for implementing the DDO system. It takes much less time for encoding summaries into numerical vectors than full texts. Cluster summaries are generated by the multiple text summarization as cluster scripts. It is more desirable to present summaries of organized texts for users than full texts as their previews. Therefore, this system is expected to be reinforced by installing the text summarization module.

Let us consider the addition of the text segmentation which was covered in Chap. 14 to the DDO system. Texts that are organized in the system have very variable length; it is possible that a very long text which spans over multiple topics may exist in the system. When a long text is loaded to the system, it is segmented into topic based subtexts by executing the text segmentation. Subtexts are generated from the long text and treated as independent ones in the system. In order to do so, we adopt the heuristic approaches, which were mentioned in Sect. 14.3.1, and expand them to the advanced approaches.

The virtual texts which are opposite to the actual ones mean texts which are synthesized artificially by concatenating subtexts from different texts. When adding the text summarization and the text segmentation to this system, a text is partitioned into paragraphs and topic-based subtexts. A full text may be constructed by assembling subtexts or paragraphs which are relevant to the topic. Individual sentences may be changed by replacing pronouns by their corresponding nouns. We will consider the topic analysis of individual texts using the HMM (Hidden Markov Model).

16.4   Issues of Dynamic Document Organization

In this section, we consider some issues of implementing the DDO system beyond the text categorization and clustering. In Sect. 16.4.1, we mention some issues in encoding texts into numerical vectors. In Sect. 16.4.2, we consider the problems which are caused by applying the decomposition into binary classifications in the transition from the creation mode to the maintenance mode. In Sect. 16.4.3, we cover conditions of transitioning from the maintenance mode to the creation mode as the process of reorganizing texts. In Sect. 16.4.4, we mention some hybrid tasks of the text clustering and the text categorization as alternatives to the tasks in the DDO system.

16.4.1   Text Representation

This section is concerned with the issues in encoding texts into numerical vectors. The previous works pointed out the problems of encoding texts so for executing text mining tasks [25, 61]. In previous works, there were trials of encoding texts into alternative structured ones that are numerical vectors [35, 61]. The DDO system where texts were encoded into string vectors, instead of numerical vectors, was implemented in Jo's PhD dissertations in 2006 [25]. In this section, we discuss the issues of encoding texts into numerical vectors and present solutions to them.

Let us consider the process of extracting the feature candidates and select one among them as the features. The entire corpus is indexed for extracting more than ten thousands of feature candidates. Only several hundred features are selected among them, most efficiently. The dimension of numerical vectors which represent texts is usually three hundreds. Better performance is achieved by encoding texts into smaller dimensional string vectors, instead of bigger dimensional numerical vectors [26].

Another issue is the sparse distribution in each numerical vector. The sparse distribution means the dominance of zero values over nonzero ones with more than 90%. The similarity between two sparse vectors tends to be zero by Eq. (6.1). It means the poor discrimination among numerical vectors is caused by the sparse distribution. It was solved by encoding texts into tables, instead of numerical vectors, by Jo in 2015 [35].

One more issue in numerical vectors which represent texts is poor transparency. There is no way of guessing text contents from representations, numerical vectors. The numerical vectors which consists of only numerical values lose symbolic values which reflect contents directly. It is given as an ordered list of numerical values, without presenting their features. In the previous works, the transparency was improved by encoding texts into tables or string vectors, instead of numerical vectors [35].

16.4.2   Binary Decomposition

This section is concerned with the binary decomposition which is the issue of implementing the DDO system. In Sect. 5.2.3, we explained the process of decomposing the multiple classification task into binary classifications. The decomposition into binary classifications as many as clusters is not choice in the transition from the creation mode to the maintenance mode. The benefit is the more reliability in maintaining the text organization, but the payment is much overhead of doing so. In this section, we discuss the issue of decomposition in the transition.

We need to relabel texts with positive or negative class, in decomposing into binary classifications. A classifier is allocated to each cluster, and texts which belong to the cluster are labeled with the positive class. Texts which do not belong to the cluster are labeled with the negative class, and among them, only some negative class texts are selected as many as the positive ones. All of texts with the positive class are fixed, but many subsets of texts labeled with the negative class exist. The performance of classifying texts which are added subsequently depends on which of texts labeled with the negative class are selected.

If the classification task is decomposed into binary classifications in the maintenance mode, the classifiers are allocated as many as clusters. They are named symbolically as categories and by the above process, sample texts which are labeled with the positive class or the negative class are prepared for each classifier. The classifiers which correspond to clusters are learned with their own sample texts, and wait for texts which are added subsequently. A novice text is classified into the positive class or the negative class by the classifiers, and it is arranged into clusters which correspond to positive classes into which the classifiers classify it. If the policy is adopted for executing the classification, we need an additional process for selecting among classified ones.

Let us consider the measure for evaluating the DDO performance in the maintenance mode. The clustering index is used for evaluating the creation mode of system. The F1 measure is used for evaluating the maintenance mode of system. The mode is evaluated by computing the macro-averaged and micro-averaged value, following the process which was described in Sect. 8.3. When operating the DDO system as a real-time system, in evaluating the maintenance mode, the balance between the learning speed and the classification performance should be kept.

Decision of the crisp classification or the fuzzy classification is the issue in operating the maintenance mode of the DDO system. If the crisp classification is set as the policy of operating the system, we don't need the process of decomposing the classification into binary classifications. If the fuzzy classification is decided, we absolutely need the decomposition. We consider swap between the crisp organization and the fuzzy organization, while the DDO system is executed. Selecting either of the two types or both of them is the issue of operating the system.

16.4.3 Transition into Creation Mode

This section is concerned with the decision on text reorganizations in the system. The reorganization is viewed as the transition from the maintenance mode to the creation mode. For deciding the transition, we need to observe and evaluate the quality of the current organization. The assumption that is underlying in operating the system is the quality of the current organization which is degraded gradually by adding and deleting texts continually. In this section, we propose some schemes of reorganizing texts, in order solve the problems.

The clustering index was mentioned by Jo in his PhD dissertation for defining the condition of transiting from the maintenance mode to the creation mode [25]. The process of adding or deleting one text in the DDO system is called transaction. The clustering index of the current text clusters is computed, whenever the transaction happens. If the clustering index after doing the transaction is less than the threshold, the texts should be reorganized as the transition to the creation mode. The reorganization of entire texts, ignoring the previous text organization, is called hard organization.

Touching only some clusters is the alternative scheme of improving the quality of text organization. Here, only clusters where transaction is needed will be observed. In deleting texts, the system decides whether a cluster is merged into another, and in adding texts, it decides whether a cluster is divided into two clusters. The clustering index is computed before and after the action and they are compared with each other. When the clustering index is improved after doing it, we decide the results after doing so.

The outlier in the context of the DDO system is a small text group whose contents are very far from the current organization and becomes an important factor for deciding the reorganization. As time passes, news topics may be created in the world, and texts about the created topic may be added. If texts are arranged into one of the existing clusters, they may exist as outliers. In this case, we need to decide the text reorganization by the outlier portion and the clustering index. The outlier detection is added to this system as a module, and texts which are added are detected as outliers or not, before classifying them.

We may add the manual editing module of text organization in the creation mode in implementing the DDO system as its commercial version. Nobody uses the system with the perfect trust of automating the text organization. We need to implement the text organization editor for enabling users for editing it manually. The human intervention is needed for rearranging subsequent texts even to the maintenance mode. We add the graphic user interface for editing the text organization to the system.

16.4.4   Variants of DDO System

This section is concerned with variants which are derived from the DDO system. It was intended for managing the text organization continually. From the tasks that are involved in the DDO system, we derive other tasks, such as novice detection, taxonomy generation, and virtual text generation. The derived tasks are applied to reinforce the performance and the functions of the system. In this section, we describe the tasks that are derived from the tasks as variants.

Let us mention the novice detection and novice classification as derived tasks. The novice detection means the process of finding novice texts which deal with emergent topics in the given corpus. The novice classification is the process of classifying texts which are added subsequently, into novice or not. The novice detection becomes a specialized text clustering where texts are clustered into the familiar group and the novice group, and the classifier for the novice classification learns texts which are clustered. Novice texts are used for building the new taxonomies of texts in the current organization, in the soft reorganization process.

Combining the taxonomy generation and the text classification with each other is similar as doing the two tasks in implementing the DDO system. The taxonomy generation is the process of generating and organizing taxonomies as topics or categories, as mentioned in Chap. 15. The taxonomy list is built and texts are linked to each taxonomy and used as the sample texts for learning classifiers. Texts which are added subsequently are classified into one or some of taxonomies. The schemes of generating and maintaining taxonomies were mentioned in Sects. 15.3 and 15.4.

In Sect. 14.4.4, we mentioned the virtual texts that are derived from actual ones. It is assumed that only actual texts are available in the current version of the DDO system. We add the module for generating virtual texts from existing and added ones. It reinforces the process of retrieving texts which are relevant to the query. Refer Sect. 14.4.4, for getting detailed description about virtual texts.

We need to nominate a particular text from each cluster as its representative one. The representative text plays the role of a script or a preview of its own cluster. The scheme of nominating the representative text was already mentioned in Sect. 10.3.4. There is a possibility of nominating a virtual text, instead of actual text, as a representative one. We need further research for doing a virtual text.

16.5   Summary and Further Discussions

We described the system of managing full texts automatically in this chapter. The online clustering where only the clustering algorithm is used is the simple scheme for organizing and managing texts. Combining the clustering with the classification is a more reliable scheme of doing so. We pointed out some issues in implementing the system beyond both the text clustering and the text categorization. In this section, we make some further discussion from what we studied in this chapter.

The DDO system is intended for managing texts automatically by compounding the text categorization and the text clustering. The combination is the cooperation of more than two tasks as independent ones, whereas the compound is to map more than two tasks into a single task. The taxonomy generation is the task which is mapped by compounding the tasks which is mentioned in Sect. 16.1. The process of synthesizing virtual texts is also the compound task of the three tasks: the text segmentation, the text clustering, and the text classification. The task which is viewed as a single task is derived as the compound of several tasks.

Text contents in the DDO system may be updated, continually. The assumption underlying in the DDO system is that text contents are fixed as long as they stay in the system. Updating texts continually causes the text reorganization: the transition from the maintenance mode to the creation mode. The soft action such as movement of text into another cluster becomes an alternative to the reorganization. If many texts are updated at almost the same time, we need to take a decision of hard or soft action.

The version of DDO system which was developed by Jo in 2006 supports only plain texts which are given as files whose extensions are "txt". The XML format is known as the standard one, and the Java class libraries for processing the XML format are available. The document is given as a PDF image and it is possible to convert it into a plain text. It is considered to process an MS Word file whose extension is "doc" as a text. HTML is another text format and tags are considered as well as contents.

The DDO system needs to be upgraded for challenging against the big data processing. The big data is characterized as the 4Vs: Volume, Velocity, Variance, and Veracity. We need to reorganize texts and manipulate clusters assuming the higher velocity of texts; addition, deletion, and updates of texts are very frequent. By the higher veracity, we must alert against noisy texts, erroneous texts, and texts with their broken characters, in maintaining the system. For managing almost infinite volume of texts, multiple DDO programs should be executed in the cloud computing environment.

References

1. Aho, A.V., Lam, M.S., Sethi, R., Ullman, J.D.: Compilers: Principles, Techniques, and Tools. Pearson, Harlow (2007)
2. Allan, J., Papka R., Lavrenko, V.: On-line news event detection and tracking. In: Proceedings of the 21st Annual International ACM SIGIR Conference on Research and Development in Information Retrieval, pp. 37–34 (1998)
3. Alpaydin, E.: Introduction to Machine Learning. MIT Press, Cambridge (2014)
4. Baeza-Yates, R., Ribeiro-Neto, B.: Information Retrieval: Concepts and Technology Behind Search. Pearson (2011)
5. Beaver, M.: Introduction to Probability and Statistics. PWS-KENT Publishing Company (1991)
6. Boiy, E., Moens, M.F.: A machine learning approach to sentiment analysis in multilingual Web texts. Inf. Retr. **12**, 525–558 (2009)
7. Brun, M., Sima, C., Hua, J., Lowey, J., Carroll, B., Suha, E., Doughertya, E.R.: Model-based evaluation of clustering validation measures. Pattern Recogn. **40**, 807–824 (2007)
8. Can, F.I., Kocbrtber, S., Baglioglu, O., Kardas, S., Ocalan, H.C., Uyar, E.: New event detection and topic tracking in Turkish. J. Am. Soc. Inf. Sci. Technol. **61**, 802–819 (2010)
9. Cai, D., He, X.: Manifold adaptive experimental design for text categorization. IEEE Trans. Knowl. Data Eng. **24**, 707–719 (2011)
10. Clark, J., Koprinska, I., Poon, J.: A neural network based approach to automated e-mail classification. In: Proceedings of IEEE/WIC International Conference on Web Intelligence, pp. 702–705 (2003)
11. Connolly, T., Begg, C.: Database Systems: A Practical Approach to Design, Implementation, and Management. Addison Wesley, Reading, MA (2005)
12. Cortes, C., Vapnik, V.: Support vector networks. Mach. Learn. **20**, 273–297 (1995)
13. Dong, Y., Han, K.: Boosting SVM classifiers by ensemble. In: Proceeding WWW '05 Special Interest Tracks and Posters of the 14th International Conference on World Wide Web, pp. 1072–1073 (2005)
14. Duda, R.O., Hart, P.E., Stork, D.G.: Pattern Classification. Wiley, New York (2000)
15. Dumais, S., Chen, H.: Hierarchical classification of Web content. In: Proceedings of the 23rd Annual International ACM SIGIR Conference on Research and Development in Information Retrieval, pp. 256–263 (2000)
16. Eiben, A.E., Smith J.E.: Introduction to Evolutionary Computing. Springer, Berlin (1998)
17. Fall, C.J., Torcsvari, A., Benzineb, K., Karetka, G.: Automated categorization in the international patent classification. ACM SIGIR Forum **37**, 10–25 (2003)

© Springer International Publishing AG, part of Springer Nature 2019
T. Jo, *Text Mining*, Studies in Big Data 45,
https://doi.org/10.1007/978-3-319-91815-0

18. Feldman, R., Sanger, J.: The Text Mining Handbook: Advanced Approaches in Analyzing Unstructured Data. Cambridge, New York (2007)
19. Grossberg, S.: Competitive learning: from interactive activation to adaptive resonance. Cogn. Sci. **11**, 23–63 (1987)
20. Hanani, U., Shapira, B., Shoval, P.: Information filtering: overview of issues, research and systems. User Model. User-Adap. Inter. **11**, 203–259 (2001)
21. Halkidi, M., Batistakis, Y., Vazirgiannis, M.: On clustering validation techniques. J. Intell. Inf. Syst. **17**, 107–145 (2001)
22. Hyvarinen, A., Oja, E.: Independent component analysis: algorihtms and applications. Neural Netw. **4–5**, 411–430 (2000)
23. Jo, T.: Neural based approach to keyword extraction from documents. Lect. Note Comput. Sci. **2667**, 456–461 (2003)
24. Jo, T.: The application of text clustering techniques to detection of project redundancy in national R&D information system. In: The Proceedings of 2nd International Conference on Computer Science and its Applications (2003)
25. Jo, T.: The Implementation of Dynamic Document Organization Using the Integration of Text Clustering and Text Categorization, University of Ottawa (2006)
26. Jo, T.: Modified version of SVM for text categorization. Int. J. Fuzzy Log. Intell. Syst. **8**, 52–60 (2008)
27. Jo, T.: Inverted Index based modified version of KNN for text categorization. J. Inf. Process. Syst. **4**, 17–26 (2008)
28. Jo, T.: Neural text categorizer for exclusive text categorization. J. Inf. Process. Syst. **4**, 77–86 (2008)
29. Jo, T.: The effect of mid-term estimation on back propagation for time series prediction. Neural Comput. Applic. **19**, 1237–1250 (2010)
30. Jo, T.: NTC (Neural Text Categorizer): neural network for text categorization. Int. J. Inf. Stud. **2**, 83–96 (2010)
31. Jo, T.: Definition of table similarity for news article classification. In: The Proceedings of Fourth International Conference on Data Mining, pp. 202–207 (2012)
32. Jo, T.: VTG schemes for using back propagation for multivariate time series prediction. Appl. Soft Comput. **13**, 2692–2702 (2013)
33. Jo, T.: Application of table based similarity to classification of bio-medical documents. In: The Proceedings of IEEE International Conference on Granular Computing, pp. 162–166 (2013)
34. Jo, T.: Simulation of numerical semantic operations on strings in medical domain. In: The Proceedings of IEEE International Conference on Granular Computing, pp. 167–171 (2013)
35. Jo, T.: Index optimization with KNN considering similarities among features. In: The Proceedings of 14th International Conference on Advances in Information and Knowledge Engineering, pp. 120–124 (2015)
36. Jo, T.: Normalized table matching algorithm as approach to text categorization. Soft Comput. **19**, 839–849 (2015)
37. Jo, T.: Keyword extraction by KNN considering feature similarities. In: The Proceedings of The 2nd International Conference on Advances in Big Data Analysis, pp. 64–68 (2015)
38. Jo, T.: AHC based clustering considering feature similarities. In: The Proceedings of 11th International Conference on Data Mining, pp. 67–70 (2015)
39. Jo, T.: KNN based word categorization considering feature similarities. In: The Proceedings of 17th International Conference on Artificial Intelligence, pp. 343–346 (2015)
40. Jo, T.: Simulation of numerical semantic operations on string in text collection. Intl. J. App. Eng. Res. **10**, 45585–45591 (2015)
41. Jo, T.: Table based KNN for indexing optimization. In: The Proceedings of 18th International Conference on Advanced Communication Technology, pp. 701–706 (2016)
42. Jo, T., Cho, D.: Index based approach for text categorization. Int. J. Math. Comput. Simul. **2**, 127–132 (2008)

43. Jo, T., Japkowicz, N.: Text clustering using NTSO. In: The Proceedings of IJCNN, pp. 558–563 (2005)
44. Jo, T., Lee, M.: The evaluation measure of text clustering for the variable number of clusters. Lect. Notes Comput. Sci. **4492**, 871–879 (2007)
45. Jo, T., Seo, J., Kim, H.: Topic spotting on news articles with topic repository by controlled indexing. Lect. Note Comput. Sci. **1983**, 386–391 (2000)
46. Jo, T., Lee, M., Kim, Y.: String vectors as a representation of documents with numerical vectors in text categorization. J. Converg. Inf. Technol. **2** 66–73 (2007)
47. Joachims, T.: Text categorization with support vector machines: learning with many relevant features. In: Proceedings of European Conference on Machine Learning, pp. 137–142 (1998)
48. Joachims, T.: Learning to Classify Text Using Support Vector Machines: Methods, Theory and Algorithms. Kluwer Academic, Boston (2002)
49. Jones, K.S.: Automatic Summarizing: Factors and Directions. In: Advanced Automate Summarization edited by Manu, I. and Maybury M., 1–12 (1999)
50. Kaski, S., Honkela, T., Lagus, K., Kohonen, T.: WEBSOM-Self organizing maps of document collections. Neurocomputing **21**, 101–117 (1998)
51. Kohonen, T.: Correlation matrix memories. IEEE Trans. Comput. **21**, 353–359 (1972)
52. Kohonen, T., Kaski, S., Lagus, K., Salojavi, J., Honkela, J.: Self organization of massive document collection. IEEE Trans. Neural Netw. **11**, 574–585 (2000)
53. Konchady, M.: Text Mining Application Programming. Charles River Media, Boston (2006)
54. Kowalski, G.J., Maybury, M.T.: Information Storage and Retrieval Systems: Theory and Implementation. Kluwer Academic, Boston (2000)
55. Kreuzthaler, M., Bloice, M.D., Faulstich, L., Simonic, K.M., Holzinger, A.: A comparison of different retrieval strategies working on medical free texts. J. Univers. Comput. Sci. **17**, 1109–1133 (2011)
56. Kroon, H.C.M.D., Kerckhoffs, E.J.H.: Improving learning accuracy in information filtering. In: International Conference on Machine Learning-Workshop on Machine Learning Meets HCI (1996)
57. Larose, D.T.: Discovering Knowledge in Data: An Introduction to Data Mining. Wiley, New York (2005)
58. Leslie, C.S., Eskin, E., Cohen, A., Weston, J., Noble, W.S.: Mismatch String Kernels for Discriminative Protein Classification. Bioinformatics **20**, 467–476 (2004)
59. Li, T., Zhu, S., Ogihara, M.: Hierarchical document classification using automatically generated hierarchy. J. Intell. Inf. Syst. **29**, 211–230 (2007)
60. Liu, J., Chua, T.S.: Building semantic perceptron net for topic spotting. In: Proceedings of the 39th Annual Meeting on Association for Computational Linguistics, pp. 378–385 (2001)
61. Lodhi, H., Saunders, C., Shawe-Taylor, J., Cristianini, N., Watkins, C.: Text classification with string kernels. J. Mach. Learn. Res. **2**, 419–444 (2002)
62. Loredana, F., Lemnaru, C., Potolea, R.: Spam detection filter using KNN algorithm and resampling. In: Proceedings of IEEE International Conference on Intelligent Computer Communication and Processing, pp. 27–33 (2010)
63. Luhn, H.: Statistical approach to mechanized encoding and searching of literary information. IBM J. Res. Dev. **1**, 309–317 (1957)
64. Maas, A.L., Daly, R.E., Pham, P.T., Huang, D., Ng, A.V., Potts, C.: Learning word vectors for sentiment analysis. In: Proceedings of the 49th Annual Meeting of the Association for Computational Linguistics: Human Language Technologies, vol. 1, pp. 142–150 (2011)
65. Manning, C.D., Raghavan, P., Schutze, H.: Introduction to Information Retrieval. Cambridge University Press, Cambridge (2009)
66. Markov, Z., Larose D.T.: Data Mining The Web: Uncovering Patterns in Web Content, Structure, and Usage. Wiley, New York (2007)
67. Martinetz, T., Schulten, K.: A "neural gas" network learns topologies. In: Artificial Neural Networks, pp. 397–402 (1991)
68. Marz, N., Warren, J.: Big Data: Principles and Best Practices of Scalable Real Time Data Systems. Manning, Shelter Island (2015)

69. Minsky, M., Papert, S.: Perceptrons. MIT Press, Cambridge (1969)
70. Mitchell, T.: Machine Learning. McGraw-Hill Companies, New York (1997)
71. Mullen, T., Collier, N.: Sentiment analysis using support vector machines with diverse information sources. In: Proceedings of Conference on Empirical Methods in Natural Language Processing, pp. 412–418 (2004)
72. Myers, K., Kearns, M., Singh, S., Walker M.A.: A boosting approach to topic spotting on subdialogues. Family Life **27**, 1 (2000)
73. Nigam, K., Mccallum, A.K., Thrun, S., Mitchell, T.: Text classification from labeled and unlabeled documents using EM. Mach. Learn. **39**, 103–134 (2000)
74. Noy, N.F., Hafner, C.D.: The state of the art in ontology design a survey and comparative review. AI Mag. **18**, 53–74 (1997)
75. Pang, B., Lee, L.: Opinion mining and sentiment analysis. Found. Trends Inf. Retr. **2**, 1–135 (2008)
76. Platt, J.: Sequential minimal optimization: a fast algorithm for training support vector machines. Technical Report MSR-TR-98-14, Microsoft Research (1998)
77. Poole, D.: Linear Algebra: A Modern Introduction. Brooks/Collen, Pacific Grove (2003)
78. Qiu, Y., Frei, H.P.: Concept based query expansion. In: The Proceedings of the 16th Annual International ACM SIGIR Conference on Research and Development in Information Retrieval, pp. 160–169 (1993)
79. Rennie, J.D.M., Rifkin, R.: Improving multiclass text classification with the support vector machine. Technical Report AIM-2001-026, Massachusetts Institute of Technology (2001)
80. Rosenblatt, F.: The perceptron: a probabilistic model for information sotrage and organization in the brain. Psychol. Rev. **65**, 385–408 (1958)
81. Rumelhart, D.E., McClelland, J.L. (eds.) Parallel Distributed Processing: Exploration in Microstructure of Cognition, vol. 1. MIT Press, Cambridge (1986)
82. Salton, G.: Automatic Text Processing: Transformation, Analysis, and Retrieval of Information by Computer. Addison Wesely, Reading (1988)
83. Salton, G., Yang, C.S.: On the specification of term values in automatic indexing. J. Doc. **29**, 351–372 (1973)
84. Schneider, K.M.: A comparison of event models for Naive Bayes anti-spam e-mail filtering. In: Proceedings of the Tenth Conference on European Chapter of the Association for Computational Linguistics, pp. 307–314 (2003)
85. Sebastiani, F.: Machine learning in automated text categorization. ACM Comput. Surv. **34**, 1–47 (2002)
86. Shanmugasundara, J., Shekita, E., Kiernan, J., Krishnamurthy, R., Viglas, E., Naughton, J., Tatarinov, I.: A general technique for querying XML documents using a relational database system. Newslett. ACM SIGMOD. **30**, 20–26 (2001)
87. Shardanand, U., Maes, P.: Social information filtering: algorithms for automating word of mouth. In: Proceedings of the SIGCHI Conference on Human Factors in Computing Systems, pp. 210–217 (1995)
88. Sriram, B., Fuhry, D., Demir, E.: Ferhatosmanoglu, H., Demirbas, M.: Short text classification in twitter to improve information filtering. In: Proceedings of the 33rd International ACM SIGIR Conference on Research and Development in Information Retrieval, pp. 441–442 (2010)
89. Sutanto, D., Leung H.C.: Automatic index expansion for concept based image query. Lect. Notes Comput. Sci. **1614**, 399–408 (2002)
90. Tan, S.: An effective refinement strategy for KNN text classifier. Expert Syst. Appl. **30**, 290–298 (2006)
91. Tan, P., Steinbach, M., Kumar, V.: Introduction to Data Mining. Addison Wesely, Boston (2006)
92. Tong, S., Koller, D.: Support vector machine active learning with applications to text classification. J. Mach. Learn. Res. **2**, 45–66 (2001)
93. Vega-Pons, S., Ruiz-Shulclopery, J.: A survey of clustering ensemble algorithms. Int. J. Pattern Recognit. Artif. Intell. **25**, 337–372 (2011)

94. Vendramin, L., Campello, R., Hruschka E.R.: On the comparison of relative clustering validity criteria. In: Proceedings of the 2009 SIAM International Conference on Data Mining, pp. 733–744 (2009)
95. Wiener, E.D.: A neural network approach to topic spotting in text. The Master Thesis of University of Colorado (1995)
96. Winter, R., Widrow, B.: Madaline rule II: training algorithm for neural networks. In: Proceedings of IEEE 2nd International Conference on Neural Networks, pp. 401–408 (1988)
97. Wu, X., Wu, G., Wei, D.: Data mining with big data. IEEE Trans. Knowl. Data Eng. **26**, 97–107 (2014)
98. Yang, Y.: An evaluation of statistical approaches to MEDLINE indexing. In: Proceedings of the AMIA Annual Fall Symposium, pp. 358–362 (1996)
99. Yang, Y., Liu, X.: A re-examination of text categorization methods. In: Proceedings of the 22nd Annual International ACM SIGIR Conference on Research and Development in Information Retrieval, pp. 42–49 (1999)
100. Youn, S., McLeod, D.: A comparative study for email classification. In: Advances and Innovations in Systems, Computing Sciences and Software Engineering, pp. 387–391 (2007)

Index

Printed in the United States
By Bookmasters